Diss. ETH No. 19142

Analysis and Design of Communication Techniques in Spectrally Efficient Wireless Relaying Systems

A dissertation submitted to the
SWISS FEDERAL INSTITUTE OF TECHNOLOGY (ETH)
ZURICH

for the degree of
Doctor of Sciences

presented by
JIAN ZHAO
M.Sc., TU Hamburg-Harburg
born May 7, 1979
citizen of China

accepted on the recommendation of
Prof. Dr. Armin Wittneben, examiner
Prof. Dr. Gerhard Bauch, co-examiner

2010

Reihe Series in Wireless Communications
herausgegeben von:
Prof. Dr. Armin Wittneben
Eidgenössische Technische Hochschule
Institut für Kommunikationstechnik
Sternwartstr. 7
CH-8092 Zürich

E-Mail: wittneben@nari.ee.ethz.ch
Url: http://www.nari.ee.ethz.ch/

Bibliografische Information der Deutschen Nationalbibliothek

Die Deutsche Nationalbibliothek verzeichnet diese Publikation in der Deutschen Nationalbibliografie; detaillierte bibliografische Daten sind im Internet über http://dnb.d-nb.de abrufbar.

Date of Doctoral Examination: July 9, 2010

ISBN 978-3-8325-2585-9
ISSN 1611-2970

Logos Verlag Berlin GmbH
Comeniushof, Gubener Str. 47,
10243 Berlin

Abstract

In the past few decades, wireless communication has become one of the fastest growing sectors in the communication industry. In order to satisfy the tremendous growth of demand, wireless service providers are striving to improve the communication systems so that higher data rate wireless transmission can be supported. One of the major breakthroughs for supporting high data rate wireless communications is the multiple-input multiple-output (MIMO) communication technology, i.e., multi-antenna communication. The utilization of multiple antennas provides a significant increase in the reliability and the information transmission rate in wireless systems. After their proposal, multi-antenna communication has been widely adopted as a key solution for providing high data rate throughput in wireless systems. Nowadays, multi-antenna communication techniques are indispensable to most of the wireless communication standards.

On the other hand, conventional wireless communication architecture does not seem to be feasible to offer the high data rate transmission required for future generation wireless communication systems in reasonably large areas, e.g., for the fourth generation (4G) cellular wireless networks. Relay communication was proposed as a means to extending the coverage range in wireless networks with reduced infrastructure deployment costs. However, conventional relaying protocols require two channel uses to transmit the data from the source to the destination via the relay, which leads to a loss in the spectral efficiency. The *two-way relaying protocol* was proposed to recover a significant portion of the spectral efficiency loss. Such a protocol combines the transmission of the bidirectional information flows in relay networks and is simple to implement in practical systems.

This dissertation deals with the analysis and design of communication techniques in MIMO relaying systems, especially in MIMO two-way relaying systems. The analysis of the considered system is based on information theoretical performance limits, such as capacity and achievable rates. For the design of communication techniques, we propose practical transmission strategies and show their performance results based on practical performance measures.

Firstly, we verify that relay communication can extend the coverage of wireless networks, and propose a quantitative analysis framework for determining the coverage extension by using decode-and-forward (DF) relays in cellular wireless systems. We define the coverage of cellular relaying networks based on the achievable rates, and provide theoretical calculations as well as simulations to show the improvement of coverage in cellular relaying systems.

Secondly, information-theoretic performance analysis on MIMO two-way relaying systems is proposed. For MIMO two-way DF relaying systems, two data-combining schemes, i.e., the superposition coding (SPC) scheme and the network coding scheme, are studies in detail. When the transmit channel knowledge is available at the relay, we propose the optimum transmission strategies at the multi-antenna relay from the information-theoretic perspective for the two data combining schemes. Furthermore, we propose the methods to find the optimum time-division (TD) strategies between the multiple access (MAC) phase and the broadcast (BRC) phase in two-way DF relaying systems, considering both peak and average power constraints.

Next, we consider the design of practical transmission schemes in MIMO two-way relaying systems. We propose a novel channel estimation method for the BRC phase of MIMO two-way DF relaying systems when the SPC scheme is applied. The self-interference (SI), which contains the known data at the receivers in two-way relaying systems, is utilized to estimate the channel. Since the SI is inherent in the received signals in two-way relaying systems, it does not consume additional system resources as pilots do. On the other hand, the SI can also be used together with pilots to offer superior channel estimation performances than the purely pilot-aided channel estimation scheme. Furthermore, we quantify the spectral efficiency improvement of the BRC phase channel with SI-aided channel estimation by deriving its achievable rates using different codebooks.

Finally, we propose a novel method to exploit the bit-level SI so that asymmetric data rates can be transmitted in the BRC phase of two-way DF relaying systems when the network coding scheme is applied. Since the network coding scheme combines the data on the bit level, the major problem faced by the network coding scheme is how to transmit with asymmetric data rates to the user stations according to their individual link qualities in the BRC phase. In the proposed scheme, the weaker link receiver exploits the *a priori* bit information in each received data symbol, so that it only needs to decode on a subset of the signal constellation. Subject to the same bit error rate constraint, the weaker link receiver can decode at lower signal-to-noise ratio (SNR) compared to the stronger link.

Kurzfassung

In den vergangenen Jahrzehnten hat sich die drahtlose Kommunikation in eine der am schnellsten wachsenden Sektoren der Nachrichtentechnik-Industrie gewandelt. Um die enorm gestiegene Nachfrage decken zu können, setzen Wireless Service Provider darauf, zukünftige drahtlose Kommunikationssysteme so zu verbessern, dass wesentlich höhere Datenraten unterstützt werden. Eine der wichtigsten Innovationen der letzten Jahre zur Unterstützung hoher Datenraten in der drahtlosen Kommunikation ist die Einführung der Multiple-Input Multiple-Output (MIMO)-Technologie, die durch die Verwendung von Merfach-Antennen-Arrays auf Seiten von Sender und Empfänger möglich wird. Die Verwendung von mehreren Antennen bietet für Wireless-Systeme eine signifikante Erhöhung der Zuverlässigkeit und der Übertragungsrate. Mittlerweile gilt die Verwendung von MIMO-Techniken allgemein als Schlüssel zur Bereitstellung hoher Übertragungsraten in drahtlosen Systemen und findet Eingang in die aktuellen Standards der drahtlosen Kommunikation.

Auf der anderen Seite wirft dies aber auch die Frage auf, ob konventionelle drahtlose Kommunikations-Architekturen in der Lage sind, die für die künftige Generation drahtloser Kommunikationssysteme notwendigen hohen Datenraten *allen* Nutzer zur Verfügung zu stellen. Ist es z.B. im Fall von LTE Advanced, der vierten Generation (4G) zellularer Mobilfunknetze, möglich, auch Usern am Zellrand mit der gebotenen Zuverlässigkeit die erstrebten hohen Übertragungsraten zu liefern? Relay-Kommunikation wird als ein Mittel zur Erweiterung der Reichweite in drahtlosen Netzen vorgeschlagen, mit dem Vorteil geringerer Kosten für die Bereitstellung der Infrastruktur im Vergleich zu anderen Optionen, wie z.B. einer Erhöhung der Dichte von Basisstationen. Allerdings erfordern konventionelle Halb-duplex Relaying Protokolle zwei Kanalzugriffe, um die Daten von der Quelle über das Relay bis zum Bestimmungsort weiterzuleiten. Dies schlägt sich in einer reduzierten spektralen Effizienz nieder. Das *Two-Way (Zwei-Wege) Relaying Protokoll* wurde vorgeschlagen, um diese Verluste signifikant zu minimieren. Two-Way Relaying Protokolle kombinieren die Übermittlung zweier bidirektionaler Informationsflüsse in Relay-Netzen und können in vielen Fällen vergleichsweise problemlos in praktischen Systemen implementiert werden.

Diese Dissertation befasst sich mit der Analyse und dem Entwurf von Kommunikationstechniken für MIMO-Relaying-Systeme, wobei insbesondere Two-Way Relaying betrachtet wird. Die Analyse der Relay-Systeme basiert auf Informationstheoretischen Metriken wie etwa der Kanalkapazität und den erreichbaren Datenraten. Wir schlagen geeignete Übertragungsstrategien vor und bewerten die Leistungsfähigkeit dieser Strategien anhand zweckmässiger Kenngrössen.

Zunächst zeigen wir dass Relay-Kommunikation zur Erhöhung der Reichweite drahtloser Netzwerke eingesetzt werden kann und schlagen einen quantitativen Analyseansatz vor, um die Verbesserung der Reichweite in zellularen Funksystemen mittels sog. Decode-and-Forward (DF) Relays zu bestimmen. Die Reichweite zellularer Netzwerke definieren wir hierbei anhand der erzielbaren Übertragungsrate und präsentieren analytische Berechnungen sowie Simulationsergebnisse, welche die Reichweitenverbesserung der betrachteten Systeme durch Relaying belegen.

Darüber hinaus schlagen wir einen Informationstheoretischen Ansatz zur Performanceanalyse von MIMO Two-Way Relaying Systemen vor. Für diese Systeme werden unter Einsatz von DF-Relaying zwei Verfahren detailliert untersucht, die sich darin unterscheiden, wie die beiden bidirektionalen Informationsflüsse kombiniert werden: das sog. Superposition Çoding (SPC) Verfahren, sowie das Network Coding Verfahren. Für den Fall, dass der sendeseitige Kanalzustand am Relay bekannt ist, entwickeln wir die optimale Übertragungsstrategie für Mehrfach-Antennen-Relays durch die Informationstheoretische Betrachtung der beiden Verfahren. Weiterhin schlagen wir für diese Verfahren Methoden vor, um die optimale zeitliche Aufteilung zwischen der Multiple Access (MAC) Phase und der Broadcast (BRC) Phase in Two-Way-DF-Relaying-Systemen zu finden. Dabei werden systembedingte Anforderungen hinsichtlich einer Beschränkung von Spitzen- und Durchschnittsleistungen berücksichtigt.

Als nächstes betrachten wir das Design von geeigneten Übertragungsverfahren für MIMO Two-Way Relaying Systeme. Wir schlagen eine neue Methode zur Kanalschätzung vor, die für die BRC Phase von MIMO Two-Way DF Relaying Systemen mit SPC angewendet werden kann. Die Selbst-Interferenz (SI), die durch das Two-Way-Relaying Prinzip in solchen Systemen entsteht, und die aus den über den Kanal zurückgesendeten und damit den Empfängern bekannten Daten besteht, wird genutzt, um den Kanal zu schätzen. Da bei Two-Way Relaying Systemen die SI inhärent in dem Empfangssignal vorhanden ist, werden keine zusätzlichen Systemressourcen wie Pilotsignale benötigt. Andererseits kann die SI auch zusammen mit Pilotsignalen genutzt werden, um eine verbesserte Kanalschätzung im Vergleich zur rein Piloten-gestützte Kanalschätzung zu bieten. Darüber hinaus quantifizieren wir die Verbesserung der spektralen Effizienz der BRC-Phase mit SI unterstützter

Kanalschätzung durch die Berechnung der erreichbaren Raten mit unterschiedlichen Codebüchern.

Schließlich schlagen wir eine neuartige Methode vor, die für ein Network-Coding basiertes Two-Way-Relaying auf Bit-Ebene die SI nutzt, um asymmetrische Datenraten in der BRC Phase von Two-Way Relaying Systemen übertragen zu können. Da das Network-Coding Schema die Daten auf der Bit-Ebene verarbeitet, ist das hauptsächliche Problem dieses Schemas, wie asymmetrische Datenraten in der BRC Phase zu den Nutzerstationen entsprechend der individuellen Verbindungsqualität übertragen werden können. In dem vorgeschlagenen Schema nutzt die schwächere Verbindung die *a priori* Bitinformation in jedem empfangenen Datensymbol, so dass nur eine Teilmenge der Signalkonstellation dekodiert werden muss. Unter der gleichen Bitfehlerratenbedingung kann der Empfänger der schwächeren Verbindung so bei einem geringeren Signal-zu-Rausch-Verhältnis (SNR) dekodieren im Vergleich zu dem SNR der stärkeren Verbindung.

Acknowledgements

First and foremost, I am deeply grateful to my advisor Prof. Armin Wittneben for his great guidance, endless support and sustained encouragement throughout my PhD journey. Working with Armin and learning from him has been an unparalleled experience. This dissertation benefits a lot from his extraordinary intuition and technical insight. Meanwhile, his humor and enthusiasm has deeply influenced me. I greatly appreciate his care and concern for me as an individual.

I would like to thank Prof. Gerhard Bauch and other colleagues in DOCOMO Euro-Labs for their far-sighted support on our research collaboration and for the insightful discussions and feedbacks during our joint meetings. Their continuous support has enabled our joint collaboration to become exceptionally successful.

My thanks also extend to Dr. Marc Kuhn for the numerous discussions and invaluable assistance for my research. It is a great pleasure for me to work with Marc in our research projects. He is always ready to help when I need his support, and his patience has greatly facilitated the progress of our research.

Special thanks go to my former and current colleagues in the Communication Technology Laboratory in ETH Zurich, who make it such a wonderful place to work and to learn.

Finally, I would like to thank my parents for their unconditional love and support.

Jian Zhao
May 2010

Contents

Chapter 1

Introduction

This dissertation discusses the multiple-input multiple-output (MIMO) relaying systems. Such systems play a central role in the next-generation wireless communication networks. In particular, we investigate the performance limits and develop transmission strategies for the systems employing a spectrally efficient relaying protocol – the two-way relaying protocol.

1.1 Motivation

The last few decades saw tremendous developments in communication technologies, especially in wireless communication. Nowadays, wireless communication has penetrated into nearly every aspect of our life and has become indispensable to the modern society. For example, cellular mobile communication has enabled people to keep contact anywhere, and mobile phones have turned into an important business and social communicating tool; wireless local area networks (WLAN) are supplanting wired Ethernet in campuses, hotels and airports. The next-generation wireless communication systems are expected to offer high speed multimedia services, such as high definition television broadcasting and high quality video conferencing. Those services need a significant boost in the system capacity. However, the performance of the system is subject to the adverse physical factors such as the limited transmit power, the scarcity of the electromagnetic spectrum and the signal attenuation and fluctuation in wireless links. In order to deliver the performance necessary for supporting the emerging applications, it is essential for future wireless systems to utilize the available spectrum in a more efficient way. Recent developments of the multiple-input multiple-output (MIMO) transmission technology, i.e., using multiple collocated antennas at both the transmitter and the receiver sides, provides a promising solution for increasing the system capacity. Theoretical studies [224] and practical implementations [67] have shown that the capacity of MIMO systems can be dramatically increased compared to that of the conventional single-input single-output (SISO) transmission systems without extra consumption of transmit power and bandwidth. In rich-scattering environments and when the signal-to-noise ratio (SNR) ρ is high, the MIMO channel capacity C scales *linearly* with the minimum of the number of the transmit antennas M and the number of the receive antennas N, i.e.,

$$C \propto \min(M, N) \log(\rho). \tag{1.1}$$

In addition, the *spatial diversity* provided by MIMO systems offers an effective way of combating the signal fluctuation due to the fading effects in wireless channels. Nowadays, MIMO has become a key communication technology on the physical layer (PHY) in many wireless standards, such as IEEE 802.11n (WLAN) [2], IEEE 802.16e/m (WiMAX) [3] and 3GPP Long Term Evolution (LTE) [6].

Relay communication is an effective approach for combating the signal attenuation and the shadowing effects in wireless channels. It is a well-known technique for transmitting signals over very long distance. Research on relay channels can be dated back to the 1970's [48,234]. Recently, due to the research on the fourth generation (4G) cellular mobile communication systems in academia and industry, relay communication was again brought to the mainstream of the wireless communication community [87, 204]. That is because of the following reasons: first, the envisioned transmission data rate of 4G systems is several orders of magnitude higher than that of the current 2G and 3G mobile systems. With the same transmit power level, the energy per bit will be just a small fraction of that in current systems. It is unlikely that 4G mobile systems can cover the same service area without additional infrastructure. Second, the 4G systems are expected to be operating in the frequency spectrum range that is much higher than current mobile systems. This leads to higher pathloss that prohibits the base station from communicating with users far away [174]. One way to overcome those problems is to deploy relay stations in the cellular system so that the coverage area can be extended and the overall data rate in the whole network can be improved.

Besides extending the coverage and enhancing the capacity in wireless networks, relay communication has also been shown by recent research to be able to improve the stability of the radio links by introducing a new form of diversity – *cooperative diversity*. The incorporation of relay stations into the system creates additional paths between the transmitter and the receiver. Signals carrying the same information are therefore transmitted over independently fading channels and can be combined at the receiver [112, 134]. Relay communication can be considered as a technique that realizes the benefits of MIMO communication in systems where multiple antennas are spatially distributed. Setting up relay stations in cellular sys-

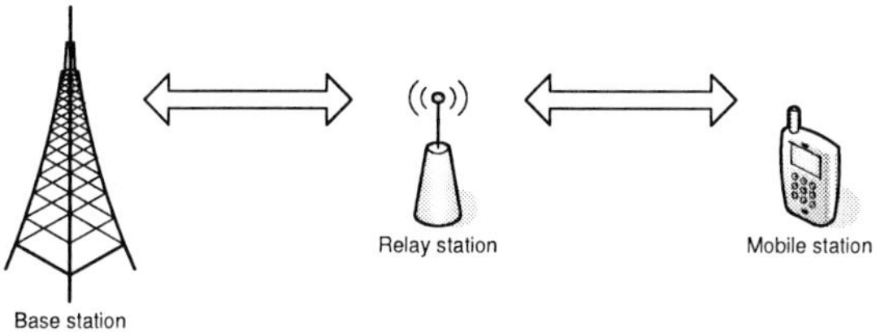

Fig. 1.1: A cellular relaying system with bidirectional information flow

tems is rather simple. For example, fixed relay nodes can be placed on the top of lampposts or high buildings where good wireless connection with the base station is possible. Those dedicated relay stations can have sufficient power supply and can be equipped with powerful signal processing hardware as well as multiple antennas. Another possible relay deployment strategy is for some users to act as relays for their neighbors, which may be useful in, e.g., WLAN hotspots. The integration of relay communication into cellular mobile networks and WLAN is envisioned to be one of the most promising wireless architecture in the years to come. Relay communication has also been incorporated into wireless standards, such as IEEE 802.11s (mesh networking) [4] and IEEE 802.11j (wireless multihop relay) [5].

Current practical wireless stations operate in half-duplex transmission mode. That is, they cannot transmit and receive data simultaneously using the same frequency channel due to the coupling between the transmit and receive circuitry [134]. For the simplest two-hop relaying scenario, where a half-duplex relay station assists the communication between the source and destination stations, the transmission of each information symbol from the source to the destination occupies two channel uses. This leads to a loss in spectral efficiency. One protocol that recovers a significant portion of the half-duplex loss is the *two-way relaying protocol* proposed in [193]. Observing that the information flows in the real world are often bidirectional, this relaying protocol considers a common scenario that both stations want to transmit data to each other via a relay station. This scenario can be considered as, e.g., the base station communicates with a mobile user via a dedicated relay in a cellular system as shown in Fig. 1.1, or two mobile clients exchange data via the access point in a WLAN. The idea of the two-way relaying protocol is to combine the "uplink" and "downlink" data transmission together by ingeniously mixing the data from different sources at the relay, instead of simply forwarding the received data from one source at a time. While traditional relaying protocols require four phases (in time or frequency) to achieve bidirectional communication between the two stations, the two-way relaying protocol only needs two phases, namely, the multiple access (MAC) phase and the broadcast (BRC) phase.

From all that has been presented above, the combination of MIMO transmission technologies and relay communication will play an important role in the next-generation wireless communication systems. This dissertation is devoted to the analysis and design in MIMO relaying systems, and we especially focus on the MIMO two-way relaying systems. We present the analysis and results from information theoretic aspects on the one hand, which serve as the fundamental performance limits for the considered system. On the other hand, we present practical transmission strategies, which are crucial for the real-world implementation of the system.

1.2 Outline of Dissertation

This dissertation studies MIMO relay transmission technologies in modern wireless communication. We focus on the analysis of the fundamental performance limits and the design of practical transmission strategies for the systems employing a spectrally efficient relaying protocol – the two-way relaying protocol. Fig. 1.2 shows the organization of this dissertation. The outline of each chapter in the dissertation is as follows.

Chapter 1 summarizes the motivation, outline and contributions of the dissertation.

Chapter 2 reviews the state of the art in modern wireless communication technologies. This chapter is divided into three parts: MIMO communications, relay communications and two-way relay communications. We present the basic system models and the important transmission schemes for the communication systems. Recent research developments and important achievements are summarized. For relay communication and two-way relay communication, we present the transmission scenarios and focus on the commonly used relaying strategies.

Chapter 3 presents the coverage analysis for MIMO decode-and-forward (DF) relaying systems. We define the coverage range of cellular relaying networks according to the quality of service (QoS) criterion based on the information theoretic outage capacity. In order to quantitatively determine the advantages of incorporating MIMO and relay communication techniques into cellular mobile systems, we present an analytical framework for analyzing the coverage range of MIMO DF relaying networks for its given system setup. Furthermore, the analytical results are verified by simulations based on the IST WINNER channel model. Both the analysis and the simulations show that the coverage of cellular mobile communication systems can be significantly enhanced by introducing multi-antenna and relay communication.

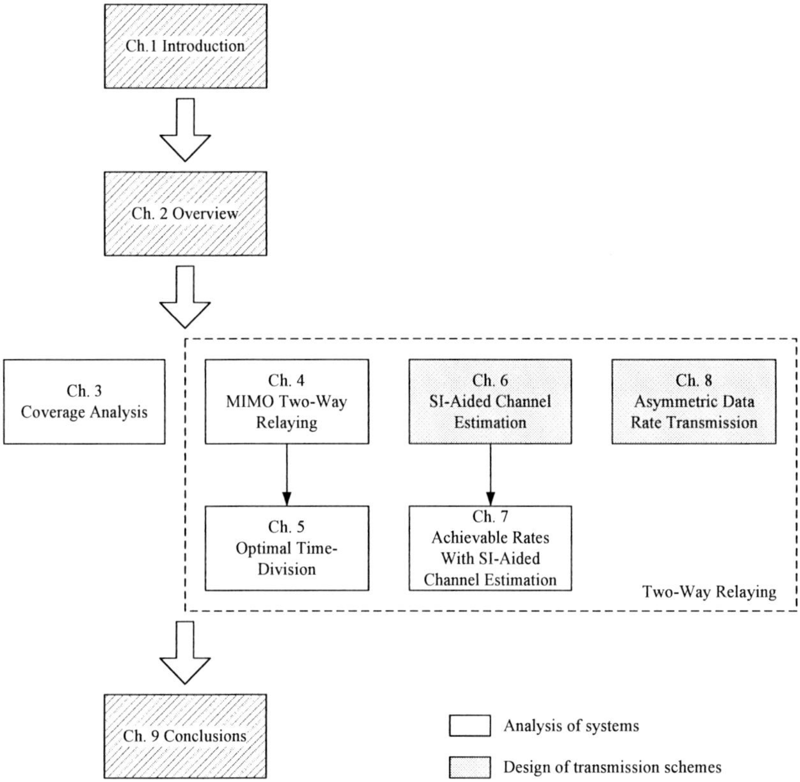

Fig. 1.2: Organization of this dissertation

Chapter 4 presents the system models and data transmission strategies for MIMO two-way relaying systems. We discuss the fundamental transmit-receive signal models and the basic relaying strategies in MIMO two-way relaying systems. The two practical data combining schemes in two-way DF relay systems, i.e., the superposition coding (SPC) scheme and the network coding scheme, are studied in detail. When the channel state information at the transmitter (CSIT) is available at the relay for the BRC phase of MIMO two-way DF relaying systems, we propose methods for calculating the achievable sum rates when the two data combining schemes are applied. Both schemes show significant advantages over the traditional one-way relaying protocol.

Chapter 5 discusses the optimum time-division (TD) strategies between the MAC and the BRC phases in MIMO two-way DF relaying systems. We show that the achievable rate region can be greatly improved by optimally allocating the spectral-temporal resources on the two transmission phases when the two-way DF relaying protocol is applied. We propose the methods to find the optimum TD strategies for the given channel, considering both peak and average power constraints. Furthermore, the achievable ergodic sum rates of the system and the average achievable user rates subject to QoS requirements are compared for different antenna configurations using the equal TD and the optimum TD strategies.

Chapter 6 proposes a novel and practical channel estimation approach for the BRC phase of MIMO two-way DF relaying systems when the SPC scheme is applied. The proposed approach exploits the self-interference (SI), which contains the known data symbols at the receivers, to estimate the BRC phase channel. The proposed SI-aided channel estimation can be applied without pilots, which achieves higher bandwidth efficiency, or the SI can also be used together with pilots to offer superior channel estimation performances than schemes that are purely based on pilots. We propose the structure of SI-aided iterative receivers and compare its performances with that of pilot-aided iterative receivers. Even when the channel estimates are solely based on the SI, the proposed receiver can still significantly outperform the pilot-aided receiver in realistic scenarios.

Chapter 7 quantifies the spectral efficiency improvement of the BRC phase channel in MIMO two-way DF relaying systems from the information theoretic perspective when the SI-aided channel estimation scheme is applied. We consider block-fading channels in the BRC phase and derive the achievable rates for systems employing the SI-aided and the pilot-aided channel estimation schemes. The idea is to consider the time slots of a coherence interval with different channel estimation qualities to be one use of a set of parallel channels with different SNRs. The spectral efficiency improvement for systems employing the SI-aided channel estimation scheme is quantified by comparing its achievable rates to that of the traditional pilot-aided channel estimation scheme.

Chapter 8 proposes a novel and practical transmission scheme for the BRC phase of two-way relaying systems when the network coding scheme is applied. The proposed scheme exploits the bit-level SI and is able to transmit with asymmetric data rates to the receivers according to their individual link qualities. The idea is that the weaker link receiver exploits the *a priori* bit information in each transmit symbol, so that it only needs to decode on a subset of the transmit symbol constellation. Subject to the same bit error rate (BER) constraint, the weaker link receiver can decode at lower SNR compared to the stronger link. The signal labeling used for mapping bits to symbols at the relay is shown to be crucial for the performance at the receivers, and we provide the criteria and methods for finding the optimized labeling schemes. Simulations show that the proposed transmission scheme can be applied to practical scenarios with asymmetric channel qualities, and the optimized labeling greatly outperforms conventional ones at both receivers.

Chapter 9 concludes the dissertation summarizing the main results and presents topics for future research.

1.3 Research Contributions

The main contributions of this dissertation are the information theoretic analysis and the proposal of practical transmission schemes for MIMO relaying systems, especially for MIMO two-way relaying systems. In the following, a detailed list of the research contributions in each chapter is presented.

Chapter 3

The main contributions of this chapter are the novel analytical framework for analyzing the coverage of cellular DF relaying networks with multiple antennas. Part of those results has been published in one conference paper.

- J. Zhao, I. Hammerström, M. Kuhn, A. Wittneben, M. Herdin, and G. Bauch, "Coverage analysis for cellular systems with multiple antennas using decode-and-forward relays," *IEEE Vehicular Technology Conference (VTC-Spring'07)*, Dublin, Ireland, April 22–25, 2007.

Chapter 4

The main contributions of this chapter are the analysis and comparisons of the relaying schemes and the data combining schemes from information theoretic perspectives in MIMO two-way relaying systems. In particular, we characterize the optimal relay transmit covariance matrices when the BRC phase CSIT is available at the relay for MIMO two-way DF relaying systems. Part of those results has been published in one conference paper.

- I. Hammerström, M. Kuhn, C. Esli, J. Zhao, A. Wittneben, and G. Bauch, "MIMO two-way relaying with transmit CSI at the relay," *Proc. IEEE International Workshop on Signal Processing Advances in Wireless Communications (SPAWC'07)*, Helsinki, Finland, Jun. 17–20, 2007.

Chapter 5

The main contributions of this chapter are the characterization of the optimal TD strategies in MIMO two-way DF relaying systems using convex optimization methods. The optimal transmit covariance matrices at the MIMO user stations and the relay station under the optimal TD strategies are determined as well. Part of those results has been published in one conference paper.

- J. Zhao, M. Kuhn, A. Wittneben, and G. Bauch, "Optimum time-division in MIMO two-way decode-and-forward relaying systems," in *Proc. 42nd Annual Asilomar Conference on Signals, Systems, and Computers*, Pacific Grove, CA, Oct. 26–Oct. 29, 2008.

Chapter 6

The main contributions of this chapter are the proposal of the novel SI-aided channel estimation scheme for the BRC phase of MIMO two-way relaying systems, which is the first scheme that *utilize* SI instead of simply *canceling* it out in two-way relaying systems. Part of those results has been published in one conference paper and a patent has been granted.

- J. Zhao, M. Kuhn, A. Wittneben, and G. Bauch, "Self-interference aided channel estimation in two-way relaying systems," in *Proc. IEEE Global Communications Conference (GLOBECOM)*, New Orleans, LA, Nov. 30–Dec. 4, 2008.

- J. Zhao, M. Kuhn, A. Wittneben, and G. Bauch, "Method, apparatus and system for channel estimation in two-way relaying networks," *Patent number(s): EP2079209-A1; CN101483622-A; JP2009171576-A; US2009190634-A1; EP2079209-B1.*

Chapter 7

The main contributions of this chapter are the derivation of the achievable rates and the quantification of the spectral efficiency improvement of the BRC phase channel in MIMO two-way DF relaying systems from the information theoretic perspective when the SI-aided channel estimation scheme is applied. Those results have been published in one conference paper.

- J. Zhao, M. Kuhn, A. Wittneben, and G. Bauch, "Achievable rates of MIMO bidirectional broadcast channels with self-interference aided channel estimation," in *Proc. IEEE Wireless Communications & Networking Conference (WCNC)*, Budapest, Hungary, Apr. 5–8, 2009.

Chapter 8

The main contributions of this chapter are the proposal of a novel asymmetric data rate transmission scheme for the BRC phase of two-way DF relaying systems when the network coding scheme is applied. The proposed approach represents a new method of exploiting the bit-level SI for network coding schemes. Furthermore, the methods and results for finding the optimized signal labeling are presented. Part of those results has been published in one conference paper and a patent application has been filed.

- J. Zhao, M. Kuhn, A. Wittneben, and G. Bauch, "Asymmetric data rate transmission in two-way relaying systems with network coding," in *Proc. IEEE International Conference on Communications (ICC)*, Cape Town, South Africa, May 23–27, 2010.
- J. Zhao, M. Kuhn, A. Wittneben, and G. Bauch, "Coding and modulation techniques for two-way relaying systems with asymmetric channel quality," *Patent pending.*

Other contributions not presented in this dissertation

Some research works not directly related to this dissertation have been published in the following four conference papers.

The proposal and the analysis of two cooperative relaying transmission schemes for a low mobility cellular relaying system downlink where two mobile users are served by two neighboring DF relays has been published in one conference paper:

- J. Zhao, M. Kuhn, A. Wittneben, and G. Bauch, "Cooperative transmission schemes for decode-and-forward relaying," *Proc. 18th Annual IEEE International Symposium on Personal, Indoor and Mobile Radio Communications (PIMRC'07)*, Athens, Greece, Sept. 3-7, 2007.

The summary of the state of the art and open issues in cellular relaying networks has been published in one conference paper:

- J. Zhao and A. Wittneben, "Cellular relaying networks: state of the art and open issues," *Proc. 2nd COST 289 Workshop*, Kemer, Antalya, Turkey, July 6–8, 2005.

Two joint papers on joint cooperative diversity and scheduling have been published in:

- I. Hammerström, J. Zhao, and A. Wittneben, "Temporal fairness enhanced scheduling for cooperative relaying networks in low mobility fading environments," *Proc. IEEE International Workshop on Signal Processing Advances in Wireless Communications (SPAWC'05)*, New York, NY, June 5–8, 2005.
- I. Hammerström, J. Zhao, S. Berger, and A. Wittneben, "Experimental performance evaluation of joint cooperative diversity and scheduling," *IEEE Vehicular Technology Conference (VTC-Fall'05)*, Dallas, TX, Sept. 25–28, 2005.

Chapter 2

Overview of Modern Wireless Communication Systems

Wireless communications have dramatically revolutionized the life of people in the past few decades. Modern wireless communication technology has become an integral part of our everyday life and also a driven force for the social and economical development. For example, the twelve-month wireless revenues in the United States alone have reached more than 151 billion dollars till June 2009 [50]. The demand for wireless communication has evolved from the original low data rate voice transmission to recent high data rate multimedia services. Such demand requires modern wireless data transmission to be *fast, stable and ubiquitous*, and propelled the invention of new wireless transmission techniques. This chapter summarizes the development of modern wireless communication technologies, where we focus on the major research achievements in the physical layer (PHY) in recent years.

Wireless communication refers to the transfer of information over a distance without the use of electrical conductors. Although its origin can be traced back to the use of beacons for defensive and navigation purposes many centuries ago, e.g., in China, the first wireless

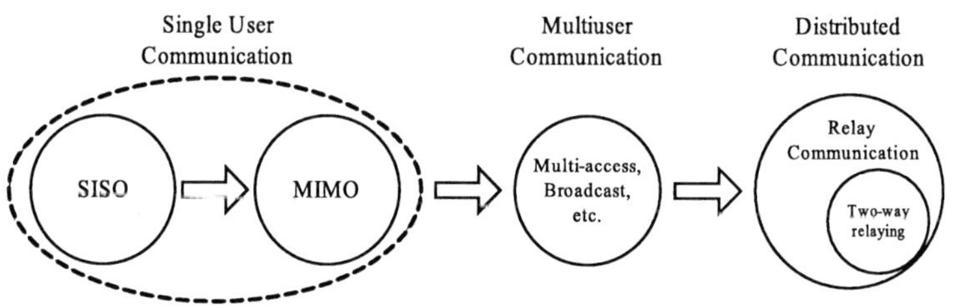

Fig. 2.1: Outline of wireless communication categories

communication system was invented by Marconi in 1895 for his radiotelegraph. Civilian wireless communication systems were established much later. For example, Motorola operated the first commercial mobile telephone service (MTS) in conjunction with the Bell System in the United States in 1946. At the early stage, the research on wireless communication focused on the implementation issues of designing real transmission systems. After the development for more than one century, wireless communication has evolved into a large research subject in electrical engineering. The transmission scenarios in wireless communications can be roughly described in Fig. 2.1. *Single user communication*, sometimes called *Point-to-point communication*, refers to the transmission scenario that one source station transmits data to one destination station. It acts as the building block for the research on larger wireless networks. When the transmitter and the receiver are each equipped with a single antenna, this communication system is called "single-input single-output" (SISO) system. "Multiple-input multiple-output" (MIMO) systems distinguishes itself from the SISO systems by equipping multiple co-located antennas at the transmitter and the receiver sides. The invention of MIMO communication systems is a significant research breakthrough in the 1990s, which improved the system performance, such as the transmission data rate and the link reliability, considerably compared to SISO systems. *Multiuser communication* refers to the communication scenario comprising multiple sources or multiple destinations. It includes the multiple-access transmission scenario, where multiple sources transmit signals to one destination, and the broadcast transmission scenario, where one source sends information to multiple destinations. Many present-day wireless systems are centralized, where one central unit is responsible for the communication with many users. For example, the base station (BS) controls the data transmission to and from multiple mobile stations (MS) in cellular systems. Multiple access transmission and broadcast transmission scenarios are important research topics for those networks. Another type of wireless systems, such as some of the mobile *ad hoc* networks, do not rely one a preexisting infrastructure and do not have centralized control. The source node relies on its neighboring nodes to forward information to the destination. We call this kind of decentralized communication pattern *distributed communication*. In distributed communication systems, relaying techniques play an important role since the information from the source relies on multihop transmission to arrive at the destination. Finally, *two-way relaying* refers to a special kind of relaying technology where bidirectional information flow is considered. In such a scenario, two or more nodes exchange information via relays. Each node acts both as a source and as a destination of information.

This chapter is organized as follows: Section 2.1 gives a briefly summarizes the important results for the MIMO communication technology. Section 2.2 discusses relay communications, which can be considered to achieve the benefits of MIMO communications in distributed fashions. After that, current developments of a spectrally efficient relaying scheme, i.e., the two-way relaying protocol, is summarized in Section 2.3.

2.1 Multi-Antenna Communication

The multi-antenna communication technique is one of the major breakthroughs in modern wireless communication technology. The available resources in wireless systems, such as the radio spectrum and the transmit power, are limited, either by regulations or by practical system constraints. The use of multiple antennas at the transmitter and/or receiver in a wireless link opens a new dimension – space. The multi-antenna communication technique can transmit higher data rate without extra consumption of bandwidth and transmit power, which is highly desirable in wireless systems. Such a technique brings fundamental improvement over traditional single antenna communication systems. Nowadays, it is at the core of the link communication techniques in many existing and future wireless standards. In literatures, *MIMO transmission* is another term used for multi-antenna transmission, which sometimes includes the case that only one side of the transmission system is equipped with multiple antennas. In this dissertation, we use the term "multi-antenna transmission" and "MIMO transmission" interchangeably.

2.1.1 Development of Multi-Antenna Communication

Before the middle of the 1990s, research on wireless communication focused on developing coding and modulation schemes for systems equipped with single transmit and receive antenna. Splendid achievements have been made, which include the discovery of capacity-achieving turbo codes [35] and low-density parity-check (LDPC) codes [70, 153]; multi-carrier modulation schemes for wideband digital communication, i.e., orthogonal frequency-division multiplexing (OFDM) [38]; and spread-spectrum technologies for serving multiple users simultaneously, i.e., code division multiple access (CDMA) [243]. However, given that the available resources in single antenna systems, such as the transmit power and the frequency band, are limited, it became increasingly difficult for single-antenna systems to support the physical transmission of high rate wireless data. The only way to satisfy the rapidly growing demand for high data rate services is to significantly improve the spectral efficiency, whereas none of the techniques above achieve this. In order to significantly improve

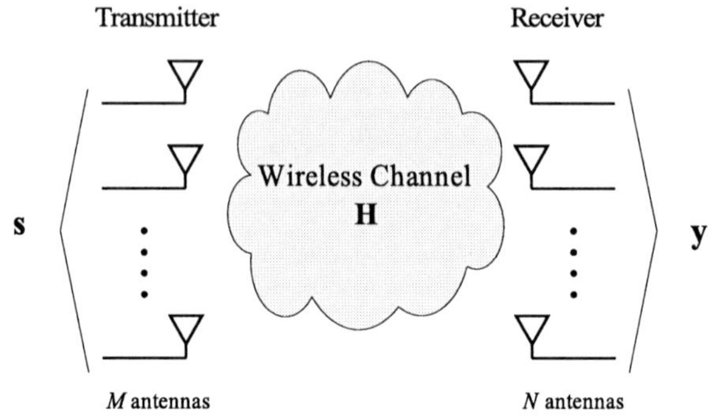

Fig. 2.2: Wireless MIMO Channel

the spectral efficiency and link reliability in wireless systems, boldly innovative techniques are called for and new frontiers must be explored.

The multi-antenna transmission technique provides such a new frontier. This technique brought a previously unexploited new dimension, i.e., space, into wireless communications. As is shown in Fig. 2.2, multi-antenna communication systems utilize multiple co-located antennas (antenna arrays) at the transmitter and the receiver side. Multiple data symbols can be transmitted or received simultaneously in the system. The use of antenna arrays has a long history and can actually be found in radar technologies during World War II. Since then, smart signal processing algorithms have been proposed for antenna arrays to identify spatial signal signatures, and to adjust the antenna array patterns by *beamforming* [133,236]. The revolution came in the 1990s. Paulraj and Kailath were among the first to realize that the wireless channel capacity can be dramatically improved by using multiple *co-located* antennas at both the transmitter and the receiver. They proposed the scheme of *spatial multiplexing* for increasing the capacity of wireless link using multiple antennas [183] in 1994. Roy and Ottersten [194] proposed to use multiple base station antennas to support multiple co-channel users in 1996. Foschini [65] proposed a new transceiver architecture, i.e., Bell Laboratories Layered Space-Time (BLAST), for transmitting multiple data streams simultaneously over the multi-antenna wireless channel. Depending on how the data are arranged on the transmit antenna arrays, the BLAST architecture consists of two variations, i.e., the vertical-BLAST (V-BLAST) architecture [259] and the diagonal-BLAST (D-BLAST) architecture [65]. In their subsequent laboratory experiments, the BLAST architecture testbed demonstrated unprecedented spectral efficiencies [76, 259]. At about the same time, fundamental research on the multi-antenna communication theory by Telatar and Foschini [66, 224] showed that the potential gains of multi-antenna systems over single-antenna systems can be rather large. Under the assumption of independent channel entries, e.g., in rich scattering environment and when each antenna is spaced sufficiently far apart, the capacity of the multi-antenna system scales linearly with the minimum number of the transmit and receive antennas. Especially, the boost in channel capacity comes at no cost in frequency bandwidth and transmit power, which provides an ideal solution for achieving higher data rates in wireless communication systems.

Except for boosting the spectral efficiency, another important benefit of the multi-antenna transmission technique is the enhancement of link quality by using space-time coding. In the presence of fading in wireless channels, combination of multiple replicas of the transmitted signal through independent faded paths provides a more reliable reception. Such a technique is called *diversity*. The use of multiple antennas brings *spatial diversity* into wireless communications. Using multiple antennas at the receiver side to achieve *receive diversity* has been known for a long time [34]. However, exploiting *transmit diversity* when there are multiple antennas at the transmitter side requires the signals to be smartly pre-processed before transmission. One of the first schemes that realized transmit diversity was the *delay diversity* scheme proposed by Wittneben [255], where each symbol is transmitted from different antennas after being delayed. This delay diversity scheme can be considered as a repetition coding scheme over space. Alamouti [12] proposed a simple transmission scheme for systems with two transmit antennas. By ingeniously arranging the transmit data symbols on the antenna array, the Almouti scheme transmits the two symbols on two effectively orthogonal channels, and achieves the second order diversity for systems with two transmit antennas and one receive antenna. Such a scheme was generalized to orthogonal *space-time block codes* (STBC) for more than two transmit antenna by Tarokh [220, 221]. Orthogonal STBC was proposed to achieve the diversity gain with low decoding complexity. Such code only needs simple linear processing at the receiver side. The drawback of orthogonal STBC is that it cannot achieve full rate transmission for more than two transmit antennas. Another type of space-time coding technique, *space-time trellis codes* (STTC), was proposed in [222], which can generally achieve higher transmission rates compared to orthogonal STBC. However, the improvement of transmission rate in STTC is at the cost of higher decoding complexity, where the decoding complexity of its maximum-likelihood (ML) detector is much higher than the linear detector used in orthogonal STBC. Unlike the orthogonal STBC that suffers from the rate loss due to its orthogonal design, some space-time coding schemes, such as the *linear complex-field space-time coding* [152] was proposed to achieve full rate and full diversity in multi-antenna transmission systems. *Linear dispersion codes* [96, 102] by Hassibi *et al.* was proposed to subsume many earlier space-time transmission techniques, such as the V-BLAST architecture [259] and the Alamouti scheme. Linear dispersion codes were designed to be be simple to encode and decode while achieving little penalty in the mutual information. The space-time coding schemes above do not require the channel knowledge to be available at the transmitter side but still require the channel knowledge at the receiver side. When there is no channel knowledge available at either side of the multi-antenna transmission system, differential space-time codes [106, 111] were proposed for the noncoherent communication.

Multi-antenna systems can achieve spatial multiplexing gain and spatial diversity gain simultaneously. However, there exists a fundamental tradeoff between the two types of gains. This *diversity-multiplexing tradeoff* (DMT) was discovered by Zheng and Tse [281], where they showed that the spatial multiplexing gain and spatial diversity gain in a multi-antenna system cannot be increased simultaneously. The search for the space-time codes that achieve the optimal DMT has been an active area of research. Space-time codes that achieve the DMT were proposed and discussed in [61,223,229,268]. Yao *et al.* [268] proposed rotation-based codes to achieve the DMT for MIMO channels with two transmit and two receive antennas; the permutation codes for achieving the DMT for parallel channels were proposed in [223] and codes based on cyclic divisional algebras were proposed in [61]. Those codes were shown to be DMT-optimal by verifying the *approximate universality conditions* [229].

Multiuser transmission is especially useful for cellular networks. The Gaussian MIMO multiple-access channel capacity region was considered by Yu [271]. The capacity region of the Gaussian MIMO broadcast channel was characterized by Weingarten *et al.* [250], where it was proved to be achievable by the *dirty paper coding* (DPC) scheme [47]. In real-world implementations, multi-antenna systems can be used with most state-of-the-art modulation schemes. The combination of multi-antenna system and OFDM techniques can be found in, e.g., [27, 145]. Good books on multi-antenna transmission technology are abundant, e.g., [24, 184].

2.1.2 Benefits of Multi-Antenna Communication Systems

Multi-antenna communication systems provide a series of advantages over traditional single-antenna communication systems, such as the *array gain*, the *diversity gain* and the *multiplexing gain*. The array gain and the diversity gain improve the *link reliability* and can be obtained in single-input multiple-output (SIMO) and multiple-input single-output (MISO) systems, as well as in MIMO systems. However, the multiplexing gain improves the *spectral efficiency* and it is unique in MIMO systems, i.e., it requires both the transmitter and receiver sides to be equipped with multiple antennas.

The gains in multi-antenna communication systems are summarized as follows:

Array gain Array gain is the improvement in the receive signal-to-noise ratio (SNR) obtained by coherently combining the signals on multiple transmit or receive antennas [30]. It is traditionally associated with the array processing techniques in smart antennas. When the coherent combining is realized at the transmit antenna arrays, i.e., for MISO systems, the channel knowledge is usually required to be available at the transmitter side to make spatial pre-processing of the transmit signals possible. The array gain improves the resistance of the system to noise, and boosts the link reliability.

Diversity gain The diversity technique mitigates the fading effects by transmitting the same data symbol over multiple independently faded branches. With those independent replicas of the same information signal, the probability that at least one of the copies is not in deep fade increases. Diversity can be realized in time (*temporal diversity*), in frequency (*frequency diversity*) or in space (*spatial diversity*). Multi-antenna systems are ideal for realized spatial diversity, which include transmit diversity and receive diversity. Receive diversity simply requires that the received signals on different receive antennas experience independent fades. Transmit diversity is more difficult to be realized since it needs more intricate coding and modulation schemes to be applied at the transmitter side. The research on transmit diversity leads to the development of the space-time coding schemes. The diversity techniques combat channel fluctuation and thus improve link reliability.

Multiplexing gain When multiple antennas are placed at both sides of the wireless radio link, multiple independent data streams can be transmitted through the parallel sub-channels established within the MIMO channel, which leads to an increase in the channel capacity. Such an improvement in channel capacity is called *spatial multiplexing gain*. It is achieved without additional consumption of power or bandwidth. However, unlike the array gain and the diversity gain, the spatial multiplexing gain requires that both the transmitter and receiver to be equipped with multiple co-located antennas. It also requires certain channel conditions, e.g., rich scattering environment, so that the receiver can separate the multiple data streams. The multiplexing gain provides higher data rate for the MIMO system, i.e., it improves the spectral efficiency of the system.

2.1.2.1 Relation Between Different Gains in MIMO Systems

Array and Diversity Gains The array gain results from the coherent combining of multiple replicas of the same signal, and it boosts the effective SNR of the received signal. The array gain only depends on the channel realization. However, diversity gain requires each branch of the transmit signal to fade independently, which depends on the statistical properties of the channel. The two types of gains can co-exist in a multi-antenna communication system and there is no tradeoff between them [179].

Array and Multiplexing Gains The array gain aims to maximize the SNR at the receiver. For a MIMO channel, maximizing the array gain implies allocating all the transmit power on the subchannel with the maximum eigenvalue [13], i.e., transmitting the data on only one subchannel. However, the multiplexing gain aims to maximize the mutual information in the system. For a MIMO channel with multiple subchannels, the optimum way to maximize the multiplexing gain is to transmit multiple data streams with the power allocation according to the waterfilling algorithm [49].

Diversity and Multiplexing Gains The diversity gain is related to the link reliability and the multiplexing gain is related to the spectral efficiency. Both gains can be simultaneously achieved in a MIMO system. However, there exists a fundamental tradeoff between the two types of gain [281], which means that the two types of gains cannot be simultaneously *increased*. The multiplexing gain r and the diversity gain d are achievable for a coding scheme if the data rate R at high SNR ρ satisfies

$$\lim_{\rho\to\infty} \frac{R(\rho)}{\log\rho} = r \tag{2.1}$$

and the average error probability P_e at high SNR ρ satisfies

$$\lim_{\rho\to\infty} \frac{\log P_e(\rho)}{\log\rho} = -d. \tag{2.2}$$

For each r, the optimal diversity gain $d^\star(r)$ is defined to be the supremum of the diversity advantage achieved over all schemes. For a MIMO system with M transmit and N receive antennas that experiences an i.i.d. Rayleigh slow-fading channel where the channel gain is random but remains constant for a duration of $l \geq M+N-1$ symbols, it has been shown in [281] that the optimal diversity gain $d^\star(r)$ achievable by any coding scheme of block length l for a given multiplexing gain r ($r \in \mathbb{N}$) is $d^\star(r) = (M-r)(N-r)$. For example, the diversity-multiplexing tradeoff for a MIMO system ($M = 6$, $N = 5$) in an i.i.d. Rayleigh fading channel is shown in Fig. 2.3.

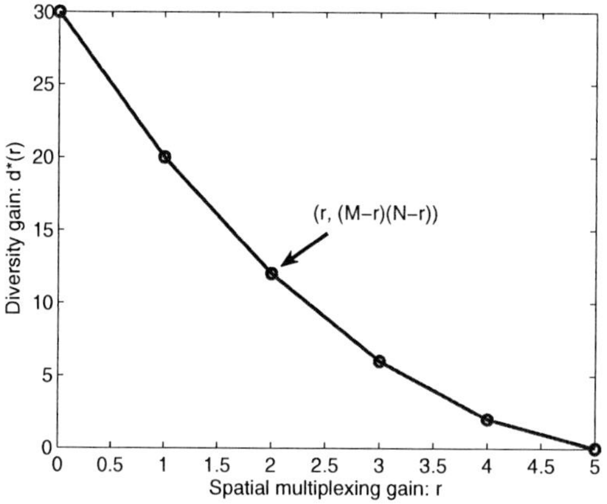

Fig. 2.3: Diversity-multiplexing tradeoff for a MIMO system (number of transmit antennas $M = 6$, number of receive antennas $N = 5$) in i.i.d. Rayleigh fading channels.

2.1.3 Single-User MIMO Channel

For a frequency flat multi-antenna transmission system with M transmit antennas and N receive antennas as shown in Fig. 2.2, the discrete-time signal model at time instance k can be expressed as

$$\mathbf{y}[k] = \mathbf{H}[k]\mathbf{s}[k] + \mathbf{n}[k] \tag{2.3}$$

$$= \sqrt{\frac{P}{M}}\mathbf{H}[k]\tilde{\mathbf{s}}[k] + \mathbf{n}[k]. \tag{2.4}$$

Each item is explained as follows:

- P is the transmit power constraint.
- $\mathbf{s}[k] \in \mathbb{C}^{M\times 1}$ denotes the transmit symbol vector and $\tilde{\mathbf{s}}[k]$ denotes the transmit symbol vector normalized with respect to the transmit power. We introduce the transmit covariance matrix as $\boldsymbol{\Omega} = \mathrm{E}\left\{\tilde{\mathbf{s}}[k]\tilde{\mathbf{s}}[k]^H\right\}$, where the expectation is taken with respect to all the time instance. In order to satisfy the transmit power constraint, we require $\operatorname{tr}\boldsymbol{\Omega} \leq M$, and it follows $\operatorname{tr}\left(\mathrm{E}\left\{\mathbf{s}[k]\mathbf{s}[k]^H\right\}\right) \leq P$.
- $\mathbf{H}[k] \in \mathbb{C}^{N\times M}$ is the channel matrix, where the (m, n)th entry in the channel matrix $\mathbf{H}[k]$ corresponds to the channel from the transmit antenna m to the receive antenna n.
- $\mathbf{y}[k] \in \mathbb{C}^{N\times 1}$ denotes the received symbol vector.

- $\mathbf{n}[k] \sim \mathcal{CN}(0, \sigma^2 \mathbf{I}_N)$ is the additive white Gaussian noise (AWGN) vector at the receiver, where σ^2 is the noise variance.

For simplicity, we will omit the discrete-time indices k without causing confusion.

Many factors impact the properties of the MIMO channel $\mathbf{H}$. Different characterizations of MIMO channels can be found in, e.g., [184]. However, one of the most widely used channel models is to assume the entries in the channel matrix $\mathbf{H}$ to be uncorrelated with each other. This is satisfied in a rich scattering environment and when the neighboring antennas in the transmit or receive antenna arrays are spaced more than the coherence distance apart. Typical coherence distance for antenna decorrelation is $\lambda/2$, where λ is the wavelength corresponds to the operation frequency.

When neither side of the transmitter and the receiver has the channel knowledge, i.e., in the noncoherent case, the capacity of the MIMO channel is still an open problem [32, 280]. For coherent communications, the channel state information (CSI) is usually required to be available at least at the receiver side in a MIMO system. Channel estimation in MIMO systems is usually performed by transmitting orthogonal training sequences from different transmit antennas. The channel state information at the transmitter (CSIT) is more difficult to obtain. it is usually acquired by channel knowledge feedback from the receiver in the frequency-division duplexing (FDD) transmission mode or assuming channel reciprocity in time-division duplexing (TDD) transmission mode. Assuming perfect CSI at the receiver but no CSIT, the channel capacity of a deterministic MIMO channel is given by

$$C = \log\det\left(\mathbf{I}_N + \frac{P}{M\sigma^2}\mathbf{H}\mathbf{H}^H\right). \tag{2.5}$$

where the capacity is achieved by transmitting M statistically independent and equal power data streams with Gaussian codebook at the transmit antennas. In a fast fading channel where the codeword length is much larger than the channel coherence interval, the capacity of the channel is replaced by the ergodic channel capacity

$$C_{\text{erg}} = \mathrm{E}\left\{\log\det\left(\mathbf{I}_N + \frac{P}{M\sigma^2}\mathbf{H}\mathbf{H}^H\right)\right\}. \tag{2.6}$$

where the expectation is taken with respective to the realizations of channel $\mathbf{H}$. For a MIMO channel with fixed number of receive antennas N and assuming each entry of the channel $\mathbf{H}$ to be i.i.d., we have the following

$$\frac{1}{M}\mathbf{H}\mathbf{H}^H \rightarrow \mathbf{I}_N, \quad \text{as } M \rightarrow \infty, \tag{2.7}$$

according to the law of large numbers [181]. Therefore the ergodic capacity of the fast fading channel approaches [30]

$$C_{\text{erg}} \to N \log\left(1 + \frac{P}{\sigma^2}\right), \quad \text{as } M \to \infty. \tag{2.8}$$

This means that the ergodic capacity of the fast fading channel increases linearly with the number of receive antennas N for the fixed SNR when the number of transmit antennas M is asymptotically large. More precisely, if the receiver perfectly knows the channel, it was shown in [66, 224] that the ergodic capacity of the i.i.d. Rayleigh fading MIMO channel scales like $\min(M, N) \log(P/\sigma^2)$ no matter the transmitter knows the channel or not. Compared to the SISO channel with the same SNR, i.e., P/σ^2, the ergodic capacity of the MIMO channel is about $\min(M, N)$ times higher in high SNR regime.

When the channel $\mathbf{H}$ is perfectly known to both the transmitter and the receiver, the capacity of a deterministic MIMO channel can be expressed as

$$C_{\text{CSIT}} = \max_{\text{tr}(\boldsymbol{\Omega}) \leq M, \boldsymbol{\Omega} \succeq 0} \log\det\left(\mathbf{I_N} + \frac{P}{M\sigma^2}\mathbf{H}\boldsymbol{\Omega}\mathbf{H}^H\right). \tag{2.9}$$

where $\boldsymbol{\Omega} \succeq 0$ means that $\boldsymbol{\Omega}$ is a positive semidefinite matrix. In this case, the MIMO channel capacity in (2.9) can be calculated as follows. Using singular value decomposition (SVD) method, the channel matrix can be written as

$$\mathbf{H} = \mathbf{U}\boldsymbol{\Sigma}\mathbf{V}^H, \tag{2.10}$$

where $\mathbf{U} \in \mathbb{C}^{N \times N}$ and $\mathbf{V} \in \mathbb{C}^{M \times M}$ are unitary matrices and $\boldsymbol{\Sigma} \in \mathbb{C}^{N \times M}$ is a diagonal matrix whose diagonal elements are made of the singular values of $\mathbf{H}$. We denote the rank of the channel matrix $\mathbf{H}$ as r, where $r \leq \min(M, N)$. The singular values of $\mathbf{H}$ are denoted as $\sqrt{\lambda_i}$, $i = 1, 2, \ldots, r$. So the MIMO channel channel can be decoupled into r parallel single-input single-output (SISO) channels, and the channel capacity can be calculated as

$$C_{\text{CSIT}} = \sum_{i=1}^{r} \log\left(1 + \frac{P\gamma_i}{M\sigma^2}\lambda_i\right). \tag{2.11}$$

γ_i reflects the transmit power allocation in the ith subchannel. The optimum power allocation in (2.11) can be calculated using the *waterfilling algorithm* [49], which is given by

$$\gamma_i = \left(\mu - \frac{M\sigma^2}{P\lambda_i}\right)^+, \quad \text{for } i = 1, 2, \ldots, r \tag{2.12}$$

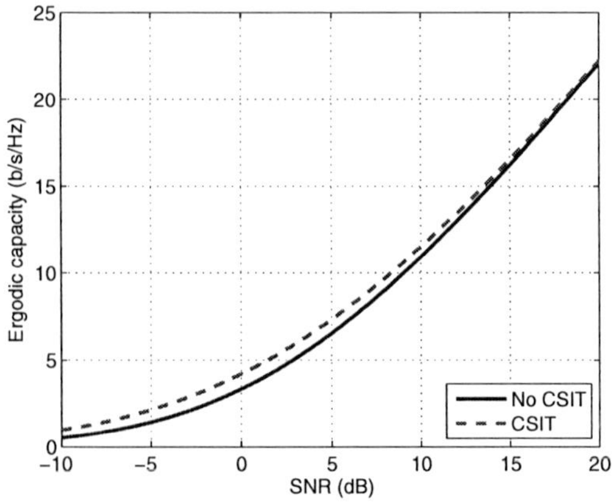

Fig. 2.4: Ergodic Capacity for $M = N = 4$ MIMO channels with and without CSIT.

where the function $(\cdot)^+$ is defined as

$$(x)^+ = \begin{cases} x & \text{if } x \geq 0 \\ 0 & \text{if } x < 0. \end{cases} \tag{2.13}$$

μ is a constant that denotes the "waterlevel" so that the power allocation satisfies

$$\sum_{i=1}^{r} \gamma_i = M. \tag{2.14}$$

Compared to the channel capacity without CSIT C in (2.5), the capacity gain of the perfect CSIT case in (2.11) is significant in low SNR regime but only marginal in high SNR regime. The ergodic capacity with and without CSIT for a MIMO system with $M = N = 4$ antennas in i.i.d. Rayleigh fading channels is shown in Fig. 2.4. With only average power constraints, a two-dimensional water-filling in both the temporal and spatial domains has been shown to be optimal in fast fading MIMO channels if the channel knowledge is available at both the transmitter and receiver sides [119, 211].

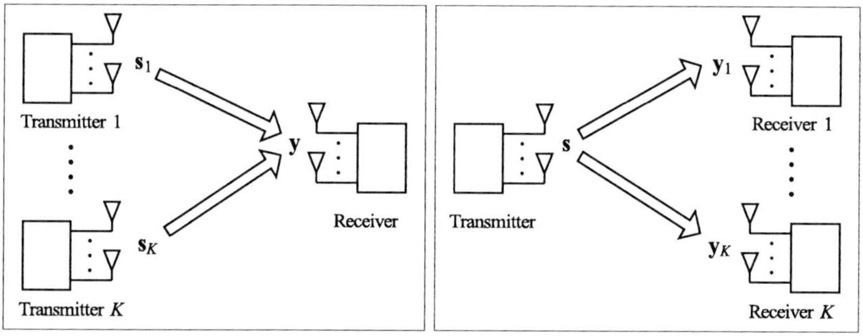

(a) MIMO multiple access channel (b) MIMO broadcast channel

Fig. 2.5: MIMO multiple access channel and MIMO broadcast channel.

2.1.4 Multiuser MIMO Channel

Multiuser communication refers to the scenario that multiple users simultaneously transmit (or receive) signals in the same frequency channel. It includes the multiple access scenario (many-to-one) and the broadcast scenario (one-to-many). The multiple access channel and the broadcast channel can respectively model the downlink and uplink transmission in cellular mobile communication systems. When some of the stations in the multiuser communication systems are equipped with multiple antennas, this channel model is called multiuser MIMO channel [77,78,150].

The MIMO multiple access channel is shown in Fig. 2.5(a). For a multiple access channel with multiple antennas, the input-output relation of the system can be described as

$$\mathbf{y} = \sum_{k=1}^{K} \mathbf{H}_k \mathbf{s}_k + \mathbf{n} \tag{2.15}$$

where the kth user is equipped with $M_{\mathrm{T}}^{(k)}$ antennas and the receiver has M_{R} antennas. $\mathbf{H}_k \in \mathbb{C}^{M_{\mathrm{R}} \times M_{\mathrm{T}}^{(k)}}$ denotes the channel from kth transmitter to the receiver. $\mathbf{s}_k \in \mathbb{C}^{M_{\mathrm{T}}^{(k)} \times 1}$ is the transmit data symbol vector at user k, and $\mathbf{Q}_k = \mathrm{E}\left\{\mathbf{s}_k \mathbf{s}_k^H\right\}$ denotes the transmit covariance matrix at the kth user without normalization with respect to the transmit power. We require $\mathrm{tr}(\mathbf{Q}_k) \leq P_k$ to satisfy the transmit power constraint P_k at the kth user. $\mathbf{n} \sim \mathcal{CN}(0, \sigma^2 \mathbf{I}_{M_{\mathrm{R}}})$ is the AWGN vector at the receiver. Furthermore, we define $\mathbf{H} = [\mathbf{H}_1, \cdots, \mathbf{H}_K]$.

Assuming the channel $\mathbf{H}$ to be static and perfectly known at both the transmitters and the receiver, the capacity region of the Gaussian multiple access channel for the given transmit power constraint $P_1, \cdots, P_K$ can be expressed as [224, 271]

$$\mathcal{C}_{\text{MA}}(P_1, \cdots, P_K; \mathbf{H}) = \bigcup_{\mathbf{Q}_k \succeq 0, \text{tr}(\mathbf{Q}_k) \leq P_k}$$

$$\left\{ (R_1, \cdots, R_k) \mid \sum_{k \in S} R_k \leq \log\det\left(\mathbf{I}_{M_\text{R}} + \sum_{k \in S} \mathbf{H}_k \mathbf{Q}_k \mathbf{H}_k^H \right), \forall S \subseteq \{1, \cdots, K\} \right\} \quad (2.16)$$

The corner points of the capacity region can be achieved by successive decoding, i.e., the receiver successively decodes users' signals and subtracts the decoded signal from the received signal. The MIMO multiple access channel capacity region in (2.16) is convex can be calculated by using convex optimization methods.

A K-user broadcast channel is shown in Fig. 2.5(b). The transmitter has M_T antennas and the kth receiver is equipped with $M_\text{R}^{(k)}$ antennas. Denoting the channel matrix between the transmitter and the user k as $\mathbf{G}_k \in \mathbb{C}^{M_\text{R}^{(k)} \times M_\text{T}}$, the received signal at the kth user is

$$\mathbf{y}_k = \mathbf{G}_k \mathbf{s} + \mathbf{n}_k \quad (2.17)$$

where $\mathbf{s} \in \mathbb{C}^{M_\text{T} \times 1}$ is the transmit vector at the transmitter, and $\text{tr}\left(\text{E}\left\{\mathbf{s}\mathbf{s}^H\right\}\right) \leq P$ satisfies the transmit power constraint. $\mathbf{n}_k \sim \mathcal{CN}(0, \mathbf{I}_{M_\text{R}^{(k)}})$ is the AWGN at the receiver k. Let $\pi(\cdot)$ denote a permutation of the user indices $1, \cdots, K$ and let $\mathbf{Q} = [\mathbf{Q}_1, \cdots, \mathbf{Q}_K]$ denote a set of positive semidefinite covariance matrices with $\text{tr}(\mathbf{Q}_1 + \cdots + \mathbf{Q}_K) \leq P$. The capacity region of the broadcast channel has been shown by Weingarten *et al.* [250] to be achievable by the DPC scheme [47]. For the given permutation π and the set of covariance matrices $\mathbf{Q}$, the following set of rate vectors is achievable

$$\mathbf{R}(\pi, \mathbf{Q}) = (R_{\pi(1)}, \cdots, R_{\pi(K)}) \quad (2.18)$$

where

$$R_{\pi(k)} = \log \frac{\det\left[\mathbf{I} + \mathbf{G}_{\pi(k)} (\sum_{j \geq k} \mathbf{Q}_{\pi(j)}) \mathbf{G}_{\pi(k)}^H\right]}{\det\left[\mathbf{I} + \mathbf{G}_{\pi(k)} (\sum_{j > k} \mathbf{Q}_{\pi(j)}) \mathbf{G}_{\pi(k)}^H\right]}, \quad k = 1, \cdots, K. \quad (2.19)$$

We define $\mathbf{G} = [\mathbf{G}_1^T, \cdots, \mathbf{G}_K^T]^T$. Assuming the channel to be deterministic and perfect channel knowledge is available at both the transmitter and the receivers, the capacity region [250] $\mathcal{C}_{\text{BC}}$ is the convex hull of the union of all such rate vectors over all permutations and all positive semidefinite covariance matrices satisfying the average power constraint P, i.e.,

$$\mathcal{C}_{\mathrm{BC}}(P, \mathbf{G}) = \mathrm{conv}\left(\bigcup_{\pi, \mathbf{Q}} \mathbf{R}(\pi, \mathbf{Q})\right). \tag{2.20}$$

The transmit signal vector is $\mathrm{s} = \mathrm{s}_1 + \cdots + \mathrm{s}_K$ using Gaussian codebooks, and the input covariance matrices are of the form $\mathbf{Q}_k = \mathrm{E}(\mathrm{s}_k \mathrm{s}_k^H)$. The DPC scheme implies that $\mathrm{s}_1, \cdots, \mathrm{s}_K$ are uncorrelated, and thus $\mathrm{tr}(\mathbf{Q}) = \mathrm{tr}(\mathbf{Q}_1 + \cdots, \mathbf{Q}_K) \leq P$. The calculation of the MIMO broadcast channel capacity region is a non-convex optimization problem due to the rate region expression (2.19) and the necessity of considering the ordering $\pi(\cdot)$ of all different users.

The capacity region of the MIMO broadcast channel in (2.20) cannot be directly solved by convex optimization methods and it is much more difficult to be characterized than the MIMO multiple access channel capacity region (2.16). Fortunately, there exists a duality between the two capacity regions [241], which greatly simplifies the problem. The duality between the multiple access channel and the broadcast channel states that: the MIMO broadcast channel capacity region $\mathcal{C}_{\mathrm{BC}}(P, \mathbf{G})$ with power constraint P and channel gain matrix $\mathbf{G}$ is equal to the union of the capacity regions of the multiple access channel with channel gain matrix $\mathbf{G}^H$, where the union is taken over all individual power constraints that sum to P. That is, we have

$$\mathcal{C}_{\mathrm{BC}}(P; \mathbf{G}) = \bigcup_{(P_1, \cdots, P_K): \sum_{k=1}^{K} P_k = P} \mathcal{C}_{\mathrm{MA}}(P_1, \cdots, P_K; \mathbf{G}^H) \tag{2.21}$$

where $\mathcal{C}_{\mathrm{BC}}(P; \mathbf{G})$ and $\mathcal{C}_{\mathrm{MA}}(P_1, \cdots, P_K; \mathbf{G}^H)$ are given in (2.20) and (2.16), respectively. The duality relation can convert the problem of determining the broadcast channel capacity region into the problem of determining its dual multiple access channel capacity region, which can be solved efficiently using convex optimization methods.

For a MIMO Gaussian broadcast channel with M transmit antennas and K users each with N antennas, the sum rate of the system scales like $M \log\log(KN)$ for large K if perfect channel knowledge is available at the transmitter (CSI is required at the receivers for decoding anyway) [210]. However, if the transmitter does not have the CSIT, the sum rate of the system in high SNR regime is approximately $\min(M, N) \log(\mathrm{SNR})$, which is essentially the single-user capacity achievable in high SNR regime [75]. For systems with a large number of users, obtaining full channel knowledge at the transmitter may be infeasible. One solution that only requires limited feedback from the users is proposed by Sharif and Hassibi in [210]. Their scheme is to construct M random orthogonal beams and transmit to users with the highest signal-to-interference-plus-noise ratios (SINRs), which dramatically reduces the feedback overhead of the full channel knowledge to only that of the SINRs. Furthermore, the authors showed that such a scheme achieves the same sum rate scaling as the perfect CSIT case for fixed M and large K.

2.1.5 Multi-Antenna Communication Schemes

2.1.5.1 Single-User Transmission Schemes Without CSIT

Even with no CSIT, multi-antenna communication systems can achieve the multiplexing and diversity gains. Not only is this promised by the theoretical results such as capacity, but also has it been demonstrated by practical communication schemes.

Spatial multiplexing architectures were proposed to achieve the multiplexing gain. It requires both the transmitter and the receiver to be equipped with multiple antennas and the multi-dimensional signals are transmitted simultaneously on different spatial substreams. The most famous spatial multiplexing architecture is the BLAST architecture. The original diagonal layered space-time architecture was proposed in [65] and was later referred to as D-BLAST. In D-BLAST, the coding sequence was dispersed across the diagonals in space and time. Such a diagonally layered coding scheme can achieve the full MN diversity promised by the MIMO channel with M transmit antennas and N receive antennas if combined with Gaussian codebooks and infinite block length. It can achieve the spatial multiplexing gain as well. However, the receiver structure of the D-BLAST architecture is complex for practical applications. In [67, 259], a less complex architecture, V-BLAST, was proposed. In the V-BLAST architecture, the transmitter simply demultiplexes independent data substreams onto the transmit antennas. No inter-substream coding is applied as done in the D-BLAST architecture. The V-BLAST architecture does not achieve the full diversity gain offered by the MIMO channel. However, the spatial multiplexing gain provided by the MIMO channel can be achieved with lower implementation complexity compared to the D-BLAST architecture.

Although the V-BLAST transmitter only needs to send independent data symbols on its transmit antennas, its receiver structure may have different complexity and may lead to different performance. The V-BLAST detector structures can be sorted as follows in the order of increasing complexity and usually better performance.

Linear detectors In MIMO communication systems, the main problem for the receiver is how to detect symbols in the presence of multi-stream interference. Linear detectors are desirable due to its low complexity. However, for spatial multiplexing schemes like V-BLAST, linear detection usually leads to suboptimal performance. Linear detectors consist of the following two types:

- Zero-forcing (ZF) detector: If the channel matrix in (2.4) is nonsingular, the ZF detector that separates the multistream from the received signal is given by

$$\hat{\mathbf{s}}_{\mathrm{ZF}} = \mathbf{H}^{\dagger}\mathbf{y} \tag{2.22}$$

  where $\mathbf{H}^{\dagger}$ denotes the pseudo-inverse of the channel matrix $\mathbf{H}$. The ZF detector is simple, yet the noise enhancement causes large performance degradation compared to other detectors [82, 154, 253].

- Minimum mean square error (MMSE) detector: The MMSE detector tries to balance the multi-stream interference mitigation and noise enhancement by minimizing the total detection errors $\mathrm{E}\{\|\hat{\mathbf{s}} - \mathbf{s}\|^2\}$, where $\hat{\mathbf{s}}$ denotes the estimated symbol using linear operations. The MMSE detector is given by $\mathbf{G}_{\mathrm{MMSE}} = \arg\min_{\mathbf{G}} \mathrm{E}\{\|\mathbf{G}\mathbf{y} - \mathbf{s}\|^2\}$, and the detected data symbol can be expressed as

$$\hat{\mathbf{s}}_{\mathrm{MMSE}} = \frac{P}{M}\mathbf{H}^H\left(\sigma^2\mathbf{I}_N + \frac{P}{M}\mathbf{H}\mathbf{H}^H\right)^{-1}\mathbf{y} \tag{2.23}$$

  Compared to the ZF detector, the MMSE detector needs to estimate the noise variance σ^2 at the receiver. The MMSE detector has much better performance compared to the ZF detector in the low SNR regime [1]. In the high SNR regime, its performance approaches that of the ZF detector.

Successive cancellation detector The successive cancellation detector can be considered as a kind of decision-feedback detector, which has been proposed to be used in the BLAST architecture [76, 259]. Such detector starts decoding from the stream with the highest SINR using the ZF or MMSE algorithm. After a stream of symbols is decoded, it is subtracted from the received signals. The advantage comes from the inherent *selection diversity* in the ordered successive cancellation. An upper bound on the performance of such a detector is provided in [81].

Maximum-likelihood (ML) detector Assuming equally likely transmitted symbols, the optimum detector for the transmit symbol $\mathbf{s}$ in (2.4) is the ML detector, which is given by $\hat{\mathbf{s}} = \arg\min_{\mathbf{s}} \|\mathbf{y} - \mathbf{H}\mathbf{s}\|_F^2$. However, the decoding complexity of the ML detector grows exponentially with the number of transmit antennas M. Sphere decoding algorithm [54, 244] was proposed to reduce the decoding complexity of ML detectors, which has worst-case exponential decoding complexity but expected polynomial decoding complexity. Soft-output sphere decoding algorithms generate not only the symbol decision but also the metrics. Its implementation can be found in [217].

Iterative detector Iterative MIMO detectors are usually combined with channel coding and use soft-input soft-output decoding algorithms [14]. In each iteration, the estimate of the data bits are refined [107, 202].

Space-time coding was proposed to achieve the transmit diversity gain of the channel. No channel knowledge is required at the transmitter. Redundancy is introduced by coding on both the spatial and the temporal dimensions. In fact, the orthogonal STBC maps different symbols onto orthogonal channels and only simple linear operation is required on the receiver side. The ML detector of orthogonal STBC symbols happens to be linear. Space-time codes that achieve the DMT were considered and proposed in [61,223,229,268]. When there is no channel knowledge available at either side of the MIMO system, differential space-time codes [106, 111] can be applied for noncoherent communications.

2.1.5.2 Single-User Transmission Schemes With CSIT

It is generally more difficult to obtain channel knowledge at the transmitter than at the receiver side. In TDD systems, channel reciprocity can be assumed for the transmission channel. So the channel knowledge during reception can be used for the transmission if the coherence time is sufficiently long. Feedback of channel knowledge from the receiver is used for FDD systems to obtain CSIT.

Single-user transmission schemes with CSIT are usually considered for linear transceiver design [247]. One of the first transceiver structure in the presence of CSIT was considered in [266], where the design objective was the minimization of mean square error (MSE) at the receiver. Several linear transceiver structures with perfect CSI at both the transmitter and the receiver were proposed in [198]. All those structures convert the MIMO channel with memory into a set of parallel flat fading subchannels. The authors of [177] generalized the existing results by developing a unified framework by considering two families of objective functions that embrace a set of reasonable criteria to design a MIMO communication system: Schur-concave and Schur-convex functions. When the performance measure is a Schur-concave function, such as the mutual information and weighted arithmetic mean of the MSE at the receiver, the optimal transceiver structure diagonalizes the effective system channel. When the performance measure is a Schur-convex function, such as the maximization of the mean bit error rate (BER) with the same transmit constellation for each substream, or the maximization of the harmonic mean of the SNR, the optimal transceiver structure diagonalizes the effective system channel up to a rotational transformation at the transmitter side and uniformly distributes the transmitted symbols among the different substreams. When the performance measure is the average BER, the design of linear transceiver structures with different constellations at each substream was considered in [176], and the authors proposed to solve the problem via primal decomposition. Joint transceiver design that combines the geometric mean decomposition with either the conventional ZF V-BLAST decoder or the ZF dirty paper precoder as proposed in [121].

When only second order statistical channel knowledge is available at the transmitter, the authors of [114] proposed to transmit signals on the eigenmodes based on the statistical channel knowledge. The authors of [162] considered the scenario that only noisy or quantized version of the channel matrix is available at the transmitter, and proposed transceiver structures to optimize the SNR or the mutual information of the system. The proposed structures transmit different data streams through the eigenmodes with different power allocation. When the transmitter is aware of the uncertain region of CSIT, another design criterion is to maximize the worst-case performance according to some performance measure. Such a maxmin design philosophy is called *robust optimization*, and transceiver structures exploiting robust optimization can be found in [180, 245].

2.1.5.3 Multiuser MIMO Transmission Schemes

Multiuser MIMO transmission schemes are active research areas due to its important applications in real-world systems, especially cellular mobile communication systems. For example, the uplink transmission in cellular systems can be typical modeled as a multiple access channel, and the downlink transmission can be modeled as a broadcast channel. Although using multiple antennas at the receiver to improve the performance when there are multiple users has been known for a long time, the capacity-achieving transmission techniques for the downlink transmission with multiple antennas were newly found [75].

In the MIMO multiple access scenario, the system can be considered as a generalization of the single user MIMO transmission scheme where the channel matrix consists of all the channel gains from all the transmit antennas to the receiver and each user's symbols can only be precoded by that user's transmitter antennas. The multiple co-located receive antennas provide the degrees-of-freedom for spatial separation of the data streams transmitted by those multiple transmitters. Linear transmitter and receiver structures for the MIMO multiple access channel have been considered in [206], where the authors assume that both the transmitters and the receiver are equipped with multiple antennas. The performance measure is the sum of MSE of all users in the system. The channel knowledge is assumed to be known at both the transmitters and the receiver, i.e., an error-free and low-delay feedback channel to each user is assumed to exist. The authors proposed algorithms to find the jointly optimum linear precoders at each transmitter and linear decoders at the receiver. They also investigated how the symbol rate should be chosen for each user with the optimum linear precoders.

Although multiple access transmission schemes do not require channel knowledge at transmitters in general, the availability of channel knowledge at the transmitter is vital for the broadcast transmission schemes. When the transmitter has enough antennas, linear precoding schemes can be applied at the transmitter. One of the simplest approaches is to pre-multiply the transmitted symbol vector by the inverse of the multiuser channel matrix. However, if the selected users are not sufficiently separable, this approach may result in inefficient use of transmit power, causing a large rate loss with respect to the optimum sum capacity solution. The reason is that while the sum capacity grows linearly with the minimum of the number of antennas and users, the sum rate of channel inversion does not. This poor performance is due to the large spread in the singular values of the channel matrix [185]. A generalization for the channel inversion scheme, known as the *block diagonalization*, was proposed in [215], which incorporates the power control for each user. The authors of [212,213] studied the problem of transceiver design in MIMO broadcast systems with the MSE as the performance measure. In [212], the authors proposed transceiver structures that minimize the downlink sum-MSE of all users under a sum power constraint, and showed that this problem can be solved efficiently by exploiting a duality between the downlink and uplink MSE feasible regions. In [213], the authors addressed the transceiver design problem for MIMO broadcast channels to a broader category, which includes, for example, minimizing the total transmit power subject to per-user MSE requirements.

Nonlinear transceiver design schemes for MIMO broadcast channels have been proposed in [64, 105, 185]. The authors of [105, 185] introduced an encoding scheme to improve the condition of the multi-user channel inverse and maximize the SINR at the receivers by regularizing the channel inverse, i.e., the transmit signal is modified to

$$\mathbf{x} = \mathbf{H}^H(\mathbf{H}\mathbf{H}^H + \alpha\mathbf{I})^{-1}\mathbf{s} \tag{2.24}$$

where α is the regularization constant, and $\mathbf{s}$ is the data symbol vector before precoding. $\mathbf{H}$

is the multi-user channel matrix. After the regularization of the channel inverse, a certain perturbation of the data using a sphere encoder was performed to further reduce the power of the transmitted signal, which enables the proposed scheme to achieve near-capacity performance at all SNRs. Another nonlinear transceiver structure for MIMO broadcast channel is based on the nonlinear precoding scheme proposed by Thomlinson [227] and Harashima [91] for intersymbol interference equalization. The authors of [64, 252] proposed the transceiver structure applying the Thomlinson-Harashima precoding scheme for MIMO broadcast systems. The proposed structure moves the decision feedback equalizer (DFE) structure, which is usually located at the receiver side, to the transmitter side. Since the information symbols are perfectly known to the transmitter, the problem of error propagation in DFE systems is avoided. This nonlinear pre-equalization scheme outperforms linear precoding schemes that increases average transmit power. In [251], a precoding structure that fills the gap between the Thomlinson-Harashima precoding scheme and the sphere decoder based scheme [105, 185] was proposed based on lattice reduction, which improves the diversity gain compared to the Thomlinson-Harashima precoding scheme.

2.2 Relay Communication

Relay communication refers to the technology that the communication between the source and the destination is established or enhanced by one or more than one relays. The relays can be dedicated relay stations that are built to support the wireless link, or other mobile users that are selected to facilitate the data transmission from the source to the destination. Relay communication is also termed *cooperative communication*, which we use in this dissertation without distinction. From the network aspects, traditional wireless communication systems, e.g., cellular mobile communication networks, are centralized. There, the transmission scenarios are point-to-point (single user), one-to-many (broadcast) or many-to-one (multi-access), which can all be categorized as *single-hop transmission*. However, in relay communication, the transmission of information from the source to the destination consists of at least two-hops. This multi-hop transmission model makes the relay communication scenarios more versatile and the research on it more difficult. Although relay communication is still a young research topic, many important results have been achieved, which makes it a fertile field of research. We summarize the important relaying strategies and the state-of-the-art research results in this section.

2.2.1 Development of Relay Communication

The first information theoretical study of relaying technique can be dated back to Van der Meulen in 1971 [234,235], where upper and lower bounds for a three-terminal relay channel were investigated. Here the source transmits information to both the relay and the destination. The received information at the relay is retransmitted to the destination. Inspired by the ALOHA system, Sato [197] also studied the relay channel in 1976. In 1979, Cover and El Gamal [48] significantly improved the upper and lower bounds proposed by Van der Meulen for such a relay channel, and they also considered various other scenarios, including relay channels with feedback from the destination and the relay. However, the capacity of the relay channel remains a open problem until now.

In the 1990s, the Third Generation Partnership Project (3GPP) Concept Group Epsilon proposed the concept of Opportunity Driven Multiple Access (ODMA) that aims to increase the range of high data rate services using relays for Universal Mobile Telecommunications System (UMTS) TDD transmission mode. Although ODMA was finally dropped in 1999 due to concerns over routing, complexity and signaling overhead, the research on relaying was revived at the beginning of the 21st century. In 2000, Gupta and Kumar studied the capacity of large-scale relay networks with randomly located but fixed nodes [87], where the neighboring nodes can assist the data transmission by forwarding the information to its destination. The authors proposed a new approach for research on network capacity, i.e., finding the scaling laws for networks with many nodes. They showed that as the number of nodes n increases in a fixed geographic area, the throughput per source-destination pair decreases approximately like $1/\sqrt{n}$. When mobility is introduced into the model and considering the scenario that users move independently around the network, Grossglauser and Tse [83] showed that the average long-term throughput per source-destination pair can be kept constant even as the number of nodes per unit area increases. This improvement is achieved by exploiting *multiuser diversity* via packet relaying. In [74], Gastpar considered the same physical model of a wireless network as [87] but under a different traffic pattern, where there is only one active source-destination pair and all other nodes act as relays to assist its transmission. They showed that as the number of nodes n in the network goes to infinity, the capacity of such a network behaves like $O(\log n)$ bits per second. Recently, Ozgur [173] showed that the total capacity in the wireless network of [87] can scale linearly with the number of nodes n as n goes to infinity. The performance gain is achieved by using a hierarchical architecture and distributed space-time coding where different nodes exchange information to realize cooperation between them.

Seminal work that drew people's attention on cooperative communication was the contribution by Sendonaris *et al.* [203–205]. In their work, the authors considered the uplink transmission scenario and proposed user cooperative strategies to improve its performance. In the proposed transmission strategies, each mobile user transmits its data to both the base station and another mobile user nearby, which helps to retransmit the data to the base station. That is, each mobile user acts both as a information source and as a relay. The authors showed that the achievable rate region can be increased by cooperation between the two transmitting users. Moreover, higher orders of diversity, i.e., cooperative diversity, can improve the stability of wireless links. Cooperative diversity was further analyzed by Laneman [134, 135]. In [134], the authors developed and analyzed some cooperative diversity protocols that combat fading induced by multipath propagations in wireless networks. They summarized the important relaying strategies, such as the amplify-and-forward (AF) and decode-and-forward (DF) relaying strategies. They also addressed the major problems for implementing practical relays, such as the half-duplex constraint. The distributed space-time coding schemes for cooperative wireless networks were proposed in [135], which generalized the space-time coding techniques for the point-to-point MIMO systems to distributed relay networks where each station is equipped with a single antenna. A summary of the distributed space-time coding schemes can be found in [146].

Capacity scaling laws in MIMO relay networks were discovered in [28], where multiple relays assist the transmission from a multi-antenna source to a multi-antenna destination. The authors showed that the point-to-point MIMO link capacity in high SNR can be achieved when the number of relays is asymptotically large. The authors of [256] proposed a *multiuser zero-forcing* (MUZF) scheme for relay networks with multiple source-destination pairs and multiple AF relays. Each node is equipped with a single antenna. The authors showed that the relay gain can be ingeniously chosen so that the multiuser interference can be perfectly canceled at the destinations nodes when the number of relay nodes is large.

2.2.2 Advantages and Challenges of Relay Communication

Relay communication can provide the benefits that traditional single-hop communication cannot achieve in many practical scenarios. Relay communication has drawn wide interests from both academia and industry [58]. For practical systems, we summarize the advantages of relay communication over single-hop communication as follows.

Combating signal attenuation The adverse effects of wireless channels include pathloss, shadowing and fading effects. The signal strength decays exponentially with the distance between the source and the destination. When the distance between the source and destination is too large, the signal attenuation becomes too high due to pathloss, which makes it impossible for the source and destination to communicate. By placing relays between the source and the destination, the distance between the source and the relay and the distance between the relay and destination is shortened. As a result, the signal strength can be boosted a lot. Moreover, due to the signal strength improvement, the source can use higher modulation symbol alphabets to transmit more data in each channel use. In this way, relaying technology not only increases the coverage of the system, but also improves the data rate transmitted to the users.

Combating shadowing effects In large cities and hilly areas, tall buildings and mountains typically block signals transmitted from the source to the destination. Such effect is called *shadowing*. Relays provide another path to circumvent the obstruction. In those scenarios, relaying is maybe the only way to provide services in shadowing environments.

Combating fading effects The fading effects arise due to the multipath propagations that lead to the fluctuations in received signals. Diversity is an effective way to combat the signal fluctuation due to the fading effects. The *cooperative diversity* introduced by the cooperative communication brings higher link reliability to the users [204,205], where multiple independently faded signals from the source and the relay are combined at the destination.

Low cost Future cellular communication systems will move to higher frequency. As a result, the coverage of each cell will shrink a lot compared to present cellular communication systems. Building more base stations can be the solution, but the cost of building those base stations will be very high. A low-cost alternative will be building relays to extend the coverage of each cell. Thus relay communication provides low-cost solutions for future generation wireless communication systems.

Infrastructure-less network In traditional cellular networks, the whole system operation depends on the centralized control, e.g., from the base station. However, in military services or due to the disasters like earthquakes, infrastructure-less networks such as *ad hoc* networks are preferable. Such networks do not rely on a preexisting infrastructure such as dedicated routers or base stations. Instead, each node participates in the routing by forwarding data for other nodes. That is each node can act as a relay, and the choice of relay nodes are determined dynamically based on the network connectivity.

Despite all those benefits that may be available by incorporating cooperative communication into future wireless communication systems, there are also challenges for implementing cooperative communications. Those challenges include:

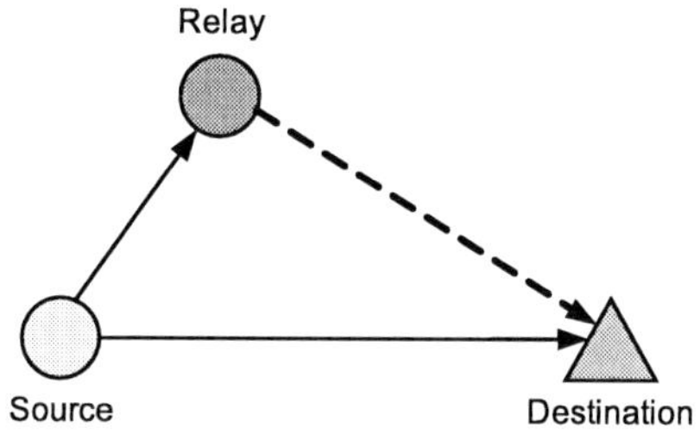

Fig. 2.6: A basic relaying system setup

Increased overhead Compared to point-to-point communication, the data in relay communication must traverse multiple links. Each link may introduce some overheads which must be considered in the implementation of relay communications. Such overheads include synchronization and channel estimation. Moreover, in some cooperative transmission schemes with multiple relays, each relay is also required to have the channel knowledge of other relays, i.e., *global CSI* of the system, which may require the relays to exchange CSI between them. The resource consumed by CSI exchange is usually not negligible.

Resource consumption Cooperative communication requires extra links between mobile users to be established. This requires extra resource consumption, which may include power (indicated by the battery life), frequency and time resources.

Increased interference and traffic In cooperative communications, each mobile user sends its data first to it neighboring user. This may cause interference to other users. Furthermore, sending data to the destination via relays leads to increased traffic for the whole system.

Spectral efficiency loss The major problem of current relays is that they cannot transmit and receive data using the same time-frequency channel. This half-duplex constraint leads to the spectral efficiency loss compare to direct transmissions. The two-way relaying protocol to be discussed in Section 2.3 was proposed as a solution to make up for the spectral efficiency loss.

2.2.3 Basic Relaying System Model

A basic relaying system setup is depicted in Fig. 2.6. The source and destination can be mobile users or one of them can be the base station for cellular systems. A wireless station's

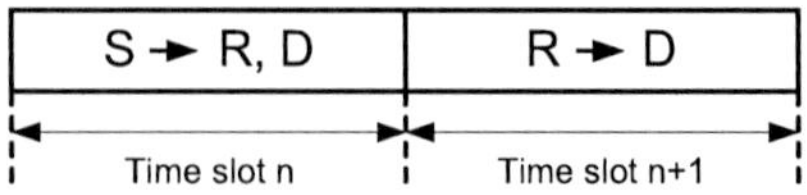

Fig. 2.7: Half-duplex Transmission

transmit signal is typically 100-150dB above its received signal and it is difficult to have sufficient electrical isolation between the transmit and receive circuitry up to now. Technologies for designing electronic circuits with duplex isolation, which enables simultaneous two-way information transfer over a common radio channel, are under research [43, 270]. However, current limitations in radio implementation make it difficult for wireless stations to achieve full-duplex communication, i.e., transmitting and receiving signals at the same time using the same frequency channel [134]. Such a constraint is called the *half-duplex constraint*, and orthogonal channels are required for the transmission and reception at wireless stations. In practice, the transmit and receive signals are usually separated in time (time-division duplexing, TDD) or in frequency (frequency-division duplexing, FDD).

For simplicity, we assume the two-hop transmission is separated in time, i.e., in time division multiple access (TDMA) fashion. We just consider half-duplex operations as depicted in Fig. 2.7, and assume equal time allocation for the two time slots and each node is equipped with a single antenna. In the first time slot n, the source sends its signal $x_{\mathrm{s}}[n]$. For a flat fading channel, the received signal at the relay and the destination can be expressed as

$$\begin{aligned} y_{\mathrm{r}}[n] &= h_{\mathrm{s,r}} x_{\mathrm{s}}[n] + z_{\mathrm{r}}[n] \\ y_{\mathrm{d}}[n] &= h_{\mathrm{s,d}} x_{\mathrm{s}}[n] + z_{\mathrm{d}}[n], \end{aligned} \tag{2.25}$$

where $y_{\mathrm{r}}[n]$ and $y_{\mathrm{d}}[n]$ are the relay and destination received signals, respectively. In the second time slot $n+1$, the source keeps silent, while the relay transmits signal $x_{\mathrm{r}}[n+1]$ to the destination.

$$y_{\mathrm{d}}[n+1] = h_{\mathrm{r,d}} x_{\mathrm{r}}[n+1] + z_{\mathrm{d}}[n+1], \tag{2.26}$$

where $y_{\mathrm{d}}[n+1]$ denotes the received signal at the destination in the time slot $n+1$. In the above equations, $h_{i,j}$ is the channel coefficient between the transmitter and receiver and z_j is the additive noise at the receiver, where $i \in \{\mathrm{s}, \mathrm{r}\}$ and $j \in \{\mathrm{r}, \mathrm{d}\}$. For Rayleigh fading channel, we can model z_j as independent $\mathcal{CN}(0, N_0)$ random variables, where N_0 denotes the noise variance.

2.2.4 Relaying Strategies

Relaying strategies refer to the processing schemes at the relay, i.e., how $x_{\mathrm{r}}[n+1]$ in (2.26) is generated at the relay. Different relaying strategies have different implementation complexities and lead to different performance at the destination. The following relaying schemes are the most commonly used processing strategies at present-day relays.

2.2.4.1 Amplify-and-Forward

The amplify-and-forward (AF) relays resemble the traditional analog relays, which transmit an amplified version of the previously received signal, i.e. [134]

$$x_{\mathrm{r}}[n+1] = \beta y_{\mathrm{r}}[n], \tag{2.27}$$

with the power constraint

$$\beta \leq \sqrt{\frac{P_{\mathrm{R}}}{|h_{\mathrm{s,r}}|^2 P_{\mathrm{S}} + N_0}}, \tag{2.28}$$

where P_{S} and P_{R} are the transmit power constraints for the source and the relay, respectively. In order to achieve the highest capacity available, the amplification factor β in (2.28) should be met with equality. Since the destination receives the same information from the source and the relay at different time slots, the two versions of the received signal can be decoded using the maximum-ratio combining technique at the destination. This leads to low-complexity relay transceivers and lower processing power consumption since there is no need of signal processing for the decoding procedures. Moreover, AF relays are transparent to adaptive modulation techniques which may be employed by the source.

Assuming that equal time is allocated for the two relaying phases as shown in Fig. 2.7, the maximum average mutual information and the outage behavior has been analyzed in [134]. The result shows that the outage probability declines proportional to SNR^{-2} at high SNR. Thus AF relaying achieves full second order diversity for the considered system in Fig. 2.6.

One of the most important question for AF relaying is how to choose the relaying function. For fading channels, when the CSI is available at the relay, the gain allocation (2.28) was proposed in [134], and the average BER and outage probability analysis for some modulation schemes in a two-hop AF relaying system without the direct link has been considered in [93, 94] over frequency flat Rayleigh-fading channels. A fixed gain allocation in such a system, which benefits from the knowledge of the first hop's average fading power, was proposed in [95]. Such a fixed gain allocation AF scheme requires less channel knowledge and has lower complexity since it does not change the gain allocation in the process of fading.

The gain allocation in (2.28) calculates the gain allocation based on the average received power of the signal. A gain allocation scheme that is based on the received signal itself was proposed in [7]. The considered scenario is a two-hop relaying system without the directlink. The channel model is AWGN and the modulation scheme is binary phase-shift keying (BPSK). The system is uncoded and memoryless, i.e., during each relay transmission, the signal transmitted by the relay only depends on its last received symbol. The authors showed that the optimal amplification function is a Lambert W function whose parameters vary with the noise variance and the input signal. Here the objective is to minimize the average probability of detection error. Furthermore, they showed that for low SNR regimes, the optimal amplification function resembles a hard limiter, and for high SNR regimes it resembles an linear amplifier. A similar problem was considered in [80], where the two-hop channel remains memoryless but the objective is to maximize the SNR at the destination. Both the single relay case and the multi-relay case are considered.

The basic relaying system where each node is equipped with a single antenna as shown in Fig. 2.6 was considered in [159], and the authors unified previous results on the protocols of such a system. Three different TDMA-based cooperative protocols that vary the degree of broadcasting and receive collision were proposed. The three protocols that the authors considered are depicted in Table 2.1. For each protocol, the authors study the ergodic and outage capacity behavior assuming Gaussian code books under the AF and DF modes of relaying. The authors analyzed the spatial diversity performance of the various protocols and find that the full spatial diversity (second-order in this case) is achieved by certain protocols provided that appropriate power control is employed. The authors also discussed the distributed space-time code design for fading relay channels operating in the AF mode. They showed that the corresponding code design criteria consist of the traditional rank and determinant criteria for the case of collocated antennas, as well as appropriate power control rules. For the AF mode, the comparison of mutual information in different scheme is

$$I_{\mathrm{I}}^{\mathrm{AF}} \geq I_{\mathrm{II}}^{\mathrm{AF}} \geq I_{\mathrm{III}}^{\mathrm{AF}}. \tag{2.29}$$

Protocol I is the only protocol that can realize a multiplexing gain in the classical sense and, hence, recover (to a certain extent) from this 50% loss in spectral efficiency due to the half-duplex constraint. All the three schemes can achieve full second order diversity for single antenna nodes if power control is applied.

Table 2.1: Three different TDMA-based protocols. S, R, and, D stand for the source, relay, and destination terminals, respectively. A $\rightarrow$ B signifies the transmission from terminal A to terminal B.

Time slot/Protocol	I	II	III
1	$\mathrm{S} \rightarrow \mathrm{R}, \mathrm{D}$	$\mathrm{S} \rightarrow \mathrm{R}, \mathrm{D}$	$\mathrm{S} \rightarrow \mathrm{R}$
2	$\mathrm{S} \rightarrow \mathrm{D}, \mathrm{R} \rightarrow \mathrm{D}$	$\mathrm{R} \rightarrow \mathrm{D}$	$\mathrm{S} \rightarrow \mathrm{D}, \mathrm{R} \rightarrow \mathrm{D}$

The relay linear processing schemes in MIMO AF relaying systems were considered in [158,219]. The authors proposed the relay linear processing scheme to maximize the mutual information between the source and destination using MIMO transmission with channel knowledge at the relays. Three different levels of CSI can be considered at the relay station:

- only first hop channel knowledge, i.e., between the source and the relay;
- first hop channel and second hop channel knowledge, i.e., between relay and destination;
- the CSI in the whole system, including the first and second hop channels and also the direct channel between source and destination.

The optimum linear processing matrix at the relay turns out to diagonalize the equivalent channels from the source to the destination, and the gain allocation on each subchannel is chosen according to the waterfilling technique.

A power allocation scheme for AF MIMO-OFDM was proposed in [89]. The authors proposed power allocation schemes over the subchannels in frequency and space domain to maximize the instantaneous rate of this link if channel state information at the transmitter (CSIT) is available. Furthermore, a heuristic scheme was proposed that pairs the subcarriers in the first and second hop channels in OFDM systems.

2.2.4.2 Decode-and-Forward

Although the AF relaying scheme is simple to implement, it forwards the signal as well as the noise to the destination and may be subject to noise enhancement. The *decode-and-forward* (DF) scheme tries to regenerate the data information at the relay and forwards a "clean" version of the data to the destination. Error-detecting codes, such as cyclic redundancy check (CRC) codes, and error-correcting codes, such as turbo codes, are usually applied in the DF relaying systems to detect and correct the errors in the received data at the relay. Compared to the AF schemes, decoding the channel codes at the relay is a distinct characteristic of the DF schemes.

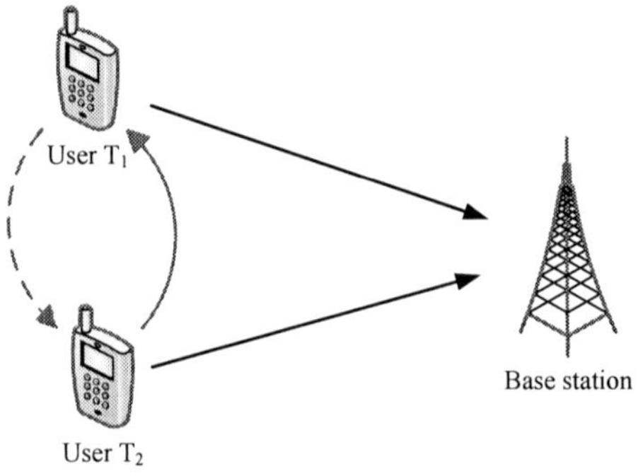

Fig. 2.8: Cooperation Transmission Scenario in [204]

The DF scheme was one of the first relaying strategy applied in the *user cooperation* scenario [204, 205], where the authors considered an uplink transmission scenario as shown in Fig. 2.8. Here the two terminal T_1 and T_2 represent two independent mobile users that both have their own data to be transmitted to the base station. The two terminals considered in [204,205] are full-duplex, i.e., they can transmit and receive signals in the same frequency channel simultaneously. Each terminal splits its transmit information into two parts: the first part is to be sent directly to the base station and the second part is to be sent to the base station via the other terminal. The power allocation of the two parts of signals enables the other terminal to perfectly decode the second part of the signal. Due to the broadcast nature of wireless channels, each terminal overhears the transmit signal when the other terminal transmits. The authors showed that, even though the interuser channel is noisy, cooperation leads to an increase in the ergodic achievable rate region for both users.

Actually the user cooperation scenario considered in Fig. 2.8 can be decomposed into the basic relaying system model as shown in Fig. 2.6, where the only difference is that each terminal act both as a source of information and as a relay. The achievable diversity gain of DF relaying strategies was calculated in [134], where the basic relaying system model as shown in Fig. 2.6 was considered. Since the data symbol from the source is transmitted from the source and the relay, people may expect to achieve second-order diversity in DF schemes. However, by analyzing the outage probabilities, the results in [134] showed that such DF relaying strategy does not offer diversity gain for large SNR if we require the relay to fully decode the source message. The outage probability can only decay proportionally to SNR^{-1} at high SNR. This is because the channel between the source and the relay limits the performance for the whole system. In order to overcome the limitations, the authors also proposed *selection relaying* and *incremental relaying* which can be used with either AF or DF relays:

- In selection relaying, the source retransmits in the second time slot if the source-relay path falls below a certain threshold. It is a adaptive version of relay transmission which falls back to direct transmission if the relay cannot decode.
- In incremental relaying, the destination sends feedbacks to the source and the relay upon receiving the symbol from the source in the first time slot. If the destination can correctly decode the received symbol in the first time slot, the source can directly transmit the next message and no relaying is required.

Combined with selection relaying or incremental relaying, the authors showed that DF relaying can achieve the full diversity (second-order diversity in this case).

Coded cooperation [112, 113, 167, 216] utilizes channel coding for user cooperation. It can be considered as a practical way of using channel coding to implement the selection DF relaying strategy. In this scheme, the data from the source is divided into two parts: the first part is a punctured convolutional code which is a valid code in itself; the second part consists of parity bits that can form a stronger code with the first part. The source transmits the first part of the code. Upon reception, the relay decodes that part and checks the CRC code. If the CRC code is correct, the relay forms the second part (parity bits) and transmits it to the destination. Otherwise the second part will not be transmitted by the relay, i.e., the user resorts to the noncooperative mode. The parity bits can be considered as incremental redundancy. Furthermore, whether or not to choose the relaying mode is automatically managed through channel coding instead of the feedback. It has been shown that coded cooperation is also capable of achieving the full diversity provided by user cooperation. Based on the idea of coded cooperation, a *space-time coded cooperation* scheme was proposed in [117], where each user transmits his own as well as his partner's second set of parity bits by splitting the available power and utilizing space-time codes upon successful decoding of the first part. The analysis and simulation show that the space-time coded cooperation scheme allows the users to capture better space-time diversity in fast fading, compared to the original coded cooperation scheme.

Instead of forwarding the parity bits, the authors of [17, 144] proposed to forward the soft reliability information obtained at the relay to the destination, i.e., *soft information relaying*. In traditional DF relaying strategies, the relay retransmits the hard decision of the decoded data bits. Thus the reliability information obtained in the relay decoding process is lost. When the channel quality between the source and the relay is poor, the relay may be unable to do perfect decoding and the hard decisions at the relay may be incorrect. In order to overcome this drawback, the authors of [144] proposed to first calculate the a posteriori probabilities (APPs) of the information symbols at the relay. Based on the APPs of the information symbols, the relay calculates the parity symbol soft estimates of the source information and then forwards them to the destination. By observing that the AF scheme with binary inputs can be viewed as a way of forwarding soft reliability information without utilizing channel codes, the authors of [17] proposed the "decode-amplify-forward" (DAF) scheme, which combines the merits of AF and DF by having the relay perform soft decoding and forward the reliability information. A hybrid scheme of DAF and coded cooperation is implemented through simple time-sharing in [17]. The soft information relaying scheme achieves better performance compared to the AF and the traditional DF relaying schemes.

Applying the idea of turbo coding in relaying systems was proposed in [278], which is called "distributed turbo codes". The idea is that the relay forwards to the destination a re-interleaved version of the convolutional code from the source upon successful decoding the messages. The destination receives two sets of convolutional codes, one from the source and one from the relay. The two sets of codes are encoded from the same information bits and interleaved by different interleavers, which can be combined and decoded using the turbo decoding principle. It has been shown that such a coding strategy performs close to the theoretical outage probability bound of a relay channel. Similarly, LDPC codes can also be integrated into the relay transmission systems [37], which are shown to be able to approach the theoretical limit of the relay channel.

Another scheme trying to combine the merits of AF and DF relaying was proposed in [143], where the authors considered a system with multiple relays. The relays that fail to decode the message from the source resort to the AF scheme, and only those relays who correctly decode the transmitted data from the source use the DF scheme to forward the data to the destination. The received signals at the destination from all the relays are combined into one signal to recover the source information. This adaptive relaying protocol outperforms the pure AF and DF relaying schemes, and it was shown that the performance gain grows as the number of relays increases.

2.2.4.3 Demodulate-and-Forward

The *demodulate-and-forward* (DemF) scheme is a simpler relaying strategy compared to the DF scheme in that the DemF scheme does not perform channel decoding at the relay. On the other hand, DemF tries to avoid noise amplification, which is the main problem in the AF scheme, by demodulating the received signals on a symbol-by-symbol basis. The demodulated signals at the relay are remodulated to form the transmit symbols. When the same modulation scheme is applied at the source and the relay, the destination can coherently combine the received signals from the source and the relay. In [249], the authors considered a system consisting of a single source-destination pair with a single or multiple relays. A weighted coherent combiner at the destination was proposed, where the weights take into account the detection errors at the relays. The authors showed that the proposed scheme achieves the maximum possible diversity of the system. The authors of [42] considered a similar system setup, and proposed demodulators with piecewise-linear combining as an approximation of the nonlinear ML detectors for coherent and noncoherent detection. The proposed scheme achieves better diversity gains than the DF strategy, but loses about half of the diversity gains compared to the AF strategy.

2.2.4.4 Compress-and-Forward

The compress-and-forward (CF) relaying strategy allows the relay station to quantize the received signal from the source node and to forward it to the destination without decoding the signal [131]. In the CF relaying strategy, Wyner-Ziv coding [49] can be used for optimal compression. The CF relaying strategy is compared with the DF relaying strategy from the information theoretical aspects in [131]. The authors considered the coding strategies of the DF and CF relaying strategies for full-duplex relays with the basic system setup as shown in Fig. 2.6. For the CF relaying strategy, coding schemes that take advantage of the statistical dependence between the channel outputs of the relay and the destination are proposed. It was shown that the DF strategy is useful for relays that are close to the source and the CF strategy is useful for relays that are close to the destination.

2.2.5 Relaying Networks

The basic relaying system setup is only a building block for larger wireless networks. The Relaying network can comprise of multiple users and multiple relays, and it may have complicated structures. We summarize the important research results on relaying networks in this section.

2.2.5.1 Multiple Relays

In a dense wireless network, multiple relays are usually available to facilitate the data transmission from the source to the destination. Those relays altogether resemble a "virtual" antenna array. Since all the data that the relays forward to the destination come from the same source, space-time codes designed for multi-antenna systems can be applied to obtain diversity gain. Applying space-time codes in the relaying scenario is usually referred to as *distributed space-time coding*.

A distributed space-time coding scheme was proposed in [135] based on the orthogonal STBC. The authors considered DF relays and require the relays to fully decode the source message. In this scheme, different relays transmit different columns of the STBC code matrix. Absence of an antenna corresponds to deletion of a column in the matrix, which is analogous to that antenna experiencing a deep fade. However, the columns still remain orthogonal, which allows the code to maintain its residual diversity benefits. The authors showed that the proposed distributed space-time coding scheme achieves full spatial diversity in the number of cooperating terminals, not just the number of decoding relays, and has higher spectral efficiencies than repetition-based schemes.

Another distributed space-time coding scheme was proposed in [124], where the relay nodes encode their received signals into a distributed linear dispersion code, and then transmit the coded signals to the receive node. The authors showed that the diversity d of the system behaves as

$$d = \min(T, R)\left(1 - \frac{\log\log P}{\log P}\right), \tag{2.30}$$

where T is the coherence interval, R is the number of relay nodes, and P is the total transmit power. when $T \geq R$ and the average total transmit power is very high ($P \gg \log P$), the relay network has almost the same diversity as a multiple-antenna system with R transmit antennas, which is the same as assuming that the R relay nodes can fully cooperate and have full knowledge of the transmitted signal. Furthermore, the optimal power allocation of the system is analyzed when there is a fixed total transmit power constraint across the entire network. Besides distributed space-time block codes, distributed STTC were discussed in [272].

In addition to achieving the diversity gain, another usage of using multiple cooperative relays is to improve the capacity in a rank-deficient MIMO channel [191, 257]. It is known that full spatial multiplexing gain in a MIMO channel is achieved only when the channel has full rank. When there are not enough scatterers to provide a rich scattering environment, the MIMO channel will be rank-deficient and suffers from a loss in capacity. The authors of [257] proposed the idea of using multiple single antenna AF relays as *active scatterers* for the MIMO channel, and showed that the full spatial multiplexing gain is achievable when a sufficient number of relays are available to act as active scatterers.

Multi-antenna relay networks were considered in [19], where multiple AF relays assist the communication between a single source-destination pair. Here all stations are equipped with multiple antennas and inter-relay cooperation is allowed. Under the ZF criterion and the MMSE criterion, the authors proposed schemes to maximize the SNR at the destination subject to certain power constraints.

The channel estimation is another important issue in relaying networks. The authors of [71] considered the channel estimation in a relaying network with a single source-destination pair and multiple AF relays. Each node is equipped with a single antenna. The proposed scheme estimates the overall channel from the source to the destination where each relay node is subject to its individual power constraint. The linear least-square estimators and the MMSE estimator, together with their optimal training sequences, are proposed for the considered relaying network. The DF relaying network with the same setup has been considered in [72], where the ML and the MMSE channel estimators as well as their optimal training sequence design have been studied.

2.2.5.2 Multiple Users

Multiuser relay networks utilize relays to assist the communication between multiple source-destination pairs. The research of multiuser relay networks is related to the *interference channel* in information theory, where multiple sources transmit information to multiple destinations in the same channel. The capacity region of the interference channel is still a open problem until now.

In the research of multiuser relay networks, the authors of [115] established the duality between the multiple access and broadcast relaying channels with two users and AF relays. In a multiuser multi-access relay channel, more than one user transmits data to a common destination with the help of relays as shown in Fig. 2.9(a); in a multiuser broadcast relay channel, a common source transmits its data to different destinations as shown in Fig. 2.9(b). The sum power constraints are applied on the transmitting user(s) and the relays separately. Under this assumption, the achievable rate region of the multi-access relay channel with a sum source power constraint $P_1 + P_2 = P$ and sum relay power constraint P_R in Fig. 2.9(a) is shown to the be identical to the achievable rate region of the broadcast relay channel with transmit power constraint P_R and a sum relay power constraint P as shown in Fig. 2.9(b).

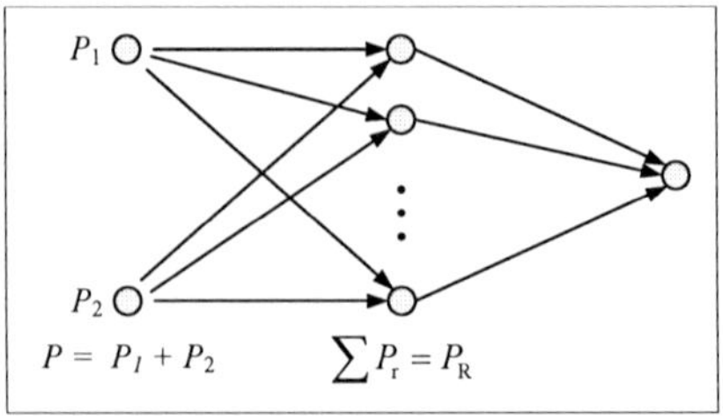

(a) Two-user multi-access AF relay channel

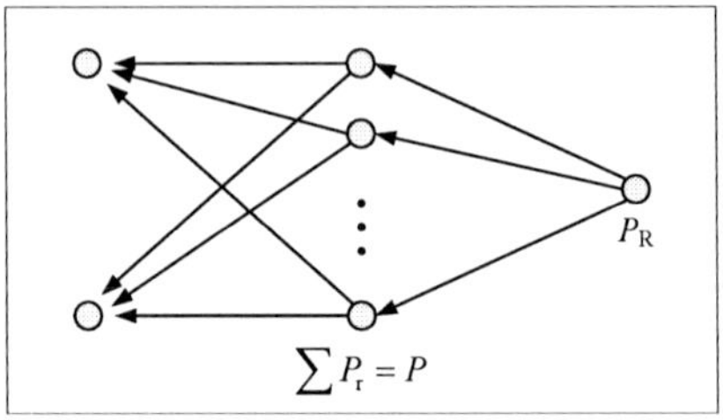

(b) Two-user broadcast AF relay channel

Fig. 2.9: Dual multi-user multi-access and broadcast relay channel

This shows that the two channel models in Fig. 2.9 are *dual*, and the achievable rate region of the broadcast relay channel can be thus characterized by determining its dual multi-access relay channel. The duality also holds for multi-hop relay channels and networks with multi-antenna nodes [79].

A multiuser zero-forcing relaying scheme was proposed in [256, 258], where multiple source nodes transmit data to their corresponding destination nodes via multiple AF relaying nodes. Each node is equipped with a single antenna. The authors proposed the gain allocations at the relays such that each destination node only receives the data symbol from its intended source node, and the inter-user interferences are canceled by those AF relays. Assuming there are N source-destination pairs, the authors showed that at least $N_{\mathrm{R}} = N(N-1)$ AF relays are required to perform this zero-forcing relaying. Multiuser zero-forcing relaying achieves distributed spatial multiplexing gain via AF relays, and this scheme has been extended into the multiuser MMSE relaying scheme in [21].

2.2.5.3 Resource Allocations

Considering a cellular network with many user, the design of such a network from a system point of view becomes challenging. The design of the system includes the optimal allocation of physical-layer resources, such as power and frequency subcarriers for each user, the optimal choice of relays and the optimal relaying strategies. The optimization of all those parameters should take into account the channel realizations and the user traffic. Joint optimization of all those parameters may sound impossible. However, it was shown in [165] that the resource allocation problem can be solved if a central unit has perfect knowledge of each link's channel conditional and user traffic. The system considered in [165] is a cellular orthogonal frequency-division multiple access (OFDMA) mobile network with one base station and many users where each user can also act as a relay to forward the data to its neighbors. The network operates in a frequency-selective slow-fading environment, and the choices of relaying strategies are AF and DF. The authors proposed a centralized utility maximization framework for the optimization of the physical-layer parameters, the choice of relays and the relaying strategies. The solution is obtained by first decomposing the Lagrangian dual of the original problem into the application-layer and physical-layer subproblems, and then solving the subproblems separately.

Another cross-layer resource allocation framework for cooperative networks was proposed in [44], where an energy-constrained cooperative network is considered. The objective is to guarantee the lifetime of each node to be equal to a target lifetime and let each node efficiently utilize its available energy to optimize its performance such as throughput and outage probability. This work considers the physical and network layers jointly. It is found that the fairness and energy constraint cannot be satisfied simultaneously if each node uses a fixed set of relays. The authors proposed a multi-state cooperation methodology to solve the problem, where the energy is allocated among the nodes state-by-state via a geometric and network decomposition approach. Given the energy allocation, the duration of each state is then optimized so as to maximize the nodes utility. It is shown that the proposed framework will not only guarantee fairness, but will also provide significant throughput and diversity gain over conventional cooperation schemes.

2.2.5.4 Capacity Scaling Laws

After the seminal work [87], the research on the network capacity scaling law became a hot research topic. The capacity scaling law in MIMO relaying networks can be found in [28]. In the considered network, a source terminal communicates with a destination terminal, with the help of K relay terminals using a half-duplex protocol. Both the source and the destination are equipped with M antennas, and each relay has N antennas. When perfect CSI is available at the destination and the relays, the network capacity C scales like

$$C \sim \frac{M \log K}{2}, \quad \text{as } K \to \infty \tag{2.31}$$

for fixed M and arbitrary, but fixed, N. This is accomplished by simply matched filtering at the relays. When the relays do not have CSI, the authors showed that the simple AF scheme turns the relay network into a point-to-point MIMO link in the high SNR regime, i.e., $C \sim (M/2)\log(\text{SNR})$, as $K \to \infty$ and for fixed M and N. This shows that the AF relays can act as active scatterers to recover spatial multiplexing gain in poor scattering environments.

2.2.6 Testbeds for Relay Communication

As a burgeoning area of research, industry soon realized the importance of deploying relays in cellular networks. Up to now, the following testbed are under development or becoming mature.

EASY-C Testbed in Dresden The research project EASY-C (Enablers for Ambient Services and Systems, Part C) is supported by the German Federal Ministry for Education and Research, and it is jointly led by Deutsche Telekom and Vodafone. Multi-cell joint signal processing as well as cooperative relaying techniques are the major research topics in this project. Those techniques are expected to provide significant improvement of user experience at the cell-edges, i.e., providing enhanced spectral efficiency and fairness in cellular systems. The testbed in Dresden, Germany, is run by the Vodafone Chair in Dresden University of Technology. In the final stage, the testbed will comprise of 10 sites with a total of 28 cells. The backhaul between the sites is built through low latency microwave links. Those links are operating in the 5 GHz frequency band and have a maximum throughput of 300 Mbit/s. This testbed will be operated in frequency division duplex (FDD) mode [118].

Berlin LTE-Advanced Testbed This testbed was set up by the Heinrich-Hertz-Institut in Berlin, and it is also part of the research project EASY-C. One study item of the testbed for LTE-Advanced is to enhance existing LTE networks by deploying DF relays. The testbed is being established with 3 sites and a total of 6 cells, where the focus is on testing applications that are enabled or improved through multi-cell joint signal processing. The sites are connected through both fiber-optic cables and laser links [254].

By indoor and outdoor field measurements using this testbed, researchers have demonstrated that relaying can have high impact on the coverage and the capacity in cellular systems.

Nokia Siemens Networks LTE Relay Testbed This testbed is built up by Nokia Siemens Networks and it operates in FDD mode. The system comprises of an LTE system supporting MIMO transmission with two transmit and two receive antennas, together with a relay station. The relaying station operates in-band, which means that the relay station inserted in the network does not need an external data backhaul. It is connected to the nearest base station by using radio resources within the operating frequency band of the base station itself. The relay station act as if it was the base station towards the user terminal, and the system offers the full functionality of LTE. The demonstration using this testbed showed that the performance at the cell edge could be increased up to 50% of the peak throughput.

RACooN Lab in ETH Zurich The Radio Access with Cooperative Nodes (RACooN) lab is a mobile simulation testbed at the Wireless Communications Group in ETH Zurich. It was set up to investigate the behavior of different transceiver schemes, including cooperative relaying. The whole testbed consists of ten single antenna relay nodes which can be combined arbitrarily, which makes RACooN an extremely adaptive and highly flexible laboratory. The operation band of the RACooN Laboratory lies between 5.1 GHz and 5.9 GHz. 34 MHz of feasible baseband user bandwidth (30 MHz with equalized phase response) correspondents to a passband user bandwidth of 68 MHz (60 MHz with equalized phase response). The carrier frequency can be set in steps of 1 MHz. The RACooN Lab has successfully demonstrated relaying transmission schemes, such as the multiuser ZF relaying [22].

2.3 Two-Way Relay Communication

Two-way relay communication can be considered as a sub-branch of relay communication. It comprises of the same system setup as conventional relay networks, with the only difference that it employs the two-way relaying protocols. The two-way relaying protocol distinguishes itself from conventional relay protocols by considering bidirectional information flows in the process of relaying. The bidirectional data traffic is prevalent in modern wireless communication systems. For example, there are both downlink data and uplink data transmissions in cellular mobile communication systems. Surprisingly, this simple introduction of a new ingredient brings significant benefits into relay communication. It provides a solution for the long-standing problem of the spectral efficiency loss due to the application of half-duplex relays. Today two-way relay communication has become an important research topic for both academia and industry. Sometimes two-way relay communication is also termed *bidirectional relaying*, and the traditional relay communication with unidirectional information flow is called *one-way relaying* or *unidirectional relaying*.

2.3.1 Development of Two-Way Relay Communication

The original study of the two-way communication problem can be traced back to Shannon's work [209] in 1961. Interestingly, in retrospect, Shannon's work in 1961 already considered an AND channel and an OR channel, which are very similar to the two-way relay channel with network coding nowadays. At the beginning of the 21st century, interests in cooperative communication and relaying systems begin to boom due to its connections to distributed MIMO and its promise to achieve coverage extension of cellular networks with low costs [174]. However, although full-duplex relays are under research, most of present-day relays are still half-duplex, i.e., they cannot transmit and receive signals using the same temporal-spectral channel due to the insufficient isolation between the transmit and receive circuitry [134]. Such half-duplex signaling requires two channel uses in two-hop networks. It causes a substantial loss in the spectral efficiency due to the pre-log factor $1/2$ in the corresponding capacity expressions. This problem pestered the research community for a long time, and researchers proposed schemes, e.g., spatial reuse of the relaying time slot as in [110, 157], to improve the spectral efficiency in relaying systems. However, those solutions are based on specific system setups. In 2005–2007, Rankov and Wittneben proposed two spectrally efficient relaying protocols, i.e., the *two-path relaying* protocol and the *two-way relaying* protocol, and summarized them in [193]. The two-path relaying protocol considers the unidirectional transmission, where two relays alternatively forward messages from a source terminal to a destination terminal. In contrast, the two-way relaying protocol considers the bidirectional data transmission and does not need to introduce additional relays into the system. Here the bidirectional data transmission means that two wireless stations exchange data via another relay station. This can be considered as, e.g., the base station communicates with a mobile user via a dedicated relay in a cellular system, or two mobile clients transmit data to each other via the access point in a WLAN. The two-way relaying protocol combines the traditional "uplink" and "downlink" data transmission together by ingeniously mixing the data from different sources at the relay. While traditional relaying protocols require four phases (in time or frequency) to achieve bidirectional communication

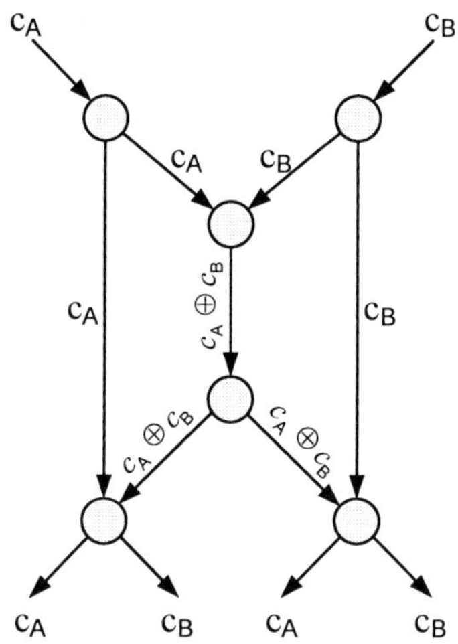

Fig. 2.10: Butterfly network

between the two stations, the two-way relaying protocol only needs two phases, namely, the multiple access (MAC) phase and the broadcast (BRC) phase.

Shortly after the proposal of the two-way relaying protocol, its connection to *network coding* was observed. Originally proposed by Ahlswede *et al.* for improving the throughput in multicast scenarios, network coding was traditionally considered as a research topic for computer and wireline communications since it operates on discrete finite fields [10]. One of the most famous example of network coding is the butterfly network as shown in Fig. 2.10. Two source nodes at the top of the figure want to send data bits c_{A} and c_{B} to the two destination nodes at the bottom of the figure. Each edge can only transmit a single data bit in each transmission. In network coding, the central line combines the data bits c_{A} and c_{B} and transmits the combined value $c_{\mathrm{A}} \oplus c_{\mathrm{B}}$ to the two destination nodes. Since the two destination nodes also receive c_{A} and c_{B} from the two sources respectively, they can subtract the known data bits from $c_{\mathrm{A}} \oplus c_{\mathrm{B}}$ and get the value of c_{B} and c_{A}, respectively. This method is better than conventional routing, where the central nodes have to send the data bit c_{A} and c_{B} separately to the two destination nodes. It was shown in [88, 136] that network coding can be integrated into two-way relaying to achieve good performance. Actually, the broadcast nature of wireless channels is more suitable for network coding than wireline channels. Two-way relaying with network coding shows a good example of cross layer design that connects the physical

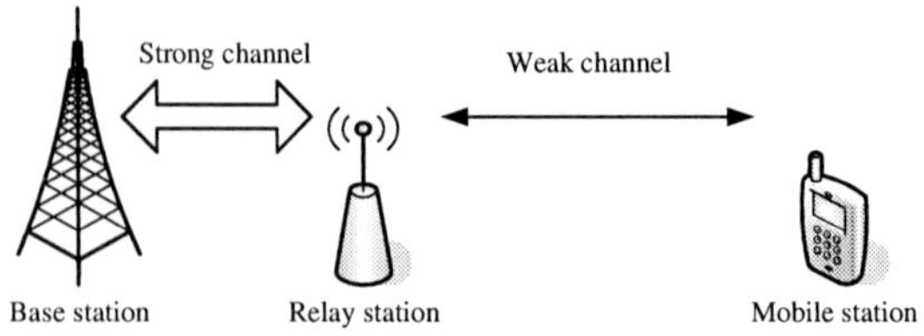

(a) Asymmetric channel conditions

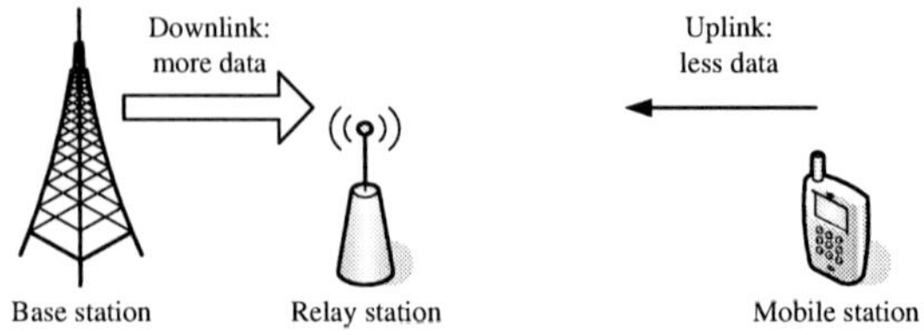

(b) Asymmetric data traffic

Fig. 2.11: Asymmetric channel conditions and asymmetric data traffic

layer (PHY) and the data link layer. Such observation brings new insights into the research of both relay communication and network coding. Nowadays, the idea of two-way relaying has been generalized to many other research areas or scenarios beyond the originally considered ones, such as secret communication [101] and uplink cooperative transmission [39].

2.3.2 Advantages and Challenges of Two-Way Relay Communication

The main advantage of applying the two-way relaying protocol in relaying networks is the improvement of the spectral efficiency. The bidirectional data transmission in a two-way relaying system is performed simultaneously. Even though the data transmission in each direction still requires two channel uses, the sum rate of the network is significantly boosted by the two-way relaying protocol due to the concurrent establishment of the bidirectional links. Another advantage of the two-way relaying protocol compared with other spectrally efficient relaying protocols is that only minor changes in the system is required to integrate the two-way relaying protocol into a conventional relaying system setup. This is highly desirable for the practical network implementation.

The two-way relaying protocol can be applied to general relay networks with bidirectional information flow. However, there is some differences between the two-way relaying systems and the conventional relaying systems. The following lists some the challenges faced by the design and implementation of practical two-way relaying systems.

Asymmetric channel conditions In wireless communication systems, the channel conditions in all the radio links cannot be guaranteed to be the same. In a typical cellular relaying network as shown in Fig. 2.11(a), the channel quality between the base station and the relay station is usually much better than that between the relay station and the mobile station if the mobile station is located near the border of the cellular coverage. This may be due to the pathloss, the different transmit power and the different number of antennas at each station. Such asymmetric channel conditions are detrimental for a two-way relaying system. This is because both the uplink and the downlink data flows have to pass through the two-hop channels. The overall data rate is limited by the hop with the worse channel quality.

Asymmetric data traffic In wireless relaying networks, the amount of data to be transmitted in one direction may be much larger than the other direction. For example, in cellular two-way relaying systems as shown in Fig. 2.11(b), the data rate to be transmitted in the downlink is usually required to be much larger than that in the uplink. How to combine different amount of data is problem especially for two-way relaying systems with network coding. This will be discussed in Chapter 8.

Synchronization Some two-way relaying protocols, e.g., the two-phase protocol, require the two user stations to transmit the data simultaneously. How to synchronize the transmission is an important problem for the practical implementation of the two-way relaying protocol.

2.3.3 Basic System Setup and Transmission Protocols

The basic system setup for two-way relaying is the same as the conventional relaying setup in Fig. 2.6. In conventional relaying protocols, the transmission of data from the source to the destination in a two-hop scenario requires two channel uses, i.e., the source first transmit the data to the relay and the relay forward the data to the destination. In real-world applications, e.g., cellular communications between the base station and the mobile station, both stations have data to transmit to the other side, i.e., each station acts both as a source and a destination. We consider such a traffic pattern, i.e., two stations A and B exchanging data via another relay R as shown in Fig. 2.12. All the stations are half-duplex. When the conventional relaying protocol is applied, it requires four phases to exchange the data as shown in Fig. 2.12(a).

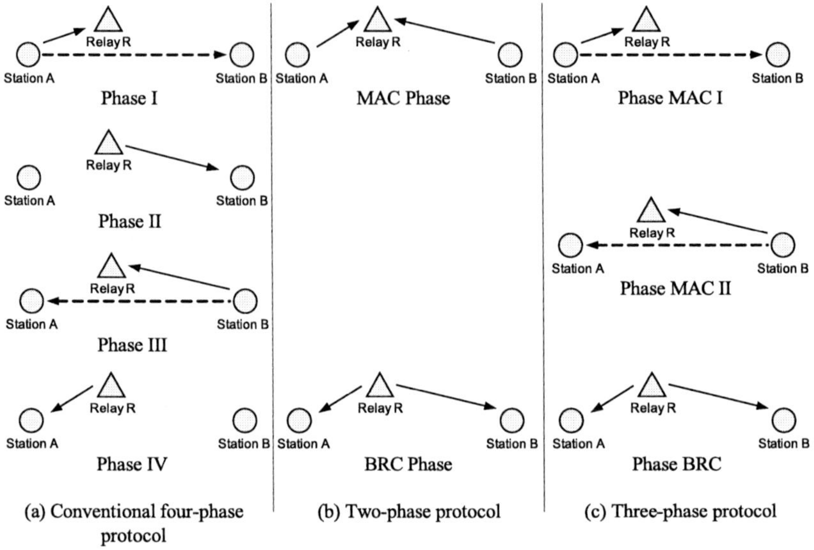

Fig. 2.12: Comparison of different protocols with bidirectional information transmission

Two-way relaying ingeniously exploits the fact that the data to be transmitted to the other side by the relay is known by its source. The idea of two-way relaying is to combine the received data from the two stations at the relay and retransmit the combined signal. Based on the data received from the relay and the previously transmitted data of the two stations, each station can extract the unknown data transmitted from the other side. Based on this transmission patterns, there are two schemes proposed for two-way relay communication: the two-phase protocol and the three-phase protocol. The core idea of the two schemes is the same, and the difference lies in how to transmit the data to the relay.

Two-phase protocol: The two-phase protocol is shown in Fig. 2.12(b). The exchange of data between the stations A and B is completed in the MAC phase and the BRC phase. In the MAC phase, the two stations transmit their data simultaneously to the relay; the relay processes the received signal and retransmits the combined signals back to the two stations. Compared to the conventional relaying schemes, the two-phase protocol saves two channel uses in exchanging the data between the stations A and B. Since both the stations and the relay are half-duplex and they cannot transmit when they receive, the directlink between the two stations A and B need not to be considered. The two-phase protocol was proposed by [193] and has been analyzed in, e.g., [85, 88, 239].

Three-phase protocol: The three-phase protocol is shown in Fig. 2.12(c), which splits the MAC phase into two phases. In this protocol, the stations A and B transmit to the relay R using two orthogonal channel uses. Then the relay combines the received data and sends the combined data back to the stations A and B. Compared to the two-phase protocol, one advantage of the three-phase protocol is that it does not require the complicated multiple access receiver at the relay when the relay has to decode. The other advantage of the three-phase protocol is that the directlink can be exploited for data transmission. When the directlink signal strength is strong, the two sets of received signals provide diversity and improve the decoding performance at the stations A and B. The drawback of the three-phase scheme is that it requires one additional channel use compared to the two-phase protocol. When the two stations are too far from each other and the directlink can be neglected, the three-phase protocol is suboptimal compared to the two-phase protocol at least from information theoretical perspective. The three-phase protocol has been considered in [136, 140, 260]. Comparison of the different protocols was performed in, e.g., [151]

2.3.4 Relaying Strategies

Only with slight modifications, the relaying strategies used in conventional relaying systems can also be applied to the two-way relaying protocol. The relaying strategies are decisive for the system performance and may differ in implementation issues, such as processing power, signal processing complexity and memory (cache) requirements. In addition, specific problems and solution, such as channel estimation, may arise for systems utilizing different relaying strategies.

2.3.4.1 Amplify-and-Forward

The two-way AF relaying strategy is sometimes called "analog network coding" [127] in literature. In this strategy, the relay simply retransmits an amplified version of the received signal in the BRC phase. It is usually applied with the two-phase two-way relaying protocol. The relay gain allocation is assumed to be known at the two receivers. The receiving stations subtract the known data symbols transmitted by itself, i.e., the *self-interference* (SI), from the received signals in the BRC phase, so the remaining signal only contains the data symbols from the other side.

The choice of the relay gain allocation is decisive for the system performance. The relay gain allocation proposed in [193] is simply based on the transmit power constraint at the relay and the received signal power in the MAC phase. Such a choice of relay gain is easy to implement but may not lead to the optimal performance at the receives. For uncoded transmission, i.e., when the relay applies an instantaneous relay function on its received signal, an optimized relay amplification function was proposed in [51] to minimize the average BER in the high SNR regime. Such an optimal relay gain is a function of the received signal value, and it is shown to be a Lambert W function, which is similar to the results obtained for one-way relaying [7].

When the relay is equipped with multiple antennas, the relay may apply linear processing on the received signal. The authors of [277] considered a two-way AF relaying system where each user station is equipped with a single antenna and the relay has multiple antennas. The linear processing strategy at the relay is reduced to the beamforming strategy. Under the given transmit power constraints, the capacity region of the system with beamforming at the relay is analyzed. The optimal relay beamforming structure, as well as an algorithm to compute the optimal beamforming matrix based on convex optimization techniques, is provided.

Space-time codes applied for two-way AF relaying with multiple relay antennas was considered in [90]. The authors of [90] considered a two-way AF half-duplex relaying system where the user stations are equipped with two antennas each and the relay has only one antenna. The source and the destination stations each transmits using the Alamouti's orthogonal STBC. Both upper and lower bounds for the average sum rate as well as an upper bound for the pairwise error probability (PEP) for the proposed two-way orthogonal STBC scheme were provided. It is showed that the average sum rate of the proposed scheme is improved compared to the single antenna case and a diversity order of two is achieved for the considered system configuration.

The availability of channel knowledge is also an important issue in the two-way AF relaying strategy. In the process of self-interference cancellation, the receiving station needs the channel knowledge between itself and the relay; in order to decode the remaining signal, the two-hop CSI from the information source is also required. The design of the training sequence to minimize the MSE of the channel estimation under ZF criterion was considered in [186]. A linear estimator that maximizes the SNR, which takes the channel estimation errors into account was proposed in [73].

2.3.4.2 Decode-and-Forward

The two-way DF relaying strategy was proposed in [193], where the relay decodes the received data bits from the two stations. Since channel decoding is performed at the relay, the two-way DF relay may have more complicated structure and require higher signal processing capability than two-way AF relays. Error-detecting codes (such as the CRC codes) can be applied in the transmitted data from the two stations to guarantee correct decoding at the relay. Unlike the two-way AF relaying strategy where the signal is actually combined on the symbol level at the relay, the decoded data at the relay in the two-way DF relaying strategy can be combined on the symbol level or on the bit level. The data combining schemes at the relay for the two-way DF relaying strategy are important for the system performance, and we summarize them as follows:

Superposition coding The superposition coding (SPC) scheme combines the data from the two user stations on the symbol level. When the SPC scheme is applied at the relay, the relay retransmits the linear sum of the two sets of *symbols* containing the decoded data from the two user stations. In the BRC phase, each user station receives a combination of the symbols containing the data from its own and its partner. By canceling its own data symbols before decoding, each station only needs to decode the data symbols from its partner. This method was proposed in [193].

Network coding The network coding schemes combines the data from the two user stations on the bit level using the XOR operation. The combined data bits are remodulated using conventional quadrature amplitude modulation (QAM) or phase-shift keying (PSK) modulations schemes and retransmitted to the two user stations. The user stations demodulate and decode the received signals, and reveal the unknown data by XOR-ing the decoded data bits with its own transmitted data bits. Network coding applied in two-way relaying systems can be found in [10, 136, 260].

Lattice coding Lattice coding uses modulo addition in multi-dimensional spaces. Like the network coding scheme, the lattice coding scheme utilizes nonlinear operations for combining the data. Applying lattice coding in two-way relaying systems was considered in [15, 160, 161]. Due to its complicated structure, most of the results obtained for the lattice coding scheme remain theoretical. How to apply the lattice coding scheme in real-world two-way relaying systems still needs to be investigated.

For the two-way DF relaying protocol, how to transmit the decoded information of each bit to the two stations is an important issue for the system performance. The authors of [140] considered forwarding signals based on the hard or soft decisions at the relay in the BRC phase of a three-phase MIMO two-way DF relaying system with the network coding scheme. When hard decisions are to be forwarded at the relay, they proposed a new decoding algorithm that takes into account the estimated bit error rate of the packets to be forwarded; when the relay forwards soft decisions, the relay retransmits estimates of the decoded signals based on the soft decisions at the relay decoder outputs, rather than their hard-decision based sliced versions, in order to retain the soft information obtained at the relay. The error in the forwarded signal is modeled as the combination of the hard decoding errors and the Gaussian soft errors, and they modified the hard decision forwarding decoders taking into account such errors. They showed that the proposed scheme achieves better decoding performance at the user receivers compared to the DF schemes that disregard the decoding errors at the relay. However, the proposed soft-decision based decoder only achieves slightly better performance compared to the hard-decision based decoder.

Two-way DF relaying systems with multiple antennas at the relay and its optimal transmission strategies in the BRC phase was considered in [172] from the information theoretical aspects. The authors of [172] considered a two-way DF relaying system where the relay is equipped with multiple antennas and the user stations each with single antenna. In the BRC phase, they showed that the relay beamforming into the subspace spanned by the channels is an optimal transmit strategy for the MISO bidirectional broadcast channel, and the correlation between the channels is advantageous. Furthermore, they presented the optimal transmit strategy at the relay that specifies the whole capacity region for this MISO bidirectional broadcast channel.

Both the two-phase protocol and the three-phase protocol as shown in Fig. 2.12 can be applied to the two-way DF relaying systems. The major difference between them is whether the directlink is exploited. The authors of [99] considered the three-phase two-way DF relaying protocol, where the coded bits from the directlink and the relay consist of different interleaved versions of the same data bits. They form distributed turbo codes and are decoded at the destination using the turbo decoding principle.

When the receivers do not have the channel knowledge, the authors of [53] considered non-coherent transmission in two-way relay systems where the CSI at receiver is not required. They proposed differential AF and DF relaying transmission strategies. Uncoded scenarios were considered and ML detectors were proposed for the AF and DF relaying strategies, where the latter can be considered as performing differential network coding at the physical layer. In addition, several suboptimal detection schemes, including decision feedback detectors and prediction based detectors, were proposed to reduce the complexity of the ML detector. Furthermore, they extended the proposed schemes to multiple-antenna cases and provided the design criterion of the differential unitary space-time modulation (DUSTM) for two-way relay channels.

Unlike the SPC and the network coding schemes, most work on the lattice coding scheme remains theoretical. The authors of [161] considered a two-way relaying system with real inputs. They showed that when the lattice coding scheme is applied at a Gaussian two-way relay channel, the achievable rates at the two stations using lattice coding are asymptotically optimal. The channels between the relay and the two user stations are average power constrained AWGN channel with the SNR of ρ, i.e., the channels between the relay and the two user stations are equally strong. Each user station, as well as the relay, is equipped with a single antenna. The upper bound of the system is shown to be $1/2\log(1+\rho)$ bits per transmitter per use of the MAC phase and the BRC phase of the bidirectional relay channel. The authors showed that the lattice coding scheme can obtain a rate of $1/2\log(1/2+\rho)$ bits per transmitter per channel use, which is asymptotically optimal when the SNR is high. The main idea is to decode the sum of the codewords modulo a lattice at the relay followed by a broadcast phase which performs Slepian-Wolf coding with structured codes. When the SNR is asymptotically low, jointly decoding of the transmitted data from the two user stations at the relay in the MAC phase is shown to be optimal. In such a scenario, the two-way AF relaying strategy is shown to be suboptimal.

A similar transmission strategy was considered in [160], where the authors considered two-way full-duplex relaying systems without the directlink between the user station pair. They proposed a transmission scheme that applies nested lattice codes in the MAC phase and use structured binning for the BRC phase. They showed that the proposed scheme achieves within $1/2$ bits from the cut-set upper bound and is asymptotically optimal when the SNR at the receiver is high. Applying the lattice coding scheme in the two-way relay channel with asymmetric channel qualities was considered in [15], where the authors proposed to apply high-dimensional lattice codes whose shaping gain is close to the optimal one for the two-way relaying communication. The authors called it "modulo-and-forward". When the transmission powers of two nodes are different, they proposed to use superposition coding and partial decoding at the relay node, which were shown to achieve better performance than the conventional AF and DF relaying strategies in certain scenarios.

2.3.4.3 Compress-and-Forward

Unlike the DF relaying strategy where the relay fully decodes the received signals and the AF relaying strategy where the relay simply transmits a linear amplified version of the received signals, the CF relaying strategy allows the relay node to quantize and compress the received signals before transmitting it to the destination node. The index of the quantized codeword is assumed to be transmitted reliably back to the two stations in the BRC phase. The quantization error is usually modeled as noise.

The authors of [190] compared the DF and CF relaying schemes in a two-way full-duplex relaying system. They showed that when the relay is near one of the two stations but far from the other, employing the pure DF or the pure CF relaying strategy at the relay results in low rate for one of the communication directions. The authors proposed to combine the two relaying strategies, which achieves better sum rates compared to the DF and CF relaying strategies. Inspired by this result, the authors of [84] proposed a two-way CF relaying scheme with two layered quantization, where one of the users receives a better description of the relay received signal. The relay operates in half-duplex transmission mode. The proposed scheme was shown to achieve rates within a half bit of the capacity region in two-way Gaussian relay channels when there is no directlink between the two user stations.

2.3.4.4 Compute-and-Forward

The idea of the two-way compute-and-forward relaying strategy is to let the relay transmit a function of the data from the two stations. The function is known to the two stations. Based on signals from the relay and the known data at the destination, the receiver can reveal the unknown data from the other side. Actually, the two-way compute-and-forward relaying strategy works similarly to the two-way DF relaying strategy. The benefit of the two-way compute-and-forward relaying strategy is that the relay does not have to fully decode the two sets of data from the two user stations. It only needs to calculate a reliable version of the function based on the received signals.

The compute-and-forward relaying strategy was originally proposed for conventional relaying networks with multiple sources [163], where multiple linear functions of the transmitted data from several sources are transmitted to the destination node such that the destination can recover the data from all the sources based on those functions. The authors of [163] proposed to use algebraically structured codes, i.e., lattice codes, to exploit the structure of the interference and protect against noise. Later, it was discovered that such a relaying strategy is more suitable for two-way relaying systems. The compute-and-forward relaying strategy applied in MIMO two-way relaying systems was considered in [276].

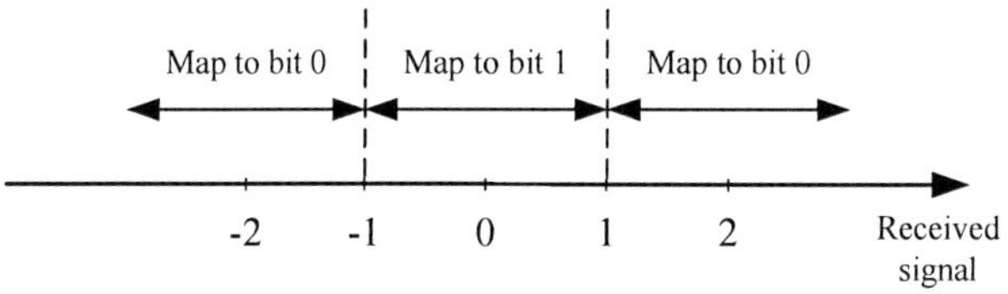

Fig. 2.13: "Denoise" mapping for the received signal at the relay

2.3.4.5 Denoise-and-Forward

The denoise-and-forward scheme was proposed in [187] for two-way relaying. Similar to the DemF scheme for one-way relaying, the denoise-and-forward scheme does not perform channel decoding at the relay. Instead, it maps the received signal to the symbol that corresponds to the combined data bits, i.e., it estimate the XOR-ed data bits transmitted from the two stations based on the received signal and maps the received signal to the XOR-ed data signal constellation. This scheme was originally proposed considering BPSK transmission from the two stations and extended to other constellations [130, 188].

An example of the "denoise" mapping is shown in Fig. 2.13. We only consider the real dimension. The two stations use BPSK modulation in the MAC phase, i.e., the bit 0 is mapped to -1 and the bit 1 is mapped to 1. We assume the channel is Gaussian, i.e., the channel gains between the user stations and the relay are both unity. The received signal at the relay has the following combinations

- Two stations transmit bit 0 and 0; relay receives signal -2, and maps it to bit $0 \oplus 0 = 0$;
- Two stations transmit bit 0 and 1; relay receives signal 0, and maps it to bit $0 \oplus 1 = 1$;
- Two stations transmit bit 1 and 0; relay receives signal 0, and maps it to bit $1 \oplus 0 = 1$;
- Two stations transmit bit 1 and 1; relay receives signal 2, and maps it to bit $1 \oplus 1 = 0$.

Based on the received signal, the relay does not have to decode the data transmitted from the two stations separately, but only needs to decode the resultant bit after the XOR operation.

2.3.5 Two-Way Relaying Networks

In real-world wireless networks, there are different network topologies. The same is true for two-way relaying network as shown in Fig. 2.14. For cellular networks, the traditional network topology is tree-structured, i.e., the base station communicates with the mobile users

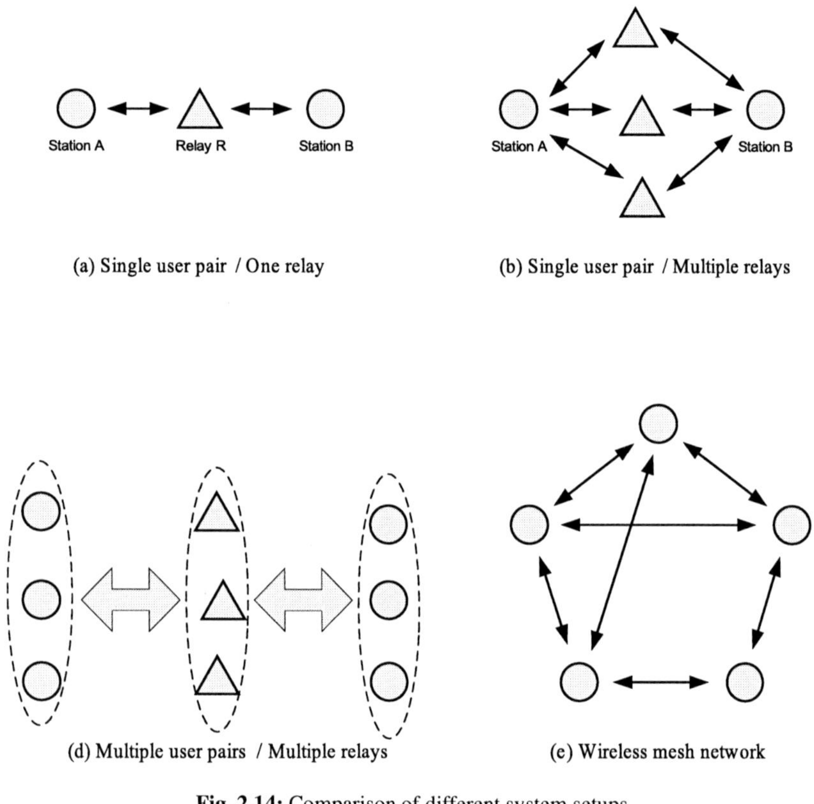

(a) Single user pair / One relay

(b) Single user pair / Multiple relays

(d) Multiple user pairs / Multiple relays

(e) Wireless mesh network

Fig. 2.14: Comparison of different system setups

via relays using multi-hop transmission. The relays forwards the received data towards the destination, and they are sometimes called the *routers*. Cellular relaying networks can be modeled as a two-hop or multi-hop network. In wireless mesh networks, the topology is more complicated since each node can act as a source or destination node of data or as a relay node. We summarize the system setups in two-way relaying networks and important results for them in this section.

2.3.5.1 Multiple Relays

The authors of [52] considered a two-way relaying system with single user pair and multiple relays, where both the two-phase protocol and the three phase protocol were considered. For the two-phase protocol, the authors considered two-way AF relaying strategies, and proposed a partial DF relaying strategy at the relay, where each relay first removes part of the noise before sending the signal to the two terminals. After processing their received signals, the relays encode the signals using a distributed linear dispersion code and retransmit them. For the proposed AF relaying strategy, it achieves the diversity order d of

$$d = \min(N, K)\left(1 - \frac{\log\log P}{\log P}\right) \tag{2.32}$$

where N is the number of relays, is P the total power of the network, and K is the number of symbols transmitted during each time slot. The proposed partial DF strategy achieves the diversity order of $\min(N, K)$ but the conventional DF can only achieve the diversity order of 1. The capacity scaling for two-way relay channel when the number of the relays goes large was considered in [239].

2.3.5.2 Multiple User Pairs

The authors of [125] studied the optimal resource allocation problem for the relay-assisted cellular system employing the orthogonal frequency-division multiple-access (OFDMA) technique. Each relay is responsible for assisting the transmission between the base station and one mobile user. The authors considered both DF and AF relaying strategies for the relays. Algorithm based on convex optimization techniques were developed for optimizing the resource allocation at the base station, the relays and mobile users, which include the subcarrier assignment and the power allocation on each transmission link. Substantial throughput gains are achieved by using two-way relaying protocols and the proposed optimal resource allocation in the considered cellular communication system.

The two-way orthogonalize-and-forward scheme was proposed in [193], where the system consists of multiple user pairs that exchange data via multiple AF relays. Each station is equipped with single antenna. The orthogonalize-and-forward scheme utilizes the scheme proposed in [256] by choosing the gain allocation at the relays to diagonalize the compound two-hop channels of the system so that each user only receives data from its corresponding transmitter.

2.3.5.3 Mesh Networks and Ad Hoc Networks

Wireless mesh networks and mobile ad hoc networks (MANET) are closed related. The main difference between them is that the nodes in mesh networks are generally fixed while the nodes in MANET are mobile. In mesh network, each node in the network may act both as a source, a destination or a relay of information. Depending on the topology, multiple connections among users within the network are possible. Any two nodes in a mesh network are connected together via multiple hops. The data transmitted from the source node in wireless mesh networks can jump around the broken or blocked paths until the destination is reached. Since there are usually more than one path between the source and destination in such networks, mesh network has the advantage of self-healing, i.e., the network can still work even when some nodes break down. Wireless mesh networks were originally developed for military applications, but are now having more and more applications in civilian communication systems. International standards such as IEEE 802.11s [4] are proposed, which defines how wireless devices can interconnect to create a WLAN mesh network. Research work considering network coding in wireless mesh networks can be found in, e.g., [60, 128].

2.3.5.4 Capacity Scaling Law and DMT in Two-Way Relaying

The authors of [239] considered the capacity scaling in MIMO two-way relay channels where all nodes work in half duplex mode. They showed that the capacity scales linear with the number of transmit antennas and logarithmically with the number of relays as the number of relays grows large. That is, for a two-way relay network with each user station equipped with M antenna, K relays assisting the communication between the two user stations, and each relay is equipped with N antennas. As $K \to \infty$, the capacity of the system C scales like

$$C \sim M \log(K). \tag{2.33}$$

This result shows that with two-way relay channels it is possible to obtain full-duplex performance asymptotically in the number of relays while using only half-duplex nodes.

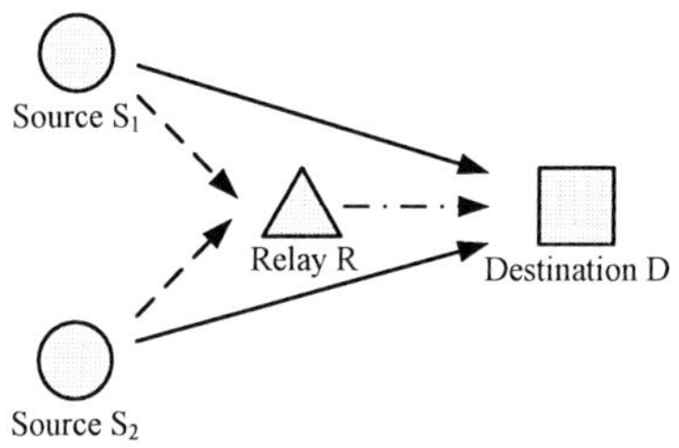

Fig. 2.15: Multi-access relay channel

The authors of [85] considered a two-way full-duplex relay channel where all the stations are equipped with multiple antennas. They characterized the optimal diversity-multiplexing gain tradeoff (DMT) curve for the system when assuming independent quasi-static Rayleigh fading channels and channel state information available at the receivers. The authors also showed that the optimal DMT can be achieved by a CF type relaying strategy in which the relay quantizes its received signal and transmits the corresponding channel codeword. With this transmission protocol, the two transmissions in opposite directions can achieve their respective single user optimal DMT performances simultaneously.

2.3.6 Related Research Topics

Nowadays the application scenarios of two-way relaying are not limited to the three-station network as shown in Fig. 2.12. The research on two-way relaying has revealed it connections to other research topics. We list some of the related research topics as follows.

Multi-access relay channel: Multi-access relay channel (MARC) was introduced in [132]. It considers the system as shown in Fig. 2.15, where two sources, S_1 and S_2 transmit information to the destination D via a half-duplex relay R. This system models the uplink transmission of two sources to the base station via a relay in between. It has been found that network coding can be applied to such a system similar to two-way relaying. Consider the following transmission scheme:

- The first phase: the source S_1 transmit its data to the relay R and the destination D,
- The second phase: the the source S_2 transmit its data to the relay R and the destination D,

- The third phase: the relay combines the data from S_1 and S_2 and transmits the combined data to the destination D.

The mathematical description of this scheme is very similar to the three-phase two-way relaying protocol. The author of [98] proposed a turbo decoding method in MARC similar to the one used in two-way relaying [99]. The authors of [274, 275] considered imperfect decoding at the relay and designed a quantizer for the soft decoding value at the relay, which outperforms the belief propagation approach in [267] where the relay sends the LLR of the network coded message to the destination. Outer bounds for discrete memoryless MARC are obtained in [195], where the authors also presented a CF strategy and an AF strategy for MARC. The DMT was analyzed in [41]. An AF relaying strategy for MARC was proposed in [269], where the authors investigated both linear and nonlinear mapping schemes at the relay.

Multicast: *Multicast* refers to the technology for delivering information to a group of destinations simultaneously, i.e, one-to-many communication. It is an important technology for streaming media and Internet television applications. In computer networks, it is typically refers to *IP multicast*. Multicast over wireless networks has been considered in, e.g, [122, 214, 238, 248]. The idea of network coding has been shown to be able to improve the multicast capacity of a network [104, 142].

Secret communication: In cryptography, the only encryption method that has peen proven to be absolutely impossible to crack if used correctly is the method of *one-time pad*. In one-time pad, the useful data is encrypted by XOR-ing it with a sequence of random data of the same length. Such a random data sequence, which acts as a random key, is only known by the transmitter and the receiver. If the key is truly random and only used once, the resulting transmitted data sequence, i.e., "ciphertext", will be impossible to be decrypted without knowing the random key. This method is called *one-time pad* and was proved by Shannon [208]. When network coding is applied at the relay in the BRC phase, the transmitted data from the relay can also be considered as a ciphertext, which can only be deciphered when the "key", i.e., the data from one station is known. Compared to direct forwarding, this property is more security for the data transmission at the relay. Works related to secrete communication in two-way relaying can be found in, e.g., [101].

Multi-way relaying channel: The authors of [86] considered the scenario that multiple users exchanges information with the help of a relay. The system is divided into multiple clusters where the users within the same cluster wish to exchange messages among themselves via the relay. Multiple interfering clusters can transmit data simultaneously. AF, DF and CF protocols are investigated for the system and the authors also considered two extreme cases: one is that every user wants to receive data from all other users and the other is that the system is composed of multiple two-way relay channels.

Chapter 3

Coverage Analysis of Multi-Antenna Decode-and-Forward Relaying Systems

Coverage extension is an important application of relay communication in cellular networks. However, the quantity of the coverage extension using relays in cellular systems is usually determined only by measurement campaigns without much theoretical analysis beforehand. Analytical frameworks for quantifying the coverage extension in cellular relay networks provide insights on relaying and serve as an important tool for cellular network planning. This chapter proposes a quantitative analysis framework of coverage extension by using decode-and-forward (DF) relays in the downlink direction of a cellular network. We consider the scenario that the relays are placed uniformly on a circle around the base station, and define the coverage of cellular relaying networks based on a 1%-outage achievable rate criterion. The considered channel model is based on the IST-WINNER model for the fourth generation (4G) cellular systems. However, the analytical method is general and can be applied to other cellular systems. To describe the coverage of a relaying system and its relation to the number of relays, we introduce the concept of *circular coverage range* and *coverage angle*. Moreover, the coverage of such a network also depends on the distance between the base and relay stations. For a fixed number of relays, there exists a certain distance between the base and relay stations such that the relaying system achieves the maximum circular coverage range. We are interested in how to determine the maximum circular coverage range given the coverage angle. Moreover, we derive upper and lower bounds for the maximum circular coverage range when the transmitters do not have the channel knowledge. We also propose a heuristic approximation for the maximum circular coverage range, which is useful for practical system design. The corresponding distance between base and relay stations that achieves this maximum circular coverage range is also calculated in the analysis. Furthermore, we optimized the transmit covariance matrices when the channel knowledge is available at the transmitters, and studied its impact on the circular coverage range extension. The comparison of coverage for cellular relaying systems and that for conventional cellular systems are shown by simulations.

3.1 Introduction

In the future fourth generation (4G) cellular wireless networks, the use of higher carrier frequencies (higher than 5 GHz) leads to smaller cell size compared to current 2G and 3G networks [156], [155]. Using relays and multihop transmission strategies to assist transmission in cellular networks has been shown to be effective in cellular coverage improvement [174]. This means higher data rates can be carried over larger distances, which reduces the base station density in the cellular network.

Currently, there are two major types of relays: amplify-and-forward (AF) relays and decode-and-forward (DF) relays. AF relays simply transmit an amplified version of the received signal, while DF relays first decode the received signal before re-encoding and retransmitting it. DF relays may be preferable in a cellular environment since they do not suffer from noise enhancement. In addition, currently relay stations cannot transmit and receive data simultaneously due to the coupling between the transmit and receive circuitry [134]. Thus half-duplexity is also a constraint for practical relays (see Chapter 2).

In cellular networks, we can either use free mobile stations (MS) as relays (called *user cooperation*) or build dedicated fixed relay stations (RS). These dedicated relay stations have sufficient power supply, higher signal processing capabilities and better antennas than the mobile stations. The positions of those dedicated relays can also be chosen intentionally to ensure stable connections to the base station (BS) of the cell. This chapter considers dedicated half-duplex DF relay stations with multiple antennas in a cellular downlink environment.

The coverage of wireless systems is an important factor in cellular network planning. The direct way to determine the coverage of a cellular system is by measurement, where one mobile test equipment is sent out to determine the signal quality as it moves around in the cell (see e.g., [92]). In this chapter, we consider such a measurement campaign scenario where this measurement equipment is assisted by one relay station at a given time.

Although measurement campaigns are effective ways of determining the coverage, the results are just valid for the specific measured system and highly depend on the environment. The aim of this chapter is to provide a theoretical framework for quantifying the coverage improvement in a cellular relaying network, and provide useful hints for cellular relaying system design.

There have been plenty of research works on the coverage of conventional cellular networks. For example, Chen *et. al* [40] proposed a scheduling scheme to replenish diversity deficient MIMO systems with multiuser diversity, and they showed that the coverage of conventional cellular networks could be significantly increased by wireless scheduling. With the advances of relaying networks, much effort has also been devoted to the research on the coverage of cellular relaying systems. Fujiwara *et. al* [68] applied the multi-hop connection scheme to a code division multiple access (CDMA) cellular system and compared its performance to that of a conventional single-hop cellular system by computer simulations. The simulation results confirmed that significant coverage area enhancement could be obtained by using multi-hop connection. Another interesting work [55] considered a similar CDMA network. The authors proposed to divide each cell into inner and outer regions, and only the users belonging to an outer region communicate with the BS using two hops. Each of the two regions is allocated a separate frequency channel. The authors analyzed the intercell interference reduction and coverage extension when the proposed method is used. Hu *et. al* [110] showed the coverage extension of a cellular relaying network by using a frequency channel reuse scheme, which uses a pre-configured relaying channel selection algorithm to minimize co-channel interference in the network. In [57], the authors investigated the cellular downlink transmission scenario where randomly positioned mobile stations are used as relays. The authors studied the coverage of such a cellular downlink scenario and considered both the pathloss only channel model and the channel model which include both pathloss and the lognormal shadowing. They shows that when the lognormal shadowing is considered, the coverage improvement using relays is more significant. Using mobile stations as relays was also considered in [232], where the mobile stations chosen as relays use DF strategies to forward the data to the destinations. The authors investigated the effects of different relay and channel selection strategies on the system performance. In [20], the authors considered using dedicated DF relay stations in 3GPP LTE-Advanced cellular networks. The authors proposed a heuristic method describing the relay transmit power, the ratio between the number of the relay and the base stations, and the performance of the system. The coverage analysis for AF relaying schemes was considered in [226]. In most existing work, they either choose a fixed cell range or choose the QoS requirement based on the signal power level. The received power alone is sometimes not a good indicator of the system throughput, especially for MIMO systems. On the other hand, the method for evaluating the coverage is usually measurement campaign or computer simulations, which is difficult to obtain insights.

In order to quantitatively analyze the coverage extension in a cellular DF relaying network, we propose the concept of *coverage angle* and *circular coverage range*. In short, the coverage angle is $360°$ divided by the number of relays, which represents the angle of the sector covered by each relay. The coverage of a relaying system is measured by circular coverage range, which is the radius of the circular coverage area of the relaying system achieved by placing those relays. When the coverage angle (or the corresponding number of relays) is given, we can find the maximum circular coverage range by varying the distance between the BS and the RS. The distance between the BS and the RS that achieves the maximum circular coverage range is called the *optimum distance* between them. The relationship between the coverage angle and the maximum circular coverage range fully characterizes the coverage improvement in the cellular relaying network. Based on these concepts, we propose analytical upper and lower bounds for the maximum circular coverage range when the power is uniformly allocated at the transmit antennas. This corresponds to the case of no channel knowledge at the transmitters. We also propose a heuristic method to approximate the maximum circular coverage range and the optimum distance between BS and RS, which may be useful for system design. The impact of channel knowledge at the transmitters on the circular coverage range will be shown by simulation results. In the analysis, we neglect the intercell and intracell interference and assume orthogonal channels are allocated for the transmission at each relay.

Our Contributions: The contributions of this chapter can be summarized as follows:

- We propose novel concepts, such as coverage angle and circular coverage range, for describing the coverage extension in cellular relaying network
- We propose a novel quantitative analytical framework for calculating the coverage extension by using MIMO DF relays, which is important for cellular network planning.

The remainder of the chapter is organized as follows: In Section 3.2, the reference channel model and metrics for describing the coverage extension in a cellular relaying network will be presented. We will first discuss how we define coverage from the outage capacity point of view. Then we will propose the idea of *circular coverage range* and its relation with *coverage angle* in detail. In Section 3.3, the achievable rate for the MIMO DF relaying systems will be summarized. The impact of channel state information at the transmitters (CSIT) will also be considered. We will present upper and lower bounds of the maximum circular coverage range for given coverage angles in Section 3.4, where we just consider uniform power allocation at the transmitters. We also propose a heuristic approximation for the maximum circular coverage range, where the corresponding optimum relay position is determined at the same time in the analysis. The comprehensive performance results are presented in Section 3.5. Conclusions are drawn in Section 3.6

3.2 Coverage of Cellular Relaying Systems

Upcoming 4G systems are expected to move to much higher carrier frequencies (> 5 GHz) than today's systems (< 2.2 GHz); this leads to higher signal attenuations. In addition, the bandwidth of those new systems will be much larger than that of today's cellular systems, but the transmit power will probably not increase. Therefore we expect that the coverage range of a 4G base station will be smaller compared to today's cellular systems. In this chapter we neglect the influence of intercell interference on the coverage by investigating a single cell scenario; this refers to a case where neighboring cells use orthogonal channels (e.g. different frequency bands). We assume different relays use different resources, e.g., in frequency, time, spreading code, space or combinations of these, to serve their mobile users. In this scenario, the intra-cell interference between the relays can be neglected.

The coverage of a cellular system is usually defined as the area that fulfills a certain quality of service (QoS) requirement. However, in a cellular relaying system, such an area may be an irregular figure, which is hard to analyze and impractical for system design. In order to quantitatively analyze the coverage of a relaying system, we propose the concept of *coverage angle* and *circular coverage range*. In the following sections, the coverage of a cellular relaying system is always measured in terms of those two notions.

3.2.1 Reference Channel Model

In order to quantitatively calculate the coverage of a cellular relaying, certain channel model has to be applied in the analysis. We use the following channel models, whose parameters were proposed in the IST WINNER project [18]. We assume the BS to be placed above rooftop level, i.e., more than 30m high. The RS is placed on rooftop level, so we can assume good channel conditions between the BS and the RS. The MS is located at street level, i.e., at the height of a person. The transmit power at BS and RS is 1W each, which is much lower than current 2G systems. We choose the same transmit power at RS and BS because the transmit power at BS is already very low. In this chapter, we assume the channels to be isotropic, i.e., the transmitted signal experiences the same fading statics in every direction. Thus the original cell (without relays) is circular.

We consider a wideband channel, but we constrict ourselves to frequency-flat fading, i.e., frequency diversity is not considered here. This restriction corresponds to the worst-case

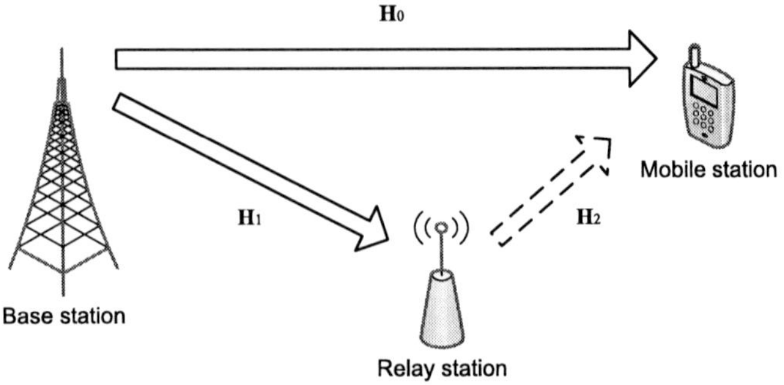

Fig. 3.1: Cellular relaying system channel model

scenario consideration because frequency diversity can also be exploited to enhance the circular coverage range in frequency-selective fading environments. As shown in Fig. 3.1, the first hop channel $\mathbf{H}_1$ is a Ricean fading channel with K-factor of 10dB, while the direct path channel $\mathbf{H}_0$ and second hop channel $\mathbf{H}_2$ are Rayleigh fading channels. The pathloss model is as follows:

$$\begin{aligned} PL_0 &= 35.0\log_{10}(d_0) + 38.4; && (3.1)\\ PL_1 &= 36.5 + 20\log_{10}(f_c/2.5) + 23.5\log_{10}(d_1); && (3.2)\\ PL_2 &= 35.0\log_{10}(d_2) + 38.4, && (3.3)\end{aligned}$$

where PL_0, PL_1 and PL_2 are the pathlosses in dB for the direct channel (BS-MS), the first hop channel (BS-RS), and the second hop channel (RS-MS), respectively. d_0, d_1 and d_2 are the distance between the transmitter and the receiver for the corresponding channels measured in meters. f_c represents the center frequency in GHz. In our simulations, $f_c =$ 5GHz. As we can see, the pathloss exponent for the first hop channel is 2.35, while it is 3.50 for the direct and second hop channels. This channel model complies with the assumption that the channel condition between BS and RS is much better than the channels to the MS. The noise variance is calculated as

$$\sigma^2 = k_B T \Delta f F = 1.3805\times10^{-23}\times290\times1\times10^{8}\times10^{(8/10)} = 2.5260\times10^{-12}\ \text{Watt} \quad (3.4)$$

where k_B is Boltzmann's constant in joules per kelvin, T is the absolute temperature in kelvins, Δf denotes the bandwidth, and F denotes the noise figure.

We summarize the parameters for the system model in Table 3.1.

Table 3.1: System Model Parameters

Bandwidth	100 MHz
Center frequency	5 GHz
Transmit power at BS and RS	1 W
Noise figure of RS and MS	8 dB
QoS requirement (1%-outage rate)	1 b/s/Hz
Pathloss exponent H_1 (BS-RS)	2.35
Pathloss exponent H_0 and H_2 (BS-MS and RS-MS)	3.50

3.2.2 QoS Requirements and SNR Regime

The choice of the QoS requirement determines the coverage area of a cellular system and lays the foundation for our following discussions. This requirement influences the minimum required SNR at the mobile stations within the coverage of the system. A common choice [137] of QoS criterion is an ε-outage data rate requirement, which means that the supported data rate for every position within coverage is guaranteed to be higher than the specified required rate C_Q with probability of at least $1 - \varepsilon$, i.e., $\mathrm{P}\{C \leq C_Q\} \leq \varepsilon$, where C is the achievable data rate of the user within coverage.

The choice of ε and C_Q influences the operational SNR regime at the border of the cell and therefore the possibilities of enhancing the range by diversity, spatial multiplexing or array gain techniques. If C_Q is chosen so large that an MS at the border of the centralized cell without relays is in the high SNR regime, spatial multiplexing techniques are efficient means for further extension of the cell. This comes from the fact that they cause a pre-log rate gain proportional to the minimum number of BS and MS antennas. Using beamforming to provide array gain only causes a logarithmic gain in achievable rate, which is less efficient.

We have to choose a QoS criterion that meets the high data rate requirement of future 4G cellular networks. On the other hand, the cell size under this QoS requirement should not be too small. So we require the 1%-outage rate for each point in the coverage area to be higher than 1b/s/Hz[1], i.e., $C_Q = 1$b/s/Hz and $\varepsilon = 0.01$. This QoS requirement makes the MS at the border of the cell to be in the low SNR regime, where spatial multiplexing gain is small and the achievable rate scales linearly with the SNR. For example, the received signal SNR of

[1]The unit b/s/Hz actually represents the spectrum efficiency. However, since the bandwidth is fixed in our system model, the achievable data rate is proportional to the spectrum efficiency. We do not distinguish the two terms unless it causes confusion in the following.

the mobile sets on the cell border is about -4dB for a 4×4 MIMO system under this QoS requirement. This makes diversity and array gain techniques much more attractive compared to spatial multiplexing. Note it does not mean every user in the system is in low SNR regime. However, those users who have high SNR do not affect the coverage because they are in the middle of the coverage area.

3.2.2.1 Coverage Range for Conventional Cellular Systems

In conventional cellular mobile systems, the BS transmits data directly to the MS, and there is no relay to facilitate the communication. We assume OFDMA technique is applied in the system and each MS is assigned with a distinct OFDM subcarrier. That is, there is no interuser interference in the system for the uplink and downlink transmission. We apply the IST WINNER C2 NLOS channel model [18] (see (3.1) in Section 3.2.1) for the BS to MS transmission and we provide numerical results on the coverage range of such a system. The QoS requirement for each MS is chosen such that the 1%-outage capacity of the border users is not smaller than 1 b/s/Hz. We discuss how the coverage range scales with the diversity order, array gain, and number of antennas in the low SNR regime.

In Fig. 3.2 the coverage range for different number of receive antennas N versus the number of transmit antennas M is depicted. The solid lines correspond to the case, that the BS has no knowledge about the channel (no CSIT), whereas the dashed lines are for the case that the BS has instantaneous CSIT and therefore can perform water-filling method to maximize the instantaneous capacity. In the left figure the absolute coverage range is shown, whereas on the right side all values are normalized with respect to the coverage range of a single-input single-output (SISO) system.

For a multiple-input single-output (MISO) system without CSIT only a diversity gain can be achieved. Compared to a SISO system, the range improvement is high for the first couple additional order of diversity. Later the increase in range saturates. It can be seen, that increasing the number of transmit antennas from $M = 1$ to $M = 2$ doubles both the diversity order and the coverage range. From $M = 4$ to $M = 8$ the range increase is only about 10%.

The blue dashed MISO curve shows that array gain pays off more in terms of range improvement than only diversity gain. When the number of transmit antennas are doubled from $M = 4$ to $M = 8$, the improvement of coverage range is about 50%.

Since the users at the border are in low SNR regime, spatial multiplexing gains plays an unimportant role in the coverage range of the system. This can also be seen from the

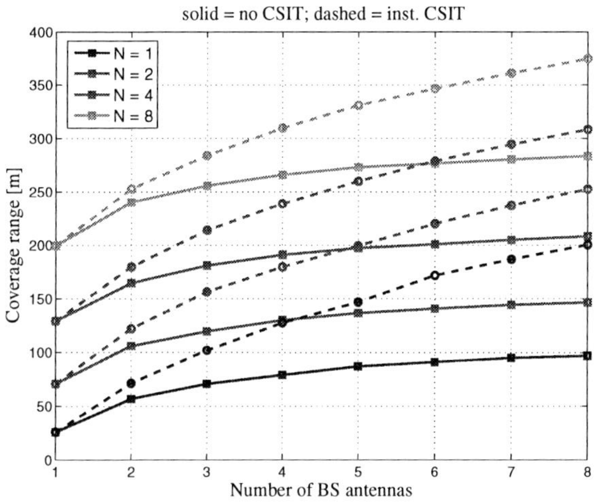

(a) absolute coverage range

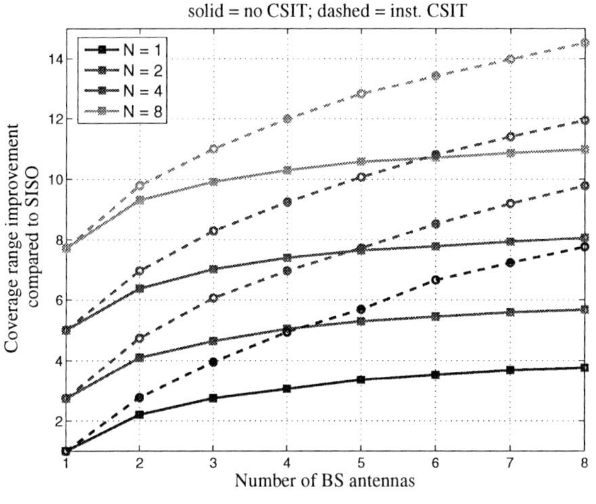

(b) relative coverage range compared to SISO

Fig. 3.2: Coverage range for direct communication

comparison of the coverage range of $(M/N) = (8/2)$ and $(M/N) = (4/4)$ with CSIT, where the coverage range for $(M/N) = (8/2)$ is larger than that for $(M/N) = (4/4)$ in this case. In the high SNR regime one would expect larger capacities for the $(M/N) = (4/4)$ case due to the spatial multiplexing gain of 4 compared to the spatial multiplexing gain of 2 for the $(M/N) = (8/2)$ case. However, in the low SNR regime spatial multiplexing gain have little impact in the capacity. In both cases $MN = 16$, so that both should have an identical diversity gain of $MN = 16$. The factor that determine the coverage range is then the array gain. When CSIT is available, it is shown in [13] that the array gain is proportional to the largest eigenvalue and is therefore upper bounded by $(\sqrt{M}+\sqrt{N})^2$. Intuitively, this is due to the fact that there are 16 channel coefficients but only $M + N$ weighting coefficients available at source and destination. In our case of $(M/N) = (8/2)$ and $(M/N) = (4/4)$ this gives us upper bounds on the array gain of 18 and 16, respectively. Thus, the $(M/N) = (8/2)$ can achieve a larger coverage range. Note that in the case of no CSIT, the coverage range of the $(M/N) = (4/4)$ system is higher than of the $(M/N) = (8/2)$ system, because the higher receive array gain of the $(M/N) = (4/4)$ system.

In Fig. 3.3 we show the CDF of the capacity for the users at the coverage border. In the left figure the CDFs for a MISO system are depicted. It can be seen, that the 1%-outage capacity is for all curves equal to the required 1 b/s/Hz. Only the mean value of the CDFs changes. In the right figure the CDFs for $N = 2$ are depicted. If one compares the curve for $(M/N) = (2/2)$ in Fig. 3.3(b) with the curve for $(M/N) = (4/1)$ in Fig. 3.3(a), it can be observed that both CDFs are almost identical. Both cases have equal diversity order.

To quantify the range improvement in a MISO system with M transmit antennas in the low SNR regime due to instantaneous CSIT compared to no CSIT, we evaluate the capacities of both schemes.

In the case of no CSIT the capacity of a MISO system is given as

$$C_{\text{noCSIT}} = \log_2\left(1 + \frac{\rho}{M} \cdot ||\mathbf{h}||^2\right),$$

where ρ is the SNR and $\mathbf{h}$ is the MISO channel vector between transmitter and receiver.

In the case of perfect instantaneous CSIT, the capacity is given as

$$C_{\text{CSIT}} = \log_2\left(1 + \rho \cdot ||\mathbf{h}||^2\right).$$

Due to the fact that we are in the low SNR regime the log-function scales linearly with the

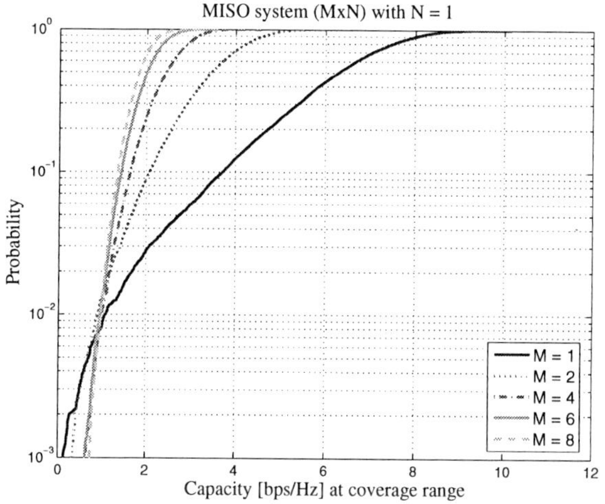

(a) MISO

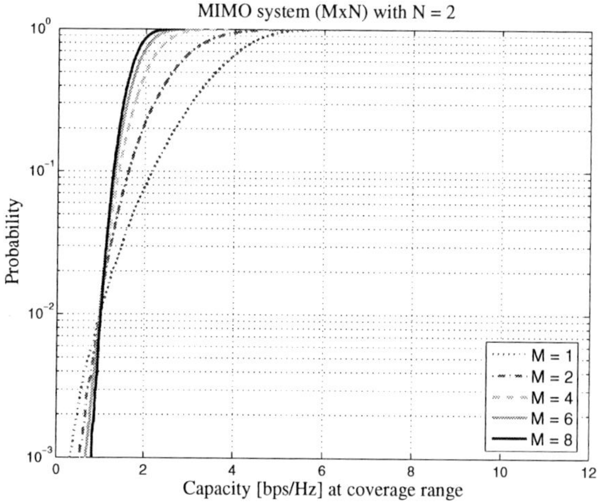

(b) MIMO

Fig. 3.3: CDFs of capacity of a direct communication MISO and MIMO system at the coverage edge

SNR, i.e.,

$$C_{\text{noCSIT}} \approx \frac{\rho}{M\log(2)} \cdot \|\mathbf{h}\|^2,$$
$$C_{\text{CSIT}} \approx \frac{\rho}{\log(2)} \cdot \|\mathbf{h}\|^2.$$

It can be seen that for the same SNR ρ, the capacity with CSIT C_{CSIT} would be M times larger than without CSIT C_{noCSIT}. Therefore, the attenuation due to the path loss in the case of CSIT can be also M times larger than the case of no CSIT. If we assume the pathloss exponent to be α, i.e., the signal strength attenuates according to $\sim r^{-\alpha}$, the distances with and without CSIT have the ratio

$$\left(\frac{r_{\text{CSIT}}}{r_{\text{noCSIT}}}\right)^{\alpha} = M.$$

Therefore, the coverage range of a MISO system with instantaneous CSIT compared to a MISO system without CSIT is given by

$$r_{\text{CSIT}} = M^{\frac{1}{\alpha}} \cdot r_{\text{noCSIT}}, \tag{3.5}$$

where α denotes the path-loss exponent. In Fig. 3.4 it can be seen that this approximation is quite tight to the simulated coverage range of a MISO system with CSIT. Here α is chosen as $\alpha = 3.5$ according to (3.1).

3.2.3 Coverage Angle vs. Circular Coverage Range

For MIMO relaying systems, the coverage range is not easily defined as single-user MIMO systems. Thus we first have to define the coverage range of a MIMO relaying system in this section. In accordance with most other papers on cellular relaying (e.g. [110]), we consider the case that the relays are placed uniformly around the BS. After placing the relays, we still require the new cell to be a *circular cell*, where the same QoS requirement is fulfilled. We refer to the cell radius achieved by the BS and the uniformly placed relays as *circular coverage range* r_{cov}. Note that we always require a circular shape for the cell. If the border of the area where the QoS requirement is fulfilled is not a circle, the circle which has the maximum radius within this shape determines r_{cov}. This is because requiring the new cell to have the same shape as the original cell provides a good basis for comparing their coverage. On the other hand, this analysis will provide the basis for analyzing cells with other shapes. For example, the circular cell considered here is the inscribed circle if we require the cell

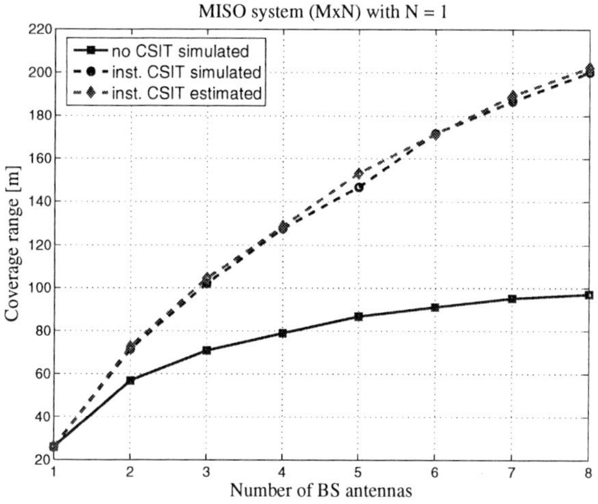

Fig. 3.4: Coverage range improvement because of CSIT for a MISO direct communication system

shape to be hexagon. If the circular coverage range r_{cov} is known, the size of the hexagon cell can be calculated accordingly.

To clarify our definition of circular coverage range, Fig. 3.5 shows the coverage region by placing eight relays. The radius r_0 is the coverage range for the BS before the relays are placed. The circular coverage range is defined by the radius r_2 which is the new r_{cov} for the system of the BS *and* relays. The circular coverage range can be extended by placing more relays around the BS. For example, in Fig. 3.5, the circular coverage range of the system can approach r_1 by placing infinitely many relays around the BS uniformly. The angle $\alpha = 360°/8 = 45°$ that determines the size of the circle sector supported by one specific relay is also shown in Fig. 3.5. We refer to this angle as *coverage angle* α_{cov}. Equivalently, when we say the coverage angle is α_{cov}, we mean that $N_{\text{r}} = \lceil 360°/\alpha_{\text{cov}} \rceil$ relays are placed uniformly in the cell. Due to symmetry, we only depict one relay in the following figures if the coverage angle is specified.

We observe that the circular coverage range also depends on the distance between the BS and the RS. Given a certain coverage angle, we are interested in the maximum achievable circular coverage range by varying the distance between the BS and the RS. This distance between the BS and the RS that achieves the maximum circular coverage range is called the *optimum distance* between them.

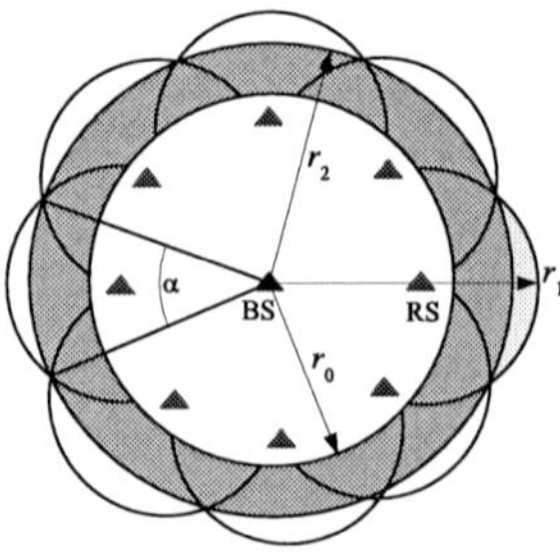

Fig. 3.5: Example for enhanced coverage region by multiple relays

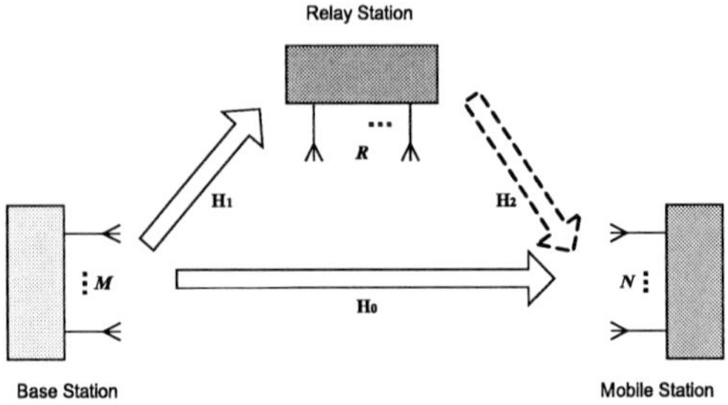

Fig. 3.6: MIMO relaying system model

3.3 Achievable Rate Expressions for MIMO DF Relaying Schemes

We consider a MIMO relaying system shown in Fig. 3.6, and we denote the number of antennas at BS, RS and MS as $M/R/N$, respectively. $\mathbf{H}_0$ is an $N \times M$ matrix representing the direct channel between the BS and MS, and $\mathbf{H}_1$ is an $R \times M$ matrix representing the first hop channel between the BS and RS. $\mathbf{H}_2$ is the second hop channel between the RS and MS, which is an $N \times R$ matrix. We denote the power constraint at BS and RS to be P_{BS} and P_{RS}, respectively. $\mathbf{n}_\text{d}^{(1)}$, $\mathbf{n}_\text{d}^{(2)}$ and $\mathbf{n}_\text{r}$ are the Gaussian noise at MS in the first time slot (direct link), the noise at MS in the second time slot (relay link) and the noise at RS, respectively. We have $\mathbf{n}_\text{d}^{(1)} \sim \mathcal{CN}(0, \sigma_\text{d}^2\mathbf{I}_N)$, $\mathbf{n}_\text{d}^{(2)} \sim \mathcal{CN}(0, \sigma_\text{d}^2\mathbf{I}_N)$ and $\mathbf{n}_\text{r} \sim \mathcal{CN}(0, \sigma_\text{r}^2\mathbf{I}_R)$, and each noise vector is i.i.d. in space and time. We only consider half-duplex DF relays.

The DF relaying scheme comprises of two time slots. In accordance with other papers (e.g. [134] and [175]), we assume the duration of the two time slots to be the same. In the first time slot, the BS first transmits the data vector $\mathbf{x}_s$. The transmit signal covariance matrix is $\mathbf{R}_1 = \mathrm{E}\{\mathbf{x}_s\mathbf{x}_s^H\}$. In order to meet the transmit power constraint, we require $\mathrm{tr}(\mathbf{R}_1) = M$. Let $\mathbf{y}_r$ and $\mathbf{y}_d^{(1)}$ represent the signal received at RS and MS in the first time slot, respectively. They can be expressed as:

$$\mathbf{y}_r = \sqrt{\frac{P_{BS}}{M}}\mathbf{H}_1\mathbf{x}_s + \mathbf{n}_r, \tag{3.6}$$

$$\mathbf{y}_d^{(1)} = \sqrt{\frac{P_{BS}}{M}}\mathbf{H}_0\mathbf{x}_s + \mathbf{n}_d^{(1)}. \tag{3.7}$$

The RS retransmits the received data to the MS if the decoding is successful. We denote the signal transmitted at the RS to be $\mathbf{x'}_s$. The signal covariance matrix $\mathbf{R}_2 = \mathrm{E}\{\mathbf{x'}_s\mathbf{x'}_s^H\}$, and we also require $\mathrm{tr}(\mathbf{R}_2) = R$ to meet the transmit power constraint. The signal received at MS in the second time slot can be expressed as:

$$\mathbf{y}_d^{(2)} = \sqrt{\frac{P_{BS}}{M}}\mathbf{H}_2\mathbf{x'}_s + \mathbf{n}_d^{(2)}. \tag{3.8}$$

The achievable rate expressions for a DF relaying system have been shown in e.g. [134] and [175]. We recapitulate the main results as follows. According to (3.6), the maximum mutual information of the first hop transmission is

$$I(\mathbf{x}_s; \mathbf{y}_r) = \log_2 \det\left(\mathbf{I}_R + \frac{P_{BS}}{M\sigma_r^2}\mathbf{H}_1\mathbf{R}_1\mathbf{H}_1^H\right). \tag{3.9}$$

Based on (3.7) and (3.8), the maximum mutual information between the transmitted data symbols and the received signals at MS can be expressed as [175]

$$I(\mathbf{x}_s, \mathbf{x'}_s; \mathbf{y}_d^{(1)}, \mathbf{y}_d^{(2)}) = \log_2 \det\left(\mathbf{I}_N + \frac{P_{BS}}{M\sigma_d^2}\mathbf{H}_0\mathbf{R}_1\mathbf{H}_0^H + \frac{P_{RS}}{R\sigma_d^2}\mathbf{H}_2\mathbf{R}_2\mathbf{H}_2^H\right). \tag{3.10}$$

Since the RS repeats the information transmitted in the first time slot, this repetition coding DF scheme only provides higher SNR but no multiplexing gain. Due to two hop transmission, the maximum transmission rate for the considered overall system can be calculated as

$$C = \max_{\substack{\mathrm{tr}(\mathbf{R}_1)=M \\ \mathrm{tr}(\mathbf{R}_2)=R}} \left\{\frac{1}{2}\min\left\{I(\mathbf{x}_s; \mathbf{y}_r), I(\mathbf{x}_s, \mathbf{x'}_s; \mathbf{y}_d^{(1)}, \mathbf{y}_d^{(2)})\right\}\right\}. \tag{3.11}$$

The first term in (3.11) represents the rate at which the relay can reliably decode the data from the BS, and the second term represents the rate the MS can reliably decode the data symbols based on the signals received from BS and RS. Since we require both RS and MS to fully decode the data symbols, the maximum rate we can choose is the minimum of the two terms. The factor $1/2$ in front of the min operator in (3.11) comes from the fact of two channel uses.

In order to find the optimum transmit covariance matrices, we can reformulate the achievable rate expression (3.11) as follows:

$$\begin{aligned} \text{maximize} \quad & \frac{1}{2}\tau \\ \text{subject to} \quad & I(\mathbf{x}_\mathrm{s};\mathbf{y}_\mathrm{r}) \geq \tau, \quad I(\mathbf{x}_\mathrm{s},\mathbf{x'}_\mathrm{s};\mathbf{y}_\mathrm{d}^{(1)},\mathbf{y}_\mathrm{d}^{(2)}) \geq \tau, \\ & \mathrm{tr}(\mathbf{R}_1) = M, \quad \mathrm{tr}(\mathbf{R}_2) = R, \end{aligned} \tag{3.12}$$

where τ is a slack variable. This problem is convex and can be solved by efficient interior-point methods [33]. Note, in order to solve the above optimization problem, all the channel matrices $\mathbf{H}_0$, $\mathbf{H}_1$ and $\mathbf{H}_2$ have to be available. One way to achieve this is to let the relay feed back the channel information about $\mathbf{H}_2$ to BS. BS feeds the optimized $\mathbf{R}_2$ to RS after it finishes the calculation of (3.12). On the other hand, if channel knowledge is not available at the transmitters, it is reasonable to allocate the power uniformly at the transmit antennas since it is known to be a robust solution under channel uncertainty [178]. That is, the signal covariance matrices are chosen to be $\mathbf{R}_1 = \mathbf{I}_M$ and $\mathbf{R}_2 = \mathbf{I}_R$.

3.4 Analysis of Circular Coverage Range of Relaying Systems with Uniform Power Allocation at Transmit Antennas

If N_r relays are placed uniformly on a circle around the BS, each relay covers a sector with coverage angle $360°/N_\mathrm{r}$. For a given coverage angle (or the corresponding number of relays), there exists an optimum position for the relays where the maximum circular coverage range is achieved. In this section, we derive analytical upper and lower bounds for the maximum circular coverage range when the coverage angle is given. The analysis is done for uniform power allocation at the transmit antennas, i.e., in the case of no channel knowledge at transmitters. Furthermore, we propose an approximation for the maximum circular coverage range, which utilizes some results in the analysis of upper and lower bounds. With the knowledge of the two bounds and the approximation, the system designers can get an idea of the achievable coverage range of the relaying system before doing measurement campaigns. We assume $P_{\text{BS}} = P_{\text{RS}}$ and $\sigma_{\text{r}} = \sigma_{\text{d}}$. Furthermore, we denote $\rho = P_{\text{BS}}/\sigma_{\text{d}}^2 = P_{\text{RS}}/\sigma_{\text{d}}^2$. Since we allocate different subcarriers in each sector and there is no intracell interference between the relays due to our assumption in Section 3.2, we can analyze the circular coverage range by considering just one sector in the system. This is because the same analysis applies to each sector independently. The analysis consists of the following steps:

- Assumptions on the BS-RS channel: Normally, dedicated relay stations are placed at those places where good connection with the BS is established. It is reasonable to assume that the first hop channel $\mathbf{H}_1$ is much better than $\mathbf{H}_0$ and $\mathbf{H}_2$, i.e., $PL_1 \ll PL_0$ and $PL_1 \ll PL_2$, at least for the users on the border. Thus for those border users, we have

$$I(\mathbf{x}_{\text{s}}; \mathbf{y}_{\text{r}}) \gg I(\mathbf{x}_{\text{s}}, \mathbf{x'}_{\text{s}}; \mathbf{y}_{\text{d}}^{(1)}, \mathbf{y}_{\text{d}}^{(2)}), \tag{3.13}$$

  where $I(\mathbf{x}_{\text{s}}; \mathbf{y}_{\text{r}})$ and $I(\mathbf{x}_{\text{s}}, \mathbf{x'}_{\text{s}}; \mathbf{y}_{\text{d}}^{(1)}, \mathbf{y}_{\text{d}}^{(2)})$ are the mutual information in (3.9) and (3.10), respectively. Since we consider the case that no channel knowledge is available at the transmitters and power is uniformly allocated at the transmit antennas, we can write the following according to (3.11)

$$\begin{aligned} C &\approx \frac{1}{2} I(\mathbf{x}_{\text{s}}, \mathbf{x'}_{\text{s}}; \mathbf{y}_{\text{d}}^{(1)}, \mathbf{y}_{\text{d}}^{(2)}) \\ &= \frac{1}{2} \log_2 \det \left(\mathbf{I}_N + \frac{P_{\text{BS}}}{M\sigma_{\text{d}}^2} \mathbf{H}_0 \mathbf{H}_0^H + \frac{P_{\text{RS}}}{R\sigma_{\text{d}}^2} \mathbf{H}_2 \mathbf{H}_2^H \right) & (3.14) \\ &= \frac{1}{2} \log_2 \det \Big(\mathbf{I}_N + \underbrace{\frac{\rho}{M} \eta_0 \cdot \bar{\mathbf{H}}_0 \bar{\mathbf{H}}_0^H}_{\text{BS contribution}} + \underbrace{\frac{\rho}{R} \eta_2 \cdot \bar{\mathbf{H}}_2 \bar{\mathbf{H}}_2^H}_{\text{RS contribution}} \Big), & (3.15) \end{aligned}$$

  where $\eta_0 = 10^{-PL_0/10}$ and $\eta_2 = 10^{-PL_2/10}$ denote the signal attenuation from the BS and the RS, respectively. $\bar{\mathbf{H}}_0$ and $\bar{\mathbf{H}}_2$ denote the normalized direct and second hop channels, respectively. Each element of $\bar{\mathbf{H}}_0$ and $\bar{\mathbf{H}}_2$ is a $\mathcal{CN}(0,1)$ random variable according to our channel model in Section 3.2.1. The achievable rate C consists of the signal contributions from the BS and from the RS. Which contribution is stronger depends on the location of the mobile user.

- Decomposition of coverage area: In general, the coverage area of the sector has an irregular shape as shown in Fig. 3.7. However, depending on which signal contribution is stronger, the whole coverage area can be considered to be composed of two independent parts: BS coverage and RS coverage. For users in BS coverage, BS signal is the major contribution to C in (3.15), i.e., $\rho\eta_0/M > \rho\eta_2/R$. It is vice versa for users in RS coverage. On average, we can characterize the strength of signal contribution

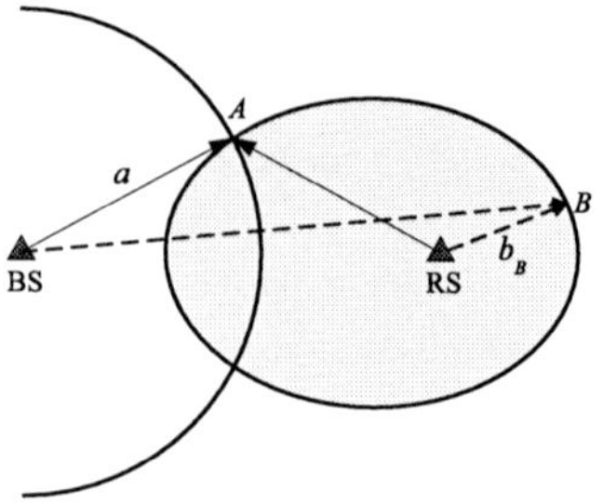

Fig. 3.7: User A and B in a cellular relaying system

just by the terms $\rho\eta_0/M$ and $\rho\eta_2/R$ because $\bar{\mathbf{H}}_0$ and $\bar{\mathbf{H}}_2$ are the normalized channels and have the same distribution.

- Approximation of the RS coverage using round disk: User A and B in Fig. 3.7 are two extreme points on the border of RS coverage. User A is the "luckiest" user on the RS coverage border in the sense that it lies on the intersection of BS and RS coverage ranges. The received signal strengths from BS and RS are equally strong, i.e., $\rho\eta_0/M = \rho\eta_2/R$. The user B is among the "most unlucky" users since the signal contribution from BS is very small. Since both user A and B have the same 1%-outage rate (1b/s/Hz), user B must receive stronger relay link signal from RS than user A. The shape of the RS coverage is like a ellipse, which has a larger distance from RS to border user A than to user B. This shape of RS coverage makes it still hard to analyze the coverage of the relaying system. In order to simplify our analysis and get upper and lower bounds for the maximum circular coverage range, we approximate the RS coverage using a round disk with the center at RS and call it "RS coverage disk". By using the distance between RS and A as radius of the disk, we overestimate the RS coverage; the RS coverage is underestimated when we use the distance between RS and B as the radius. We also use a round disk with the distance between BS and user A as the radius to approximate the BS coverage, which we call "BS coverage disk".

- Calculation of disk radiuses and maximum circular coverage range: The next step is to calculate the disk radiuses, i.e., to determine the distance between BS, RS and A, B. The exact expressions for the distances are hard to find analytically. However, we can easily calculate upper and lower bounds for them. After we get the upper and lower bounds for the radiuses, the corresponding upper and lower bounds for the maximum circular coverage range can be determined by geometry.

The last two points will be elaborated in the following subsections.

3.4.1 Upper Bound of Maximum Circular Coverage Range

Since the user A lies on the watershed separating the BS coverage and RS coverage, its signal contributions from BS and RS should be equally strong, i.e., $\rho\eta_0/M = \rho\eta_2/R$. According to (3.1) and (3.3), we can calculate an upper bound of the distance between user A and BS, RS. By using them as the BS and RS coverage disk radiuses, we get an overestimate of the relaying system coverage and can derive an upper bound of the maximum circular coverage range of the system.

We just show how to calculate the upper bound of the distance between A and RS. The upper bound of the distance between A and BS can be calculated in the same way. We denote $\bar{\mathbf{H}} = \left[\bar{\mathbf{H}}_0 \; \bar{\mathbf{H}}_2\right]$, which is a $N \times (M+R)$ matrix with each element i.i.d. $\sim \mathcal{CN}(0,1)$. The rate expression of user A can be written as

$$C \approx \frac{1}{2}\log_2 \det\left(\mathbf{I}_N + \frac{\rho}{M}\eta_0 \cdot \bar{\mathbf{H}}_0\bar{\mathbf{H}}_0^H + \frac{\rho}{R}\eta_2 \cdot \bar{\mathbf{H}}_2\bar{\mathbf{H}}_2^H\right) \tag{3.16}$$

$$= \frac{1}{2}\log_2 \det\left(\mathbf{I}_N + \frac{\rho\eta_2}{R}\bar{\mathbf{H}}\bar{\mathbf{H}}^H\right) \tag{3.17}$$

$$= \frac{1}{2}\sum_{i=1}^{N}\log_2\left(1 + \frac{\rho\eta_2}{R}\lambda_i\right), \tag{3.18}$$

where λ_i, $i = 1 \dots N$ are the eigenvalues of $\bar{\mathbf{H}}\bar{\mathbf{H}}^H$. The distribution of (3.18) is difficult to characterize. However, since we are interested in an upper bound of coverage, we can use the inequality: $\log_2(1+x) \leq x\log_2(e), \forall x \geq 0$, and get

$$C \leq C^{(u)} = \frac{\rho\eta_2}{4R}\log_2(e) \cdot \left(\sum_{i=1}^{N} 2\lambda_i\right). \tag{3.19}$$

For the i.i.d. Rayleigh fading channel, $\sum_{i=1}^{N} 2\lambda_i$ is a Chi-squared random variable with $2(M+R)N$ degrees of freedom [184]. Its cumulative distribution function (CDF) is:

$$F[x; 2(M+R)N] = G[(M+R)N, x/2] \tag{3.20}$$

$$- \frac{\gamma[(M+R)N, x/2]}{\Gamma[(M+R)N]}, \tag{3.21}$$

where $G(\cdot,\cdot)$ denotes the regularized incomplete gamma function. $\gamma(\cdot,\cdot)$ is the lower incomplete Gamma function, and $\Gamma(\cdot)$ denotes the Gamma function.

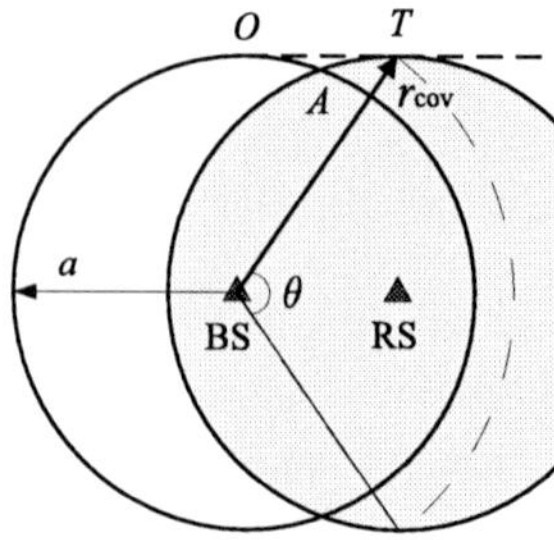

Fig. 3.8: Calculation of the upper bound for the maximum circular coverage range when $\theta \geq 60^\circ$ ($M = R$)

$C^{(u)}$ is an overestimate of the achievable rate for user A. By using $C^{(u)}$ instead of C to characterize our QoS requirement, we can get an overestimate of the distance between RS and A. That is,

$$\mathrm{P}\left\{C^{(u)} \leq C_\mathrm{Q}\right\} = F\left[\frac{4RC_\mathrm{Q}}{\rho\eta_2 \log_2(e)}; 2(M+R)N\right] = \varepsilon. \tag{3.22}$$

Here $C_\mathrm{Q} = 1\text{b/s/Hz}$ and $\varepsilon = 0.01$. We denote $F^{-1}(y;k)$ as the inverse function for the CDF $F(x;k)$. We have $F^{-1}(y;k) = 2G^{-1}(k/2, y)$, where $G^{-1}(\cdot,\cdot)$ is the inverse of the regularized incomplete gamma function. The pathloss between A and RS can be calculated as

$$PL_2 = 10\log_{10}\left\{\frac{\rho\log_2(e)}{2RC_\mathrm{Q}} \cdot G^{-1}[(M+R)N, \varepsilon]\right\} \tag{3.23}$$

$$= 35.0\log_{10}(b_A^{(u)}) + 38.4 \tag{3.24}$$

By solving the above equation, we can get the upper bound of the distance $b_A^{(u)}$ between user A and the RS, which is also an upper bound of the radius for the RS coverage disk. Similarly, we can get an upper bound $a^{(u)}$ for the radius of the BS coverage disk.

When the number of antennas at BS and RS are the same, i.e., $M = R$, the radiuses of BS and RS coverage disks are the same. Given the radiuses of the BS and RS coverage disks, the next step is to determine the optimum distance between the BS and RS, and calculate the maximum circular coverage range. For illustration purpose, we just show the $M = R$ case in the figures. Given the coverage angle, the maximum circular coverage range is the maximum radius such that every user in the *circular* sector fulfills the QoS requirement. As depicted in Fig. 3.8, the line OT is tangent to the BS disk and RS disk at O and T, respectively. When

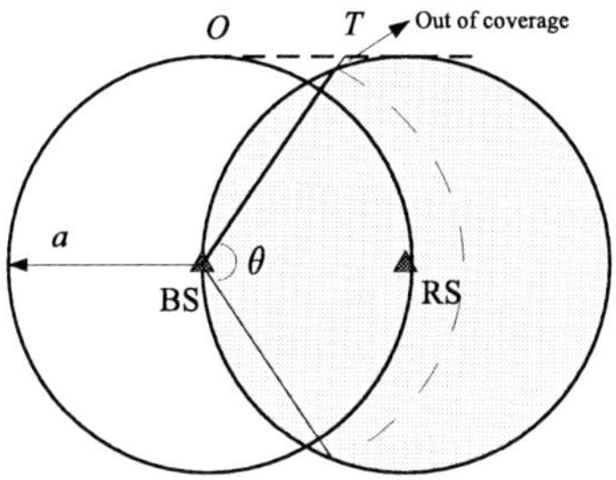

Fig. 3.9: The circular coverage range will shrink if RS moves farther away ($\theta \geq 60°$, $M = R$)

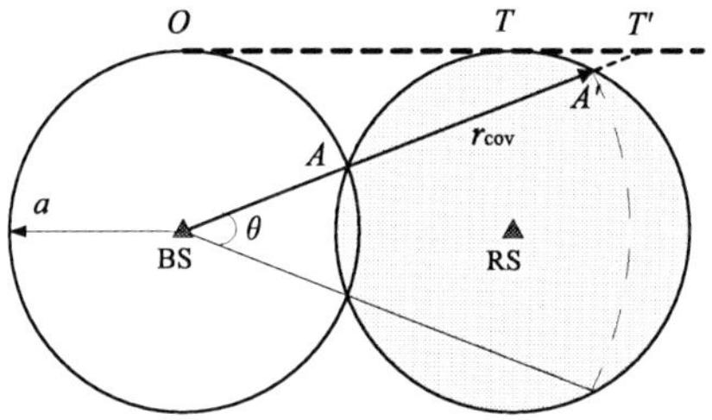

Fig. 3.10: Calculation of the upper bound for the maximum circular coverage range when $\theta < 60°$ ($M = R$)

$\theta > 60°$, the maximum circular coverage range is achieved when the point T is on the ray of angle θ. The distance between BS and T is the maximum circular coverage range. Otherwise if we move the RS further away, the circular coverage range will shrink. This is shown in Fig. 3.9.

For the coverage angle $\theta \leq 60°$ as depicted in Fig. 3.10, the maximum circular coverage range is achieved when the ray of the angle θ crosses the crossing point of the BS and RS circles. The maximum circular coverage range equals the distance from BS to the point A'. This is because if we move the RS and BS further apart, some area in the sector will not fulfill the QoS requirement and is not in the coverage of the system. This is shown in Fig. 3.11.

When $M = R$, the upper bound for the maximum circular coverage range can be expressed as:

$$r_{\mathrm{cov}} = \begin{cases} a^{(u)} + 2a^{(u)} \cos\theta, & \text{if } \theta < 60°; \\ a^{(u)} / \sin(\theta/2), & \text{if } \theta \geq 60°. \end{cases} \tag{3.25}$$

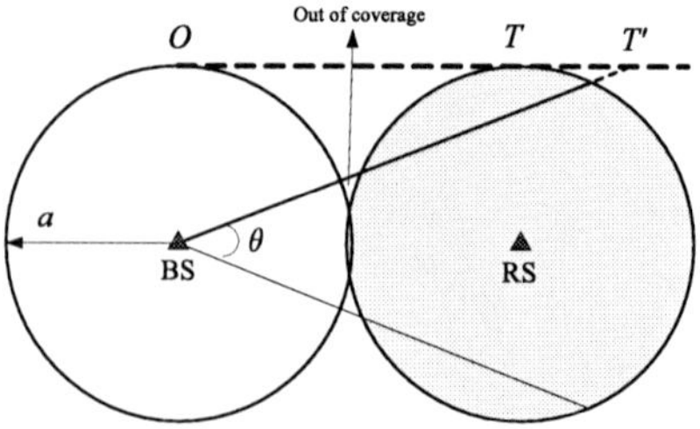

Fig. 3.11: Some area will be out of coverage if RS moves farther away ($\theta < 60°$, $M = R$)

where $a^{(u)}$ is the radius of the BS and RS coverage disks since they are the same. The corresponding distance r_{relay} between BS and RS is

$$r_{\text{relay}} = \begin{cases} 2a^{(u)} \cos(\theta/2), & \text{if } \theta < 60°; \\ a^{(u)} \cot(\theta/2), & \text{if } \theta \geq 60°. \end{cases} \tag{3.26}$$

When $M \neq R$, the upper bound for maximum circular coverage range can be calculated as follows according to coverage angle θ

$$r_{\text{cov}} = \begin{cases} a^{(u)} + 2b_A^{(u)} \cos\varphi, & \text{if } \theta < 2\arcsin\left(\frac{\sqrt{p^2+8}-p}{4}\right); \\ b_A^{(u)} / \sin(\theta/2), & \text{if } \theta \geq 2\arcsin\left(\frac{\sqrt{p^2+8}-p}{4}\right). \end{cases} \tag{3.27}$$

where φ satisfies $b_A^{(u)} \sin\varphi = (a^{(u)} + b_A^{(u)} \cos\varphi)\tan(\theta/2)$ and $p = a^{(u)}/b_A^{(u)}$. Here $a^{(u)}$ and $b_A^{(u)}$ are the upper bounds for the radiuses of the BS and RS coverage disks, respectively. The corresponding distance between BS and RS is

$$r_{\text{relay}} = \begin{cases} b_A^{(u)} \sin\varphi / \sin(\theta/2), & \text{if } \theta < 2\arcsin\left(\frac{\sqrt{p^2+8}-p}{4}\right); \\ b_A^{(u)} \cot(\theta/2), & \text{if } \theta \geq 2\arcsin\left(\frac{\sqrt{p^2+8}-p}{4}\right). \end{cases} \tag{3.28}$$

3.4.2 Lower Bound of Maximum Circular Coverage Range

In the last subsection, we first calculated an overestimate of the BS and RS coverage radiuses and then derived an upper bound for the maximum circular coverage range. Following the same idea, we can also derive an lower bound. In Fig. 3.7, user B is on the other side of the relay and is much farther away from the BS than from the RS. Thus the signal contribution from BS is much smaller than RS signal contribution at this position, i.e., $\rho\eta_0/M \ll \rho\eta_2/R$. We ignore the signal contribution from the BS and get the rate expression for user B as

$$
\begin{aligned}
C &\approx \frac{1}{2}\log_2 \det\left(\mathbf{I}_N + \frac{\rho}{M}\eta_0 \cdot \bar{\mathbf{H}}_0\bar{\mathbf{H}}_0^H + \frac{\rho}{R}\eta_2 \cdot \bar{\mathbf{H}}_2\bar{\mathbf{H}}_2^H\right) && (3.29)\\
&\geq \frac{1}{2}\log_2 \det\left(\mathbf{I}_N + \frac{\rho}{R}\eta_2\bar{\mathbf{H}}_2\bar{\mathbf{H}}_2^H\right) && (3.30)\\
&= \frac{1}{2}\sum_{i=1}^{N}\log_2\left(1 + \frac{\rho}{R}\eta_2\lambda_i'\right) && (3.31)\\
&\geq \frac{1}{2}\log_2\left(1 + \frac{\rho}{2R}\eta_2\sum_{i=1}^{N}2\lambda_i'\right) = C^{(l)}, && (3.32)
\end{aligned}
$$

where λ_i', $i = 1 \ldots N$ are the eigenvalues of $\bar{\mathbf{H}}_2\bar{\mathbf{H}}_2^H$. $\sum_{i=1}^{N} 2\lambda_i'$ is a Chi-squared random variable with $2RN$ degrees of freedom. The last inequality (3.32) follows from

$$
\sum_{i=1}^{N}\log_2(1+x_i) = \log_2\left[\prod_{i=1}^{N}(1+x_i)\right] \geq \log_2(1+\sum_{i=1}^{N}x_i), \quad \text{for } x_i \geq 0. \qquad (3.33)
$$

By using $C^{(l)}$ instead of C to characterize the QoS requirement, we can derive an underestimate of the distance between RS and B. We have

$$
\mathrm{P}\left\{C^{(l)} \leq C_{\mathrm{Q}}\right\} = F\left[\frac{2R(2^{2C_{\mathrm{Q}}}-1)}{\rho\eta_2}; 2RN\right] = \varepsilon. \qquad (3.34)
$$

Here $C_{\mathrm{Q}} = 1$b/s/Hz and $\varepsilon = 0.01$. So the pathloss between B and RS can be calculated as

$$
\begin{aligned}
PL_2 &= 10\log_{10}\left\{\frac{\rho}{R(2^{2C_{\mathrm{Q}}}-1)} \cdot G^{-1}[RN, \varepsilon]\right\} && (3.35)\\
&= 35.0\log_{10}(b_B^{(l)}) + 38.4 && (3.36)
\end{aligned}
$$

By solving the above equation, we can get the distance $b_B^{(l)}$, which is a lower bound of the radius for RS coverage disk. The inequality (3.33) should also be used to calculate a lower bound $a^{(l)}$ on the distance between A and BS, which is also a lower bound on the radius of the BS coverage disk. Similarly, we can get a lower bound $b_A^{(l)}$ on the distance between RS and user A using the inequality (3.33), which will be used in the next subsection.

After calculating lower bounds on the radius $a^{(l)}$ of BS coverage disk and the radius $b_B^{(l)}$ of RS coverage disk, the next step is to determine the optimum distance between BS and RS, and calculate the lower bounds of the maximum circular coverage range. This is depicted

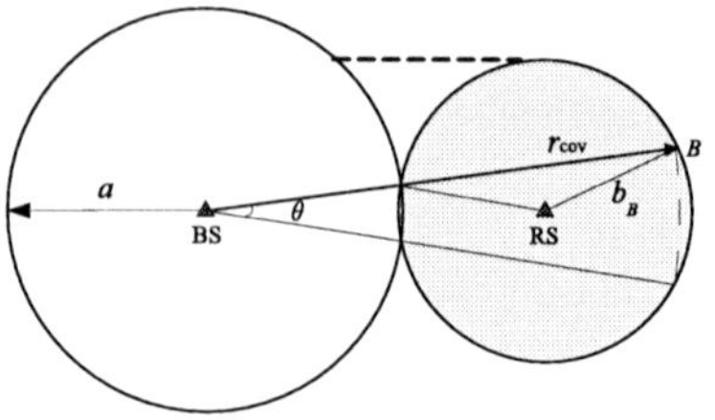

Fig. 3.12: Calculation of the lower bound for the maximum circular coverage range

in Fig. 3.12. Following the same discussion as in subsection 3.4.1, the lower bound for the maximum circular coverage range vs. coverage angle in such a case can be calculated as

$$r_{\text{cov}} = \begin{cases} a^{(l)} + 2b_B^{(l)} \cos\varphi, & \text{if } \theta < 2\arcsin\left(\frac{\sqrt{p^2+8}-p}{4}\right); \\ b_B^{(l)} / \sin(\theta/2), & \text{if } 2\arcsin\left(\frac{\sqrt{p^2+8}-p}{4}\right) \leq \theta < 2\arcsin(\frac{1}{p}); \\ a^{(l)}, & \text{if } \theta \geq 2\arcsin(\frac{1}{p}), \end{cases} \tag{3.37}$$

where $\cos\varphi = \cos(\theta/2)\sqrt{1 - p^2\sin^2(\theta/2)} - p\sin^2(\theta/2)$ and $p = a^{(l)}/b_B^{(l)}$. Here $a^{(l)}$ and $b_B^{(l)}$ are lower bounds on the radius of BS and RS coverage disks, respectively. The optimum distance r_{relay} between BS and RS can be calculated as

$$r_{\text{relay}} = \begin{cases} b_B^{(l)} \sin\varphi / \sin(\theta/2), & \text{if } \theta < 2\arcsin\left(\frac{\sqrt{p^2+8}-p}{4}\right); \\ b_B^{(l)} \cot(\theta/2), & \text{if } \theta \geq 2\arcsin\left(\frac{\sqrt{p^2+8}-p}{4}\right), \end{cases} \tag{3.38}$$

3.4.3 Heuristic Approximation for Maximum Circular Coverage Range

We derived upper and lower bounds for the maximum circular coverage range of DF relaying systems in the previous subsections. However, a designer of a real system may be more interested in an approximation for the maximum circular coverage range of the system, especially for the case of low coverage angle (e.g., below 90°). This is because usually more than four relays are expected to be used in a cell in order to get high coverage extensions. Thus we propose a heuristic approximation of the maximum circular coverage range for DF relaying systems.

As we can see from Fig. 3.7, the RS coverage is like an ellipse which has larger distance to A than to B. Because the border users are in low SNR regime, $\rho\eta_2\lambda_i'/R$ is a small number in

(3.32). We have $C \approx C^{(l)}$. As an approximation, we can average the two distances $b_A^{(l)}$ and $b_B^{(l)}$, and use it as the radius of the RS coverage disk. In addition, the radius of BS coverage disk is between the overestimate $a^{(u)}$ and the underestimate $a^{(l)}$. As an approximation, we also take the average of it. That is, we define

$$\bar{a} = \left(a^{(u)} + a^{(l)}\right)/2 \tag{3.39}$$

$$\bar{b} = \left(b_A^{(l)} + b_B^{(l)}\right)/2. \tag{3.40}$$

Then we substitute $\bar{a}$ and $\bar{b}$ for $a^{(l)}$ and $b_B^{(l)}$ in (3.37) and (3.38) to get the approximation expression for the maximum circular coverage range and the optimum distance between BS and RS for the coverage angles.

3.5 Simulation Results

Monte Carlo simulations are carried out to determine the 1%-outage rate for each point in a DF relaying system. We use the channel model discussed in Section 3.2.1, which reflects the fact that the RS is chosen to be placed at the positions where stable connections to the BS is established. Nevertheless, the outage at the RS is considered in our simulation results. We restrict ourselves to two extreme cases in the assumptions of CSIT: perfect CSIT and no CSIT. When perfect CSIT is available, we calculate the achievable rate of each point according to (3.11) and optimize the covariance matrices $\mathbf{R}_1$ and $\mathbf{R}_2$ numerically. When no CSIT is available, we choose the transmit covariance matrices to be $\mathbf{R}_1 = \mathbf{I}_M$ and $\mathbf{R}_2 = \mathbf{I}_R$. The channel knowledge at the receivers is always assumed to be perfectly available. Different antenna configurations are displayed as $M/R/N$ in the figures.

3.5.1 Analysis of Circular Coverage Range for DF Relaying with Uniform Power Allocation at Transmit Antennas

Fig. 3.13(a) depicts the upper and lower bounds for a DF relaying system with $M/R/N = 2/2/2$ antenna configuration. The simulation result and the approximation are also shown. For the coverage angle $90^\circ \leq \alpha_{\text{cov}} \leq 180^\circ$, the lower bound remains unchanged. This is because at those points, the RS coverage disk merges into the BS coverage disk. So the maximum circular coverage range equals the radius of the BS coverage disk when $90^\circ \leq \alpha_{\text{cov}} \leq 180^\circ$. As we can see, the upper and lower bounds are not tight. In a real system, it

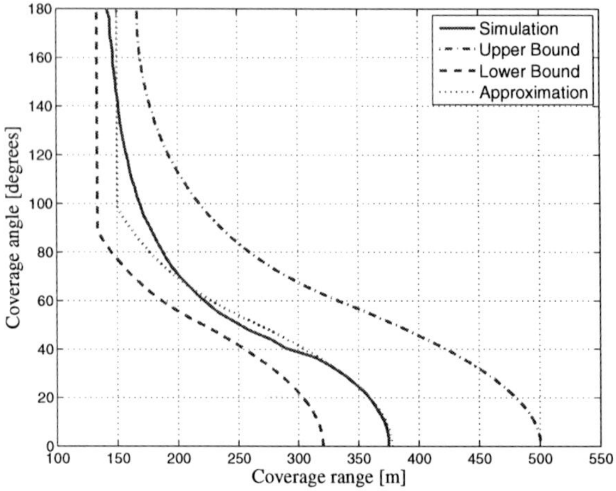

(a) $M/R/N = 2/2/2$ antenna configuration

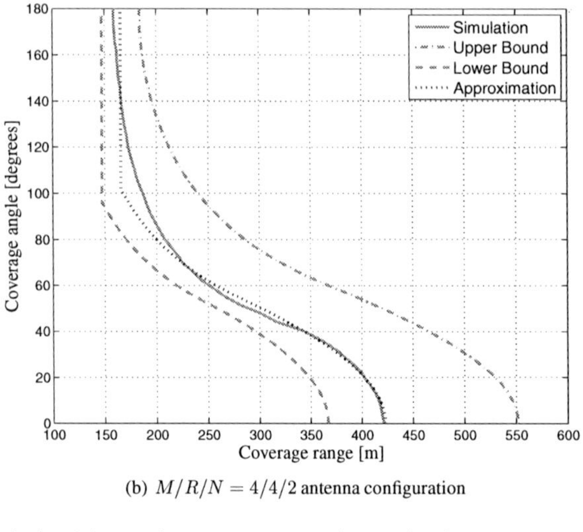

(b) $M/R/N = 4/4/2$ antenna configuration

Fig. 3.13: Analysis of the maximum coverage angle vs. circular coverage range for DF relaying systems: uniform power allocation at transmit antennas, $M/R/N = 2/2/2$ and $4/4/2$ antenna configurations

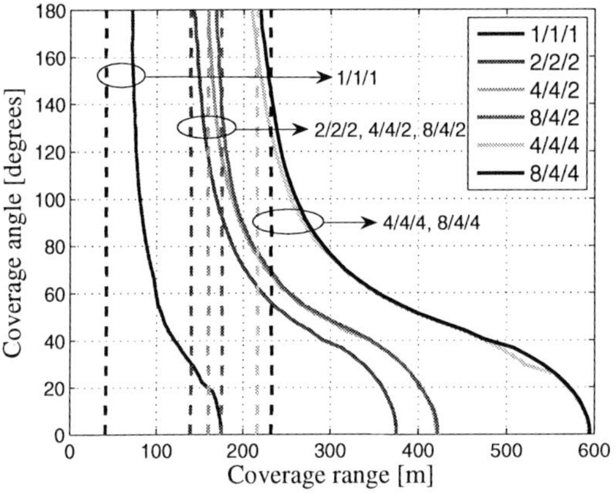

Fig. 3.14: Coverage angle vs. maximum circular coverage range for DF relaying schemes: Uniform power allocation at transmit antennas. The solid lines represent the maximum circular coverage range of relaying systems, and the dashed lines represent the circular coverage range of direct transmission.

may be more interesting to consider the approximation for the maximum circular coverage range, which fits well with the simulated results for $0^\circ \leq \alpha_{\text{cov}} \leq 70^\circ$. In this range, the maximum difference between simulated and approximated circular coverage range is less than 20m. Similar observations can be obtained for $M/R/N = 4/4/2$ antenna configuration shown in Fig. 3.13(b).

3.5.2 Circular Coverage Range Improvement for DF Relaying Without CSIT

Fig. 3.14 shows the maximum circular coverage range of DF relaying schemes compared to the coverage range of direct transmission for different types of antenna configurations with uniform power allocation at transmitter antennas. Relay outage is considered. For all relaying schemes, except the $1/1/1/$ antenna configuration, the maximum circular coverage ranges at $\alpha_{\text{cov}} = 180^\circ$ are similar as their corresponding direct transmission schemes. This is because border users at this coverage angle receive approximately equally strong signals from BS and RS. This SNR improvement at $\alpha_{\text{cov}} = 180^\circ$ compensates for the transmission rate loss due to two hop transmission.

For the considered uniform power allocation at transmitter cases, the maximum circular coverage ranges of the relaying systems are better than their corresponding direct transmission systems for coverage angle $\alpha_{\text{cov}} < 140°$. This means $N_{\text{r}} = 3$ relays per cell can guarantee the circular coverage range extension. As shown in Fig. 3.14, placing more relays (lowering coverage angle) will achieve higher circular coverage range extension. For example, for a $(4/4/4)$ configuration and at coverage angle $\alpha_{\text{cov}} = 60°$, which corresponds to 6 relays per cell, the DF relay achieves the maximum circular coverage range of $r_{\text{cov}} \approx 356$m. Compared to the coverage range of direct transmission ($r_{\text{cov}} \approx 216$m), this corresponds to an 65% increase in circular coverage range.

Using multiple antennas is another effective means for coverage extension. As shown in Fig. 3.14, the maximum circular coverage range of $2/2/2$ antenna configuration more than doubles the maximum circular coverage range of $1/1/1$ antenna configuration for any coverage angle. The maximum circular coverage range of $4/4/4$ relaying system is about 50% larger than that of the $2/2/2$ system. Furthermore we conclude that the maximum circular coverage range does not improve much by placing more antennas at BS if we have fixed antenna configurations (R/N) at RS and MS. This can be observed from the simulation result of $8/4/2$ antenna configuration as compared to that of $4/4/2$ antenna configuration. Their maximum circular coverage ranges are nearly identical. The same is also true for the simulation result of $8/4/4$ antenna configuration as compared with that of $4/4/4$ antenna configuration. This is because the first hop channel between BS and RS is already very good and the overall transmission rate is limited by the second hop. Thus placing more antennas at BS provide no more advantages.

3.5.3 Circular Coverage Range Improvement for DF Relaying with Full CSIT

Fig. 3.15 shows the maximum circular coverage range of DF relaying schemes compared to direct transmission for different types of antenna configurations when channel knowledge is available at the transmitters. Using the optimum covariance matrices as discussed in Section 3.3 at the transmitters can provide surprisingly large coverage extension for MIMO DF relaying systems. For 8×4 cellular system, waterfilling at the BS can provide an additional 87m circular coverage range extension compared to the uniform power allocation case for direct transmission. Using the optimum signal covariance matrices can provide an additional 210m circular coverage range extension compared to the uniform power allocation case for DF relaying with $8/4/4$ antenna configuration at the coverage angle $60°$ (i.e., using 6 relays).

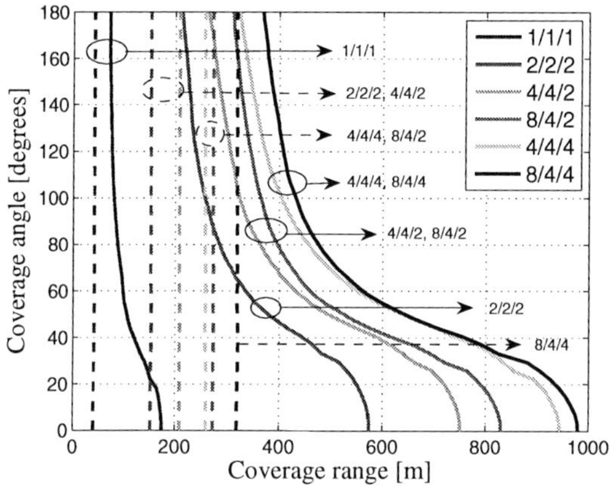

Fig. 3.15: Coverage angle vs. maximum circular coverage range for DF relaying system: full CSIT at transmitters. The solid lines represent the maximum circular coverage range of relaying systems, and the dashed lines represent the coverage range of direct transmission.

However, in order to calculate the optimum covariance matrices, the RS has to feedback the channel knowledge on $\mathbf{H}_2$ to the BS, and the BS has to inform the RS about the calculated covariance $\mathbf{R}_2$. This may introduce additional cost for a real system. From Fig. 3.15, we can also observe that a DF relaying system with $8/4/2$ antenna configuration has similar maximum circular coverage range as a $4/4/2$ system when full CSIT is available. The same is also true for $4/4/4$ and $8/4/4$ DF relaying systems. This is similar to what is observed in Fig. 3.14.

3.5.4 Circular Coverage Range Improvement by Using Fixed Number of Relays

In Fig. 3.16 and Fig. 3.17, we compare the absolute circular coverage range extension and the relative improvement of DF relaying schemes compared to direct transmission for 6 and 9 relays per cell. The corresponding coverage angles are $\alpha_{\text{cov}} = 60°$ and $\alpha_{\text{cov}} = 40°$, respectively.

In the *symmetric scenarios* in which BS and RS have the same number of antennas, i.e., for $2/2/2$, $4/4/2$ and $4/4/4$ antenna configurations, their relative improvement in the maximum

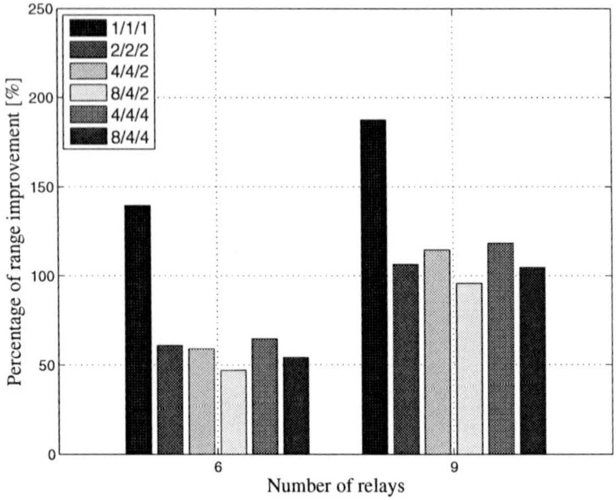

(a) DF relay; no CSIT

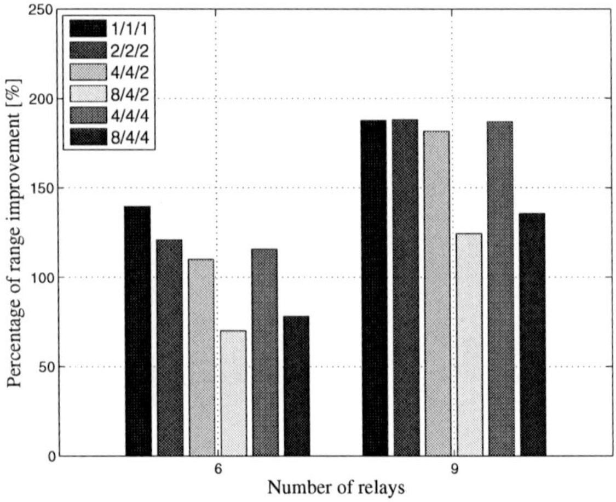

(b) DF relay; with CSIT

Fig. 3.16: Relative circular coverage range improvement compared to direct transmission for $N_r = 6$ and $N_r = 9$ DF relaying systems. Antenna configurations from left to right: $1/1/1$, $2/2/2$, $4/4/2$, $8/4/2$, $4/4/4$, $8/4/4$.

circular coverage range in the case of no CSIT (i.e., the gain relative to the coverage range of the direct communication between BS and MS) do not differ much (cf. Fig. 3.16(a)). For $1/1/1$ antenna configuration, its relative circular coverage range improvement is much higher than the others. This is because its coverage range is small in direct transmission, and small circular coverage range improvement in DF relaying leads to high relative circular coverage range improvement. For the *asymmetric scenarios*, i.e., $M > R$, the relative improvement in circular coverage range is smaller, especially when CSIT is available. This behavior is due to the fact that the direct transmission scheme improves significantly in the mentioned asymmetric scenarios, while in the DF relaying schemes the rate is limited by the second hop, and therefore the maximum circular coverage ranges in DF relaying do not improve as much as direct transmission. Note that to achieve the depicted circular coverage range extension in the case of DF without CSIT (cf. Fig. 3.17(a)) space-time coding techniques have to be used.

The use of more antennas at the BS than at the MS does not improve the performance significantly (cf. Fig. 3.17), at least in the considered scenarios, in which the link between BS and RS is stable compared to the two links BS-MS and RS-MS. Therefore, it is sufficient to use as many BS antennas as RS antennas. Additional BS antennas can be used for SDMA (space-division multiple access) techniques. For example, they can be used to transmit data to two different relays in the same cell in the same band, in order to improve the overall spectral efficiency.

To answer the question when it does pay off to use relays instead of more base stations, we calculate a rough estimate. As a fair comparison, we assume the same amount of resources and the same number of active users in both cases. Fig. 3.16 shows that DF relaying in a 4/4/4 scenario more than doubles the maximum circular coverage range if 9 relays are used (i.e., coverage angle $\alpha_{\text{cov}} = 40°$), whether CSIT is available or not. Therefore, the covered area is improved by the factor of 4, which would require 3 additional base stations if no relays would be deployed. This means relaying is profitable if $9/3 = 3$ relays are cheaper than one base station in this example.

3.6 Chapter Summary and Conclusions

We proposed the concepts of *coverage angle* and *circular coverage range* for the analysis of coverage extension in cellular relaying networks. The upper and lower bounds for the maximum circular coverage range when uniform power is allocated at the transmitters were

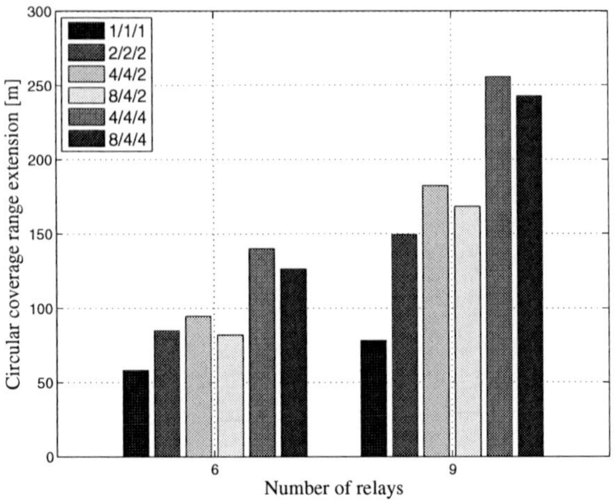

(a) DF relay; no CSIT

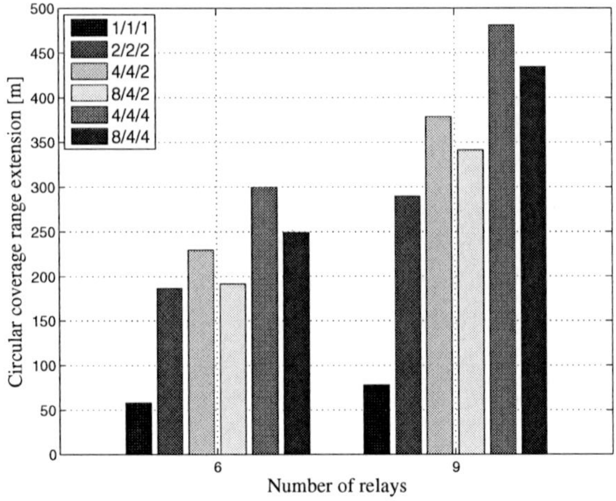

(b) DF relay; with CSIT

Fig. 3.17: Circular coverage range extensions compared to direct transmission for $N_\mathrm{r} = 6$ and $N_\mathrm{r} = 9$ DF relaying systems. Antenna configurations from left to right: $1/1/1, 2/2/2, 4/4/2, 8/4/2, 4/4/4, 8/4/4$.

presented. We also proposed an approximation for the maximum circular coverage range and verified our proposals by simulations. Our simulation results showed that the circular coverage range extension for MIMO relaying systems is significant compared to the conventioanl direct transmission, even though we have a rather high QoS requirement and require a circular cell. If CSIT is available, additional coverage range can be achieved by using the optimum transmit signal covariance matrices. We also found that if the first hop channel is already very good, as is the case for most relaying systems, placing more antennas at BS does not provide substantial additional coverage extension.

Chapter 4

Two-Way Relaying Systems With Multiple Antennas

The two-way relaying protocol provides an effective means to reduce the spectral efficiency loss in half-duplex relaying systems that is due to the two channel uses required for the transmission from the source station to the destination station. This is achieved by bidirectional simultaneous transmission of data between the two stations. In this chapter, we consider two-way relaying systems with multiple antennas equipped at each station, i.e., MIMO two-way relaying systems. Fundamental transmit and receive signal models for two-way amplify-and-forward (AF) and decode-and-forward (DF) relaying protocols are presented. In particular, we discuss in detail two practical data combining schemes at the relay for the broadcast (BRC) phase in the two-way DF relaying protocol, i.e., the *superposition coding* scheme and the *network coding* scheme. The difference of the two schemes lies in combining the decoded data at the relay on the symbol level or on the bit level. A unified view for the data combining schemes at the relay is also proposed. Furthermore, we explore how to choose the optimum transmit signal covariance matrices at the relay and characterize the achievable sum rates using the two data combining schemes for given channels. The capacity region in the BRC phase achieved by random coding approaches is also presented. In this chapter, we allocate equal time resources on the multiple access (MAC) and BRC phases. The optimum time-division strategies between the MAC and BRC phases will be discussed in Chapter 5.

4.1 Introduction

Conventional relaying schemes suffer from the half-duplex constraint, which causes the spectral efficiency loss due to the two channel uses required for the transmission from the source to the destination. This results in the pre-log factor of 1/2 in the corresponding rate expressions. In [192, 193], the *two-way relaying protocol* was proposed to recover the spectral efficiency loss in half-duplex relaying systems. The two-way relaying protocol considers the bidirectional information exchanges between two wireless stations via a relay. Conventional relaying protocols require four channel uses to exchange the data of those two stations. The two-way relaying protocol in [192, 193] only needs two phases, which represents the channel uses, to finish the information exchange. In order to distinguish the two phases, the first phase is referred to as the *multiple access* (MAC) phase, and the second phase is referred to as the *broadcast* (BRC) phase (see Chapter 2). In the MAC phase, both stations transmit their messages simultaneously to the relay. In the BRC phase, the relay combines the data from the two stations and transmits the combined data back to the two stations. Since the two stations know their own data, they can subtract the back-propagating known data, i.e., *self-interference*, and the reveal the information from its partner. Both the amplify-and-forward (AF) and decode-and-forward (DF) relaying strategies were considered for the two-way relaying protocol in [192, 193], where each station is equipped with only a single antenna. The data combining scheme at the relay proposed by [192, 193] is realized on the symbol level, and it is called the *superposition coding (SPC)* scheme.

The network coding scheme was proposed in [10, 142] originally for data communication in computer networks. In [136, 260], a network coding scheme for relaying networks was considered, where two stations communicate via an access point in three time slots. In the first and second slot, the two nodes transmit their messages via orthogonal channels to the relay. The relay combines both messages on bit-level by means of the XOR operation and retransmits it to both nodes. The nodes use the XOR operation on the decoded message and the own transmitted message to obtain the message from the other node. The basic idea of the schemes in [136, 260] is also the information combining at the relay and the known information cancellation at the receiving stations. The difference to the SPC scheme is that the information combination at the relay is realized on the bit level in the network coding scheme.

The optimum random coding schemes and the capacity region in the BRC phase of two-way DF relaying systems were proposed in [170, 264] for discrete memoryless channels, and extended to multiple-input multiple-output (MIMO) Gaussian channels in [262]. The considered scenario is that the relay wants to transmit information two two stations and each station knows perfectly the information intended for the other side. Such a channel model was termed *bidirectional broadcast channel.* The interesting results showed that independent messages can be sent out to two receivers simultaneously at their respective link capacities by the same input at the relay using random coding approaches.

The optimum linear processing matrix at the relay that maximizes the achievable rate in MIMO unidirectional AF relaying systems can be found in [158] without considering the directlink between the transmitter and the receiver. However, for MIMO two-way AF relaying systems, finding the optimum linear processing matrix at the relay to maximize the sum rate exploiting the channel state information at the transmitter (CSIT) in the BRC phase is a non-convex optimization problem, which is difficult to solve. Recently, the authors of [139] proposed an iterative scheme to find a relay linear processing matrix aiming to maximize the sum rate for MIMO two-way AF relay systems. Such a scheme iteratively identifies a local optimal solution by deriving the gradient of the sum rate and applying the gradient descent algorithm. Linear precoder and decoder which minimize the sum of mean squared error at the two stations can be found in [141].

In this chapter, we extend the two-way AF and DF relaying protocol [193] to multiple antennas at all stations, i.e., MIMO two-way AF and DF relaying systems. We present the signal models for the MIMO two-way AF and DF relaying protocols and assume further the knowledge of transmit CSI at the DF relay. In the time-division duplex (TDD) transmission mode, the relay has to estimate the MIMO channels for decoding in the MAC anyway. So, in the BRC phase this knowledge can be used for precoding if the bursts are short enough compared to the coherence time of the MIMO channels. In the frequency-division duplex (FDD) transmission mode, the relay may require feedback of channel knowledge from the two receiving stations. Equal time (or frequency) resources are allocated to the MAC and BRC phases. We compare the two practical approaches, i.e., the SPC and the network coding schemes, for combining the information at the relay in the BRC phase of two-way DF relaying protocol. Furthermore, we also present the method of characterizing the capacity region in the MIMO bidirectional broadcast channel and calculating its maximum sum rate. We show that two-way relaying achieves a quite substantial improvement in spectral efficiency compared to conventional relaying with and without transmit CSI at the relay. We show that the difference in sum-rate compared to the case where no CSIT is used, increases with increasing ratio between number of relay antennas and number of node antennas. We further show that the network coding scheme achieves nearly the optimal sum rate when CSIT is available at the relay. When CSIT is not available at the relay, the two-way DF relaying protocol always significantly outperforms the two-way AF relaying protocol.

Our Contributions: The contributions of this chapter can be summarized as follows:

- We provide the analysis and comparison of the relaying strategies and the data combining schemes from information theoretic perspectives in MIMO two-way relaying

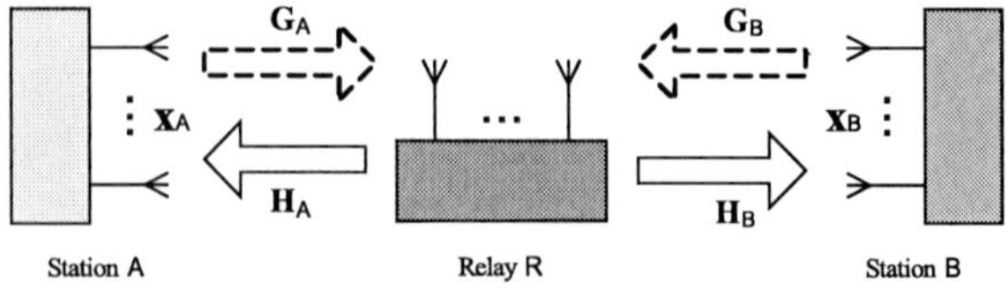

Fig. 4.1: MIMO two-way DF relaying system. The dashed arrows and solid arrows represent the transmissions in the MAC and BRC phases, respectively.

systems. In particular, we characterize the optimal relay transmit covariance matrices when the BRC phase CSIT is available at the relay for MIMO two-way DF relaying systems;

- We provide a unified view of the SPC and the network coding schemes;
- We propose an algorithm for characterizing the BRC phase capacity region.

This chapter is organized as follows: Section 4.2 presents the system setup for MIMO two-way relaying systems. Section 4.3 discusses the signal models in MIMO two-way AF relaying systems, where the relay linear processing matrix is chosen according to [193]. Section 4.4 presents the signal models in MIMO two-way DF relaying systems, where the two practical relay combining schemes, the SPC scheme and the network coding scheme, are discussed in detail. A unified view for the SPC scheme and the network coding scheme is also discussed. Furthermore, we present the method of characterizing the capacity region in the MIMO bidirectional broadcast channel and calculating its maximum sum rate. Comprehensive performance results are presented in Section 4.5, and this chapter is summarized in Section 4.6.

4.2 System Setup

A generic MIMO two-way relaying system model is shown in Fig. 4.1, where two wireless stations A and B exchange data via a relay station R. All stations are half-duplex. We refer the stations A and B collectively as *user stations*. We consider a discrete-time baseband transmission model. The channel is assumed to be flat-fading, which can be considered as one subcarrier in orthogonal frequency-division multiplexing (OFDM) systems. The number of antennas at Station A, the relay and Station B are denoted as N_A, N_R and N_B, respectively. The transmission in two-way relaying protocols is separated in *transmission phases*, which represent channel uses. Each transmission phase can be a time-slot in the time division duplex (TDD) transmission mode or a frequency subcarrier in the frequency division duplex (FDD) transmission mode. Our discussions are based on the two-phase protocol, i.e., the data from stations A and B are exchanged in the multiple access (MAC) phase and the broadcast (BRC) phase. Since all the stations are half-duplex, the directlink has no impact on the two-phase transmission and need not to be considered. In Fig. 4.1, $\mathbf{G}_k \in \mathbb{C}^{N_\mathsf{R} \times N_k}$ and $\mathbf{H}_k \in \mathbb{C}^{N_k \times N_\mathsf{R}}$ respectively denote the channel matrices between Station k and the relay in the MAC and BRC phases, where $k \in \{\mathsf{A}, \mathsf{B}\}$. All the channels remain constant during its corresponding transmission phase. We assume that the information generated by the two user stations are independent, and the transmission in the system is perfectly synchronized. R_A denotes the information rate of the data to be transmitted from Station A to B, and R_B denotes the information rate of the data to be transmitted from Station B to A.

4.3 Amplify-and-Forward Relaying

The amplify-and-forward (AF) protocol is a simple relaying protocol where the relay transmits a linearly processed version of its received signal to the destination. The two-way AF relaying protocol was proposed in [193] where each station is equipped with a single antenna. We generalized the two-way AF relaying system to the MIMO case, and present the transmission models for the scenario where multiple antennas at each station.

In the MAC phase, Station A and B transmit their data simultaneously to the relay, and the received data $\mathbf{y} \in \mathbb{C}^{N_\mathsf{R} \times 1}$ at the relay is

$$\mathbf{y} = \mathbf{G}_\mathsf{A}\mathbf{x}_\mathsf{A} + \mathbf{G}_\mathsf{B}\mathbf{x}_\mathsf{B} + \mathbf{n} \tag{4.1}$$

where $\mathbf{x}_\mathsf{A} \in \mathbb{C}^{N_\mathsf{A} \times 1}$ and $\mathbf{x}_\mathsf{B} \in \mathbb{C}^{N_\mathsf{B} \times 1}$ denote the signal vectors transmitted by Station A and B in the MAC phase under the transmit power constraint P_A and P_B, respectively. $\mathbf{n} \sim \mathcal{CN}(0, \sigma^2 \mathbf{I}_{N_\mathsf{R}})$ denotes the additive white Gaussian noise vector at the relay. We assume that stations A and B do not have channel state information at transmitters (CSIT) in the MAC phase, and equal power is allocated on each data stream. That is, we have $\mathrm{E}\left\{\mathbf{x}_\mathsf{A}\mathbf{x}_\mathsf{A}^H\right\} = P_\mathsf{A}/N_\mathsf{A} \cdot \mathbf{I}_{N_\mathsf{A}}$ and $\mathrm{E}\left\{\mathbf{x}_\mathsf{B}\mathbf{x}_\mathsf{B}^H\right\} = P_\mathsf{B}/N_\mathsf{B} \cdot \mathbf{I}_{N_\mathsf{B}}$.

The AF relay linearly process the received signal and transmits an amplified version of the signal vector $\mathbf{y}$. We assume the linear processing matrix at the AF relay to be $\mathbf{F} \in \mathbb{C}^{N_\mathsf{R} \times N_\mathsf{R}}$.

The transmitted symbol at the relay in the BRC phase is then given by

$$\mathbf{s}_\mathsf{R} = \mathbf{F}\mathbf{y} \tag{4.2}$$
$$= \mathbf{F}\mathbf{G}_\mathsf{A}\mathbf{x}_\mathsf{A} + \mathbf{F}\mathbf{G}_\mathsf{B}\mathbf{x}_\mathsf{B} + \mathbf{F}\mathbf{n}. \tag{4.3}$$

Furthermore, we require the signal vector $\mathbf{s}_\mathsf{R}$ to satisfy the transmit power constraint P_R at the relay, i.e., $\mathrm{tr}\left(\mathrm{E}\left\{\mathbf{s}_\mathsf{R}\mathbf{s}_\mathsf{R}^H\right\}\right) \leq P_\mathsf{R}$. Equivalently, the linear processing matrix $\mathbf{F}$ must satisfy

$$\mathrm{tr}\left[\mathbf{F}\left(\frac{P_\mathsf{A}}{N_\mathsf{A}}\mathbf{G}_\mathsf{A}\mathbf{G}_\mathsf{A}^H + \frac{P_\mathsf{B}}{N_\mathsf{B}}\mathbf{G}_\mathsf{B}\mathbf{G}_\mathsf{B}^H + \sigma^2\mathbf{I}_{N_\mathsf{R}}\right)\mathbf{F}^H\right] \leq P_\mathsf{R}. \tag{4.4}$$

We assume the relay knows the MAC phase channel matrices $\mathbf{G}_\mathsf{A}$ and $\mathbf{G}_\mathsf{B}$. One linear processing matrix that satisfies the transmit power constraint at the relay can be chosen as [193]

$$\mathbf{F} = \sqrt{\frac{P_\mathsf{R}}{\frac{P_\mathsf{A}}{N_\mathsf{A}}\mathrm{tr}(\mathbf{G}_\mathsf{A}\mathbf{G}_\mathsf{A}^\mathrm{H}) + \frac{P_\mathsf{B}}{N_\mathsf{B}}\mathrm{tr}(\mathbf{G}_\mathsf{B}\mathbf{G}_\mathsf{B}^\mathrm{H}) + \sigma^2 N_\mathsf{R}}} \cdot \mathbf{I}_{N_\mathsf{R}}. \tag{4.5}$$

So the received signals at stations A and B can be expressed as

$$\mathbf{y}_\mathsf{A} = \mathbf{H}_\mathsf{A}\mathbf{s}_\mathsf{R} + \mathbf{n}_\mathsf{A} \tag{4.6}$$
$$= \mathbf{H}_\mathsf{A}\mathbf{F}\mathbf{y} + \mathbf{n}_\mathsf{A} \tag{4.7}$$
$$= \underbrace{\mathbf{H}_\mathsf{A}\mathbf{F}\mathbf{G}_\mathsf{A}\mathbf{x}_\mathsf{A}}_{\text{SI for Station A}} + \mathbf{H}_\mathsf{A}\mathbf{F}\mathbf{G}_\mathsf{B}\mathbf{x}_\mathsf{B} + \mathbf{H}_\mathsf{A}\mathbf{F}\mathbf{n} + \mathbf{n}_\mathsf{A} \tag{4.8}$$
$$\mathbf{y}_\mathsf{B} = \mathbf{H}_\mathsf{B}\mathbf{s}_\mathsf{R} + \mathbf{n}_\mathsf{B} \tag{4.9}$$
$$= \mathbf{H}_\mathsf{B}\mathbf{F}\mathbf{y} + \mathbf{n}_\mathsf{B} \tag{4.10}$$
$$= \mathbf{H}_\mathsf{B}\mathbf{F}\mathbf{G}_\mathsf{A}\mathbf{x}_\mathsf{A} + \underbrace{\mathbf{H}_\mathsf{B}\mathbf{F}\mathbf{G}_\mathsf{B}\mathbf{x}_\mathsf{B}}_{\text{SI for Station B}} + \mathbf{H}_\mathsf{B}\mathbf{F}\mathbf{n} + \mathbf{n}_\mathsf{B} \tag{4.11}$$

where $\mathbf{n}_\mathsf{A} \sim \mathcal{CN}(0, \sigma_\mathsf{A}^2\mathbf{I}_{N_\mathsf{A}})$ and $\mathbf{n}_\mathsf{B} \sim \mathcal{CN}(0, \sigma_\mathsf{B}^2\mathbf{I}_{N_\mathsf{B}})$ are the additive white Gaussian noise vectors at stations A and B, respectively. The linear processing matrix $\mathbf{F}$ is also known to both stations A and B. We assume that Station A knows the channel matrices $\mathbf{H}_\mathsf{A}\mathbf{F}\mathbf{G}_\mathsf{A}$ and $\mathbf{H}_\mathsf{A}\mathbf{F}\mathbf{G}_\mathsf{B}$, while Station B knows the channel matrices $\mathbf{H}_\mathsf{B}\mathbf{F}\mathbf{G}_\mathsf{A}$ and $\mathbf{H}_\mathsf{B}\mathbf{F}\mathbf{G}_\mathsf{B}$. The signal contribution of their own transmitted data symbols is called *self-interference* (SI) and can be canceled before decoding. Assuming that the channel matrices remain constant during their transmission phases, the information rate of the data from stations A and B can be respectively expressed as

$$R_{\mathsf{A}}^{\mathrm{AF}} = \frac{1}{2}\log\det\left[\mathbf{I}_{N_{\mathsf{B}}} + \left(\frac{P_{\mathsf{A}}}{N_{\mathsf{A}}}\mathbf{H}_{\mathsf{B}}\mathbf{F}\mathbf{G}_{\mathsf{A}}\mathbf{G}_{\mathsf{A}}^{H}\mathbf{F}^{H}\mathbf{H}_{\mathsf{B}}^{H}\right)\left(\sigma_{\mathsf{B}}^{2}\mathbf{I}_{N_{\mathsf{B}}} + \sigma^{2}\mathbf{H}_{\mathsf{B}}\mathbf{F}\mathbf{F}^{H}\mathbf{H}_{\mathsf{B}}^{H}\right)^{-1}\right] \quad (4.12)$$

$$R_{\mathsf{B}}^{\mathrm{AF}} = \frac{1}{2}\log\det\left[\mathbf{I}_{N_{\mathsf{A}}} + \left(\frac{P_{\mathsf{B}}}{N_{\mathsf{B}}}\mathbf{H}_{\mathsf{A}}\mathbf{F}\mathbf{G}_{\mathsf{B}}\mathbf{G}_{\mathsf{B}}^{H}\mathbf{F}^{H}\mathbf{H}_{\mathsf{A}}^{H}\right)\left(\sigma_{\mathsf{A}}^{2}\mathbf{I}_{N_{\mathsf{A}}} + \sigma^{2}\mathbf{H}_{\mathsf{A}}\mathbf{F}\mathbf{F}^{H}\mathbf{H}_{\mathsf{A}}^{H}\right)^{-1}\right], \quad (4.13)$$

which can be achieved by using i.i.d. Gaussian codebooks. The transmission in each direction still suffers from the pre-log factor one-half due to the two-hop transmission. However, the bidirectional information flows are established simultaneously, i.e., the sum rate of $R_{\mathsf{A}}^{\mathrm{AF}}$ and $R_{\mathsf{B}}^{\mathrm{AF}}$ is achieved simultaneously which roughly doubles the spectral efficiency.

Unlike the optimum linear processing matrix design problem for MIMO unidirectional AF relaying systems [158], finding the optimum linear processing matrix $\mathbf{F}$ exploiting CSIT on $\mathbf{H}_{\mathsf{A}}$ and $\mathbf{H}_{\mathsf{B}}$, which are available at the relay, to maximize the sum rate $R_{\mathsf{A}}^{\mathrm{AF}} + R_{\mathsf{B}}^{\mathrm{AF}}$ is a non-convex optimization problem, which is difficult to solve. Simple choice of AF gain matrix can be chosen according to (4.5). Linear precoder and decoder which minimize the sum of mean squared error at the two stations can be found in [141].

4.4 Decode-and-Forward Relaying

The AF relaying protocol is a simple scheme that forwards the received signals only after linear processing. Compared to the AF relaying protocol, the DF relaying protocol requires the relay to decode the received the signals, which is more complicated. However, the DF relaying protocol has some advantages: firstly, the DF relays do not have to store the received waveforms as AF relays do. So the DF relays may require less memory compare to the AF relays. Secondly, due to the decoding process done at the DF relays, the transmitted signal from the relays is "clean" and does not contain noise. As a result, the DF relays do not suffer from the noise amplification as AF relays do. The two-way DF relaying protocol works as follows.

Same as the two-way AF relaying protocol, the stations A and B transmit their data simultaneously to the relay in the MAC phase. The received signal $\mathbf{y} \in \mathbb{C}^{N_{\mathsf{R}}\times 1}$ at the relay is given by

$$\mathbf{y} = \mathbf{G}_{\mathsf{A}}\mathbf{x}_{\mathsf{A}} + \mathbf{G}_{\mathsf{B}}\mathbf{x}_{\mathsf{B}} + \mathbf{n} \quad (4.14)$$

where $\mathbf{x}_{\mathsf{A}} \in \mathbb{C}^{N_{\mathsf{A}}\times 1}$ and $\mathbf{x}_{\mathsf{B}} \in \mathbb{C}^{N_{\mathsf{B}}\times 1}$ denote the signal vectors transmitted by Station A and B in the MAC phase under the transmit power constraint P_{A} and P_{B}, respectively. $\mathbf{n} \sim$

$\mathcal{CN}(0, \sigma^2 \mathbf{I}_{N_\mathsf{R}})$ denotes the additive white Gaussian noise vector at the relay. We assume that stations A and B do not have channel state information of $\mathbf{G}_\mathsf{A}$ and $\mathbf{G}_\mathsf{B}$ in the MAC phase. So equal power is allocated on the transmit data streams, i.e., $\mathrm{E}\left\{\mathbf{x}_\mathsf{A}\mathbf{x}_\mathsf{A}^H\right\} = P_\mathsf{A}/N_\mathsf{A} \cdot \mathbf{I}_{N_\mathsf{A}}$ and $\mathrm{E}\left\{\mathbf{x}_\mathsf{B}\mathbf{x}_\mathsf{B}^H\right\} = P_\mathsf{B}/N_\mathsf{B} \cdot \mathbf{I}_{N_\mathsf{B}}$.

Unlike the AF relay that simply stores and retransmits a linearly processed version of the received signal, the DF relay fully decodes the signal received in the MAC phase. The capacity region for the MAC phase channel is the closure of all the set of achievable rates $(R_\mathsf{A}, R_\mathsf{B})$ in the MAC phase. It can be described by

$$R_\mathsf{A} \leq I_\mathsf{A}^{\mathrm{MAC}} = \frac{1}{2}\log_2 \det\left(\mathbf{I}_{N_\mathsf{R}} + \frac{P_\mathsf{A}}{N_\mathsf{A}\sigma^2}\mathbf{G}_\mathsf{A}\mathbf{G}_\mathsf{A}^H\right), \tag{4.15}$$

$$R_\mathsf{B} \leq I_\mathsf{B}^{\mathrm{MAC}} = \frac{1}{2}\log_2 \det\left(\mathbf{I}_{N_\mathsf{R}} + \frac{P_\mathsf{B}}{N_\mathsf{B}\sigma^2}\mathbf{G}_\mathsf{B}\mathbf{G}_\mathsf{B}^H\right), \tag{4.16}$$

$$R_\mathsf{A} + R_\mathsf{B} \leq I_{\mathrm{sum}}^{\mathrm{MAC}} = \frac{1}{2}\log_2 \det\left(\mathbf{I}_{N_\mathsf{R}} + \frac{P_\mathsf{A}}{N_\mathsf{A}\sigma^2}\mathbf{G}_\mathsf{A}\mathbf{G}_\mathsf{A}^H + \frac{P_\mathsf{B}}{N_\mathsf{B}\sigma^2}\mathbf{G}_\mathsf{B}\mathbf{G}_\mathsf{B}^H\right), \tag{4.17}$$

The MAC phase capacity region can be achieved by i.i.d. Gaussian codebooks [49].

In the BRC phase, the relay re-encodes the information and broadcasts the combined information to both stations on the same time or frequency channel. We denote the data bits contained in the MAC phase transmit symbols $\{\mathbf{x}_\mathsf{A}\}$ and $\{\mathbf{x}_\mathsf{B}\}$ as $\{b_\mathsf{A}\}$ and $\{b_\mathsf{B}\}$, respectively. The transmit signal of the relay $\mathbf{s}_\mathsf{R}$ is determined by the decoded bit-sequences of the MAC phase, i.e.,

$$(\{b_\mathsf{A}\}, \{b_\mathsf{B}\}) \overset{\mathrm{mod}}{\longmapsto} \{\mathbf{s}_\mathsf{R}\}.$$

Since both stations A and B respectively knows the bit sequence information $\{b_\mathsf{A}\}$ and $\{b_\mathsf{B}\}$ that have been transmitted in the MAC phase, they can cancel this contribution and reveal the information from their partners after receiving the signals transmitted from the relay.

There are two basic approaches for combining the information from the stations A and B at the relay in the two-way DF relaying protocol. The first is based on superposition coding (SPC) scheme as in [192], whereas the second approach is based on the networking coding scheme as in [260]. Their difference lies in that the SPC scheme combines the decoded data from the stations A and B on the symbol level, while the network coding scheme combines two sets of data on the bit level.

4.4.1 Superposition Coding Scheme

In the SPC scheme, the two sets of decoded bit sequences $\{b_\mathsf{A}\}$ and $\{b_\mathsf{B}\}$ are re-encoded separately and respectively mapped to the transmit symbol vectors $\mathbf{s}_\mathsf{A}$ and $\mathbf{s}_\mathsf{B}$, i.e., $\{b_\mathsf{A}\} \overset{\text{mod}}{\longmapsto} \mathbf{s}_\mathsf{A}$ and $\{b_\mathsf{B}\} \overset{\text{mod}}{\longmapsto} \mathbf{s}_\mathsf{B}$. Note that the encoding and mapping methods of $\mathbf{s}_\mathsf{A}$ and $\mathbf{s}_\mathsf{B}$ can differ from those of $\mathbf{x}_\mathsf{A}$ and $\mathbf{x}_\mathsf{B}$ used in the MAC phase, and those methods have to be known to the receivers of the stations A and B, which may be agreed upon *a priori* within the two-way DF relaying protocol. The transmit signal vector in the BRC phase is given by

$$\mathbf{s}_\mathsf{R} = \mathbf{s}_\mathsf{A} + \mathbf{s}_\mathsf{B} \tag{4.18}$$

$$= \beta\frac{P_\mathsf{R}}{N_\mathsf{R}}\tilde{\mathbf{s}}_\mathsf{A} + (1-\beta)\frac{P_\mathsf{R}}{N_\mathsf{R}}\tilde{\mathbf{s}}_\mathsf{B}. \tag{4.19}$$

Here $\mathbf{s}_\mathsf{A}$ and $\mathbf{s}_\mathsf{B}$ contain the same data as $\mathbf{x}_\mathsf{A}$ and $\mathbf{x}_\mathsf{B}$. $\tilde{\mathbf{s}}_\mathsf{A}$ and $\tilde{\mathbf{s}}_\mathsf{B}$ denote the normalized transmit signal vectors. P_R denotes the transmit power constraint at the relay. β and $1-\beta$ respectively denote the power allocation for symbol $\mathbf{s}_\mathsf{A}$ and $\mathbf{s}_\mathsf{B}$, where $0 < \beta < 1$. We denote the transmit signal covariance matrices as $\mathbf{\Omega}_\mathsf{A} = \mathrm{E}\left\{\tilde{\mathbf{s}}_\mathsf{A}\tilde{\mathbf{s}}_\mathsf{A}^H\right\}$ and $\mathbf{\Omega}_\mathsf{B} = \mathrm{E}\left\{\tilde{\mathbf{s}}_\mathsf{B}\tilde{\mathbf{s}}_\mathsf{B}^H\right\}$. In order to satisfy the transmit power constraint, we require $\mathrm{tr}(\mathbf{\Omega}_\mathsf{A}) \leq N_\mathsf{R}$ and $\mathrm{tr}(\mathbf{\Omega}_\mathsf{B}) \leq N_\mathsf{R}$. Here we assume the relay knows the channel of $\mathbf{H}_\mathsf{A}$ and $\mathbf{H}_\mathsf{B}$ in the BRC phase. In the TDD transmission mode, the CSIT can be obtained by assuming channel reciprocity, since the channel knowledge is anyway required to be available at the relay receiver for decoding in the MAC phase. In the FDD transmission mode, the CSIT on $\mathbf{H}_\mathsf{A}$ and $\mathbf{H}_\mathsf{B}$ may be obtained by the feedback from the stations A and B.

The received signals at stations A and B can be expressed as

$$\mathbf{y}_\mathsf{A} = \mathbf{H}_\mathsf{A}\mathbf{s}_\mathsf{A} + \mathbf{H}_\mathsf{A}\mathbf{s}_\mathsf{B} + \mathbf{n}_\mathsf{A} \tag{4.20}$$

$$= \underbrace{\beta\frac{P_\mathsf{R}}{N_\mathsf{R}}\mathbf{H}_\mathsf{A}\tilde{\mathbf{s}}_\mathsf{A}}_{\text{SI for Station A}} + (1-\beta)\frac{P_\mathsf{R}}{N_\mathsf{R}}\mathbf{H}_\mathsf{A}\tilde{\mathbf{s}}_\mathsf{B} + \mathbf{n}_\mathsf{A} \tag{4.21}$$

$$\mathbf{y}_\mathsf{B} = \mathbf{H}_\mathsf{B}\mathbf{s}_\mathsf{A} + \mathbf{H}_\mathsf{B}\mathbf{s}_\mathsf{B} + \mathbf{n}_\mathsf{B} \tag{4.22}$$

$$= \beta\frac{P_\mathsf{R}}{N_\mathsf{R}}\mathbf{H}_\mathsf{B}\tilde{\mathbf{s}}_\mathsf{A} + \underbrace{(1-\beta)\frac{P_\mathsf{R}}{N_\mathsf{R}}\mathbf{H}_\mathsf{B}\tilde{\mathbf{s}}_\mathsf{B}}_{\text{SI for Station B}} + \mathbf{n}_\mathsf{B} \tag{4.23}$$

where $\mathbf{n}_\mathsf{A} \sim \mathcal{CN}(0, \sigma_\mathsf{A}^2\mathbf{I}_{N_\mathsf{A}})$ and $\mathbf{n}_\mathsf{B} \sim \mathcal{CN}(0, \sigma_\mathsf{B}^2\mathbf{I}_{N_\mathsf{A}})$ denote the additive white Gaussian noise at stations A and B, respectively. Since each station knows its own transmitted data, the back-propagating SI can be subtracted at the receivers before decoding the data symbols from the partner station. This requires the knowledge of the MIMO channel from the relay to the receivers to be available at the corresponding receivers. The coding and modulation schemes for constructing the symbol vectors $\mathbf{s}_\mathsf{A}$ and $\mathbf{s}_\mathsf{B}$ also needs to be known to both the stations A and B. With this interference cancellation technique we essentially have an interference-free reception at each receiving station, which results in two equivalent point-to-point single-user channels. It follows that the achievable rates in the BRC phase using the SPC scheme have to fulfill the constraints

$$R_\mathsf{A} \le I_\mathsf{A}^{\mathrm{SPC}} = \frac{1}{2}\log\left(\mathbf{I}_{N_\mathsf{B}} + \frac{\beta P_\mathsf{R}}{N_\mathsf{R}\sigma_\mathsf{B}^2}\mathbf{H}_\mathsf{B}\mathbf{\Omega}_\mathsf{A}\mathbf{H}_\mathsf{B}^H\right), \tag{4.24}$$

$$R_\mathsf{B} \le I_\mathsf{B}^{\mathrm{SPC}} = \frac{1}{2}\log\left(\mathbf{I}_{N_\mathsf{A}} + \frac{(1-\beta) P_\mathsf{R}}{N_\mathsf{R}\sigma_\mathsf{A}^2}\mathbf{H}_\mathsf{A}\mathbf{\Omega}_\mathsf{B}\mathbf{H}_\mathsf{A}^H\right). \tag{4.25}$$

The rate expressions in (4.24) and (4.25) are only coupled via the transmit power constraint P_R at the relay. For a given power allocation β, the transmit signal covariance matrices $\mathbf{\Omega}_\mathsf{A}$ and $\mathbf{\Omega}_\mathsf{B}$ can be chosen separately. In this case, the maximum information rate transmitted to station A and B in the BRC phase can be calculated according to the point-to-point single-user MIMO channel capacity with the transmit power constraint βP_R and $(1-\beta)P_\mathsf{R}$, respectively. They can be expressed as

$$I_\mathsf{A}^{\mathrm{SPC},\star}(\beta) = \max_{\mathrm{tr}(\mathbf{\Omega}_\mathsf{A})\le N_\mathsf{R}, \mathbf{\Omega}_\mathsf{A}\succeq 0} \frac{1}{2}\log\left(\mathbf{I}_{N_\mathsf{B}} + \frac{\beta P_\mathsf{R}}{N_\mathsf{R}\sigma_\mathsf{B}^2}\mathbf{H}_\mathsf{B}\mathbf{\Omega}_\mathsf{A}\mathbf{H}_\mathsf{B}^H\right), \tag{4.26}$$

$$I_\mathsf{B}^{\mathrm{SPC},\star}(1-\beta) = \max_{\mathrm{tr}(\mathbf{\Omega}_\mathsf{B})\le N_\mathsf{R}, \mathbf{\Omega}_\mathsf{B}\succeq 0} \frac{1}{2}\log\left(\mathbf{I}_{N_\mathsf{A}} + \frac{(1-\beta) P_\mathsf{R}}{N_\mathsf{R}\sigma_\mathsf{A}^2}\mathbf{H}_\mathsf{A}\mathbf{\Omega}_\mathsf{B}\mathbf{H}_\mathsf{A}^H\right), \tag{4.27}$$

where $\mathbf{\Omega}_\mathsf{A} \succeq 0$ and $\mathbf{\Omega}_\mathsf{B} \succeq 0$ denote that they are positive semidefinite matrices. For each given value of β, the optimum covariance matrices $\mathbf{\Omega}_\mathsf{A}$ and $\mathbf{\Omega}_\mathsf{B}$ in (4.26) and (4.27) can be calculated via the waterfilling algorithm [49] as described in Section 2.1.3.

Maximizing the Sum Rate: The set of achievable rate pairs $(R_\mathsf{A}, R_\mathsf{B})$ for the data from station A and B have to satisfy the MAC phase constraints (4.15)–(4.17) and the BRC phase constraints (4.24)–(4.25) simultaneously. Unlike the AF case, the achievable rates not only depends on the given channel but also depends on the power allocation β, i.e., the expressions of (4.24) and (4.25) are coupled by the transmit power constraint at the relay. For the given channel and the given power allocation β, the overall rate region of the two-way DF relaying system with SPC scheme can be expressed as

$$R_\mathsf{A} \le \min\left(I_\mathsf{A}^{\mathrm{MAC}}, I_\mathsf{A}^{\mathrm{SPC},\star}(\beta)\right) \tag{4.28}$$

$$R_\mathsf{B} \le \min\left(I_\mathsf{B}^{\mathrm{MAC}}, I_\mathsf{B}^{\mathrm{SPC},\star}(1-\beta)\right) \tag{4.29}$$

$$R_\mathsf{A} + R_\mathsf{B} \le I_{\mathrm{sum}}^{\mathrm{MAC}}, \tag{4.30}$$

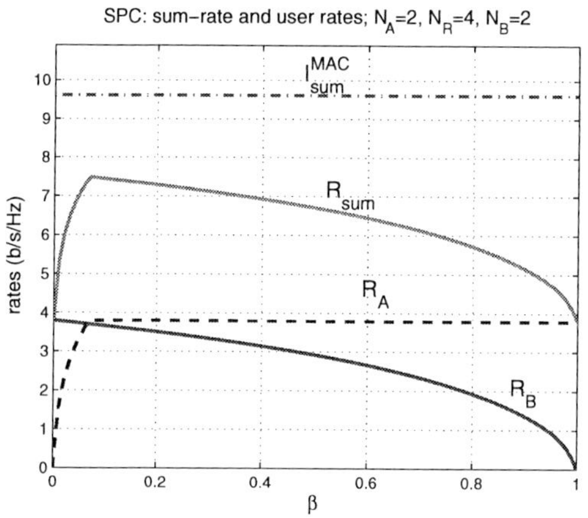

Fig. 4.2: Superposition Coding (SPC): Sum-rate and user rates for a specific realization of $\mathbf{H}_\mathsf{A}$, $\mathbf{H}_\mathsf{B}$ vs. β.

where R_A and R_B denote the achievable information rates from station A to station B and from station B to station A, respectively. The maximum sum-rate of the two-way DF relaying protocol using the SPC scheme is thus given by

$$\begin{aligned} R_\text{sum} &= R_\mathsf{A} + R_\mathsf{B} \\ &= \min\left\{\min(I_\mathsf{A}^\text{MAC}, I_\mathsf{A}^{\text{SPC},\star}(\beta)) + \min(I_\mathsf{B}^\text{MAC}, I_\mathsf{B}^{\text{SPC},\star}(1-\beta)), I_\text{sum}^\text{MAC}\right\}. \end{aligned} \tag{4.31}$$

Because the transmit signal covariance matrices $\mathbf{\Omega}_\mathsf{A}$ and $\mathbf{\Omega}_\mathsf{B}$ in (4.26) and (4.27) can be optimized individually when the power allocation β is given, the maximization of (4.31) can be reduced to the problem of finding the optimum value of β. Fig. 4.2 shows an example of the rates changing over β for a specific realization of $\mathbf{H}_\mathsf{A}$ and $\mathbf{H}_\mathsf{B}$. Here the sum-rate (4.31) is not limited by I_sum^MAC in (4.17), and there exists a unique $\beta^\star$ which maximizes (4.31), $\beta^\star = 0.07$ in Fig. 4.2. For this $\beta^\star$ only one pair of possible R_A and R_B exists. However, if the sum-rate is limited by I_sum^MAC, R_sum is constant over a certain range of β. In this case, multiple choices of the user rates are possible. Since the maximization of the sum rate in (4.31) over the whole achievable rate region of the two way DF relaying system using the SPC scheme only depends on the choice of one variable β, the maximization of R_sum can be calculated using the golden section search algorithm as described in Algorithm 1.

Algorithm 1 Maximizing R_{sum} for two-way DF relaying systems using SPC scheme

Initialize $\beta^{\min} = 0$, $\beta^{\max} = 1$,
$\beta^{\text{mid}} = \beta^{\min} + 0.618(\beta^{\max} - \beta^{\min})$ and
$\beta = \beta^{\min} + 0.618(\beta^{\text{mid}} - \beta^{\min})$.
repeat
Calculate $R_{\text{sum}}(\beta^{\min})$, $R_{\text{sum}}(\beta^{\max})$, $R_{\text{sum}}(\beta^{\text{mid}})$ and $R_{\text{sum}}(\beta)$ in (4.31) using the water-filling algorithm [49].
if $R_{\text{sum}}(\beta) < R_{\text{sum}}(\beta^{\text{mid}})$ **then**
$\beta^{\min} = \beta$;
$\beta^{\max}$ and β^{mid} unchanged;
$\beta = \beta^{\min} + 0.618(\beta^{\max} - \beta^{\min})$;
else
$\beta^{\max} = \beta^{\text{mid}}$;
$\beta^{\min}$ and β^{mid} unchanged;
$\beta = \beta^{\min} + 0.382(\beta^{\max} - \beta^{\min})$;
end if
until $|R_{\text{sum}}(\beta) - R_{\text{sum}}(\beta^{\text{mid}})| < \epsilon$ or $|\beta^{\max} - \beta^{\min}| < \zeta$
return $R_{\text{sum}}(\beta)$.

4.4.2 Network Coding Scheme

In the two-way DF relaying protocol, the binary information bit sequences $\{b_{\mathsf{A}}\}$ and $\{b_{\mathsf{B}}\}$ are both available at the relay after the decoding in the MAC phase. Another practical scheme for processing the data at the relay is the network coding scheme [260]. The basic idea is that the relay combines the decoded bit sequences $\{b_{\mathsf{A}}\}$ and $\{b_{\mathsf{B}}\}$ on the bit level prior to re-encoding. Specially the relay applies the bitwise XOR operation on the decoded bit sequences $\{b_{\mathsf{A}}\}$ and $\{b_{\mathsf{B}}\}$, and remodulates the combined bit sequence $\{b_{\mathsf{R}}\}$ into transmit symbol sequence $\mathbf{s}_{\mathsf{R}}$, i.e.,

$$\{b_{\mathsf{A}} \oplus b_{\mathsf{B}}\} = \{b_{\mathsf{R}}\} \overset{\text{mod}}{\longmapsto} \{\mathbf{s}_{\mathsf{R}}\}. \tag{4.32}$$

where $\mathbf{s}_{\mathsf{R}} \in \mathbb{C}^{N_{\mathsf{R}} \times 1}$ denotes the transmit symbol vector at the relay. The encoding and modulation schemes for the bit sequence $\{b_{\mathsf{R}}\}$ and $\mathbf{s}_{\mathsf{R}}$ can be different compared to those used at the stations A and B in the MAC phase, but they have to be known at the two stations for decoding in the BRC phase. By using the XOR operation, the decoded two bit sequences are combined into one. As a consequence, lower data rate is required to be transmitted from the relay. Compared to the SPC scheme, the transmit energy per bit is increased in the network coding scheme, and the network coding scheme may have advantages in the transmit SNR

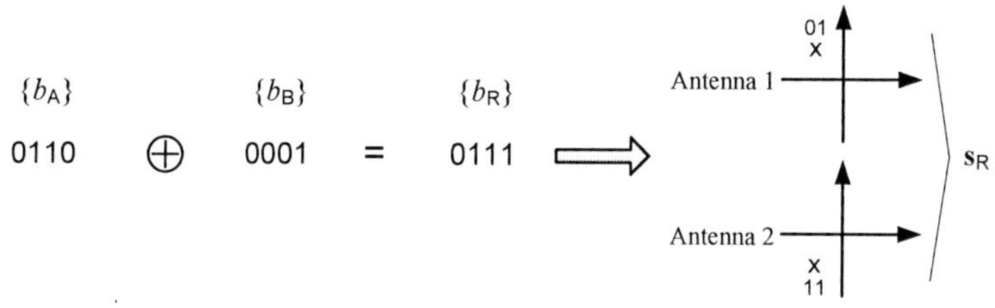

Fig. 4.3: Re-encoding at the relay

per bit. The received signals at stations A and B are

$$\mathbf{y}_{\mathsf{A}} = \mathbf{H}_{\mathsf{A}}\mathbf{s}_{\mathsf{R}} + \mathbf{n}_{\mathsf{A}} \tag{4.33}$$

$$= \sqrt{\frac{P_{\mathsf{R}}}{N_{\mathsf{R}}}}\mathbf{H}_{\mathsf{A}}\tilde{\mathbf{s}}_{\mathsf{R}} + \mathbf{n}_{\mathsf{A}}, \tag{4.34}$$

$$\mathbf{y}_{\mathsf{B}} = \mathbf{H}_{\mathsf{B}}\mathbf{s}_{\mathsf{R}} + \mathbf{n}_{\mathsf{B}} \tag{4.35}$$

$$= \sqrt{\frac{P_{\mathsf{R}}}{N_{\mathsf{R}}}}\mathbf{H}_{\mathsf{B}}\tilde{\mathbf{s}}_{\mathsf{R}} + \mathbf{n}_{\mathsf{B}} \tag{4.36}$$

where $\tilde{\mathbf{s}}_{\mathsf{R}}$ denotes the normalized transmit signal vector. The transmit signal covariance matrix of the transmit signal in (4.34) and (4.36) is defined as $\mathbf{\Omega} = \mathrm{E}\{\tilde{\mathbf{s}}_{\mathsf{R}}\tilde{\mathbf{s}}_{\mathsf{R}}^H\}$, where $\mathrm{tr}(\mathbf{\Omega}) \leq N_{\mathsf{R}}$ satisfies the transmit power constraint. $\mathbf{n}_{\mathsf{A}} \sim \mathcal{CN}(0, \sigma_{\mathsf{A}}^2)$ and $\mathbf{n}_{\mathsf{B}} \sim \mathcal{CN}(0, \sigma_{\mathsf{B}}^2)$ are the additive Gaussian noise vectors at stations A and B in the BRC phase, respectively. Upon receiving the signals, the two stations first demodulate the received signals and convert the received signals into the binary information bit sequence, i.e.,

$$\mathbf{s}_{\mathsf{R}} \overset{\text{demod}}{\longmapsto} \{\hat{b}_{\mathsf{R}}\}. \tag{4.37}$$

Then the stations reveal the unknown data bits by XOR-ing the decoded data $\{\hat{b}_{\mathsf{R}}\}$ with their own transmitted data on the bit level. Here, the SI cancellation is accomplished after demodulation using the XOR operation again. That is,

$$\{\hat{b}_{\mathsf{B}}\} = \{\hat{b}_{\mathsf{R}} \oplus b_{\mathsf{A}}\}, \quad \text{at Station A;}$$
$$\{\hat{b}_{\mathsf{A}}\} = \{\hat{b}_{\mathsf{R}} \oplus b_{\mathsf{B}}\}, \quad \text{at Station B.}$$

An example of encoding and decoding procedure for the network coding scheme is shown in Fig. 4.3 and Fig. 4.4. The decoded bit sequences $\{b_{\mathsf{A}}\}$ and $\{b_{\mathsf{B}}\}$ at the relay in the MAC phase is $\{b_{\mathsf{A}}\} = 0110$ and $\{b_{\mathsf{B}}\} = 0001$. The relay first combines the two bit sequences using

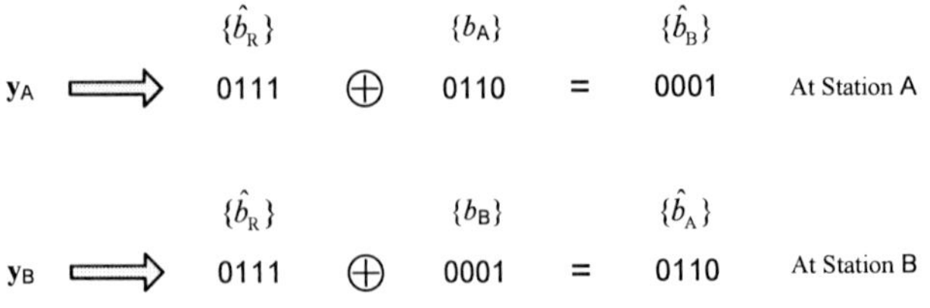

Fig. 4.4: SI cancellation at the stations A and B

the XOR operation and obtains $\{b_R\} = \{b_A \oplus b_B\} = 0111$. Then the relay modulates the combined bit sequence $\{b_R\}$ to the transmit symbol vector $\mathbf{s}_R$ on the its antennas as shown in Fig. 4.3. We assume that the relay is equipped with two antennas and we use the 4QAM constellation with Gray labeling to map the bit sequences to the transmit symbols on the transmit antennas. The first two bits of $\{b_R\}$, i.e., 01, is mapped to the symbol on the first antenna; the last two bits of $\{b_R\}$, i.e., 11, is mapped to the symbol on the second antenna, which is shown in Fig. 4.3.

After receiving the signals $\mathbf{y}_A$ and $\mathbf{y}_B$ as in (4.34) and (4.36), the stations A and B demodulate the received signals in the BRC phase and obtain the demodulated bit sequence $\{\hat{b}_R\} = 0111$. Since the bit sequence $\{b_A\}$ and $\{b_B\}$ is already respectively known at the stations A and B, station A can reveal the bit sequence $\{\hat{b}_B\}$ transmitted from station B by using the XOR operation, i.e.,

$$\begin{aligned}\{\hat{b}_B\} &= \{\hat{b}_R \oplus b_A\} \\ &= 0111 \oplus 0110 \\ &= 0001,\end{aligned}$$

and the same is done at station B, i.e.,

$$\begin{aligned}\{\hat{b}_A\} &= \{\hat{b}_R \oplus b_B\} \\ &= 0111 \oplus 0001 \\ &= 0110\end{aligned}$$

as shown in Fig. 4.4. In this way, both receivers obtain the bit sequence information from the other side.

In the network coding scheme, two different sets of information bits $\{b_A\}$ and $\{b_B\}$ are combined in the same transmit symbol $\{s_{\mathsf{R}}\}$. The relay broadcasts the common information, i.e., s_{R}, to both station A and B, and both stations have to be able to decode s_{R}. For the given transmit signal covariance matrix $\boldsymbol{\Omega}$, the information rate transmitted from the relay to the stations A and B in the BRC phase with the network coding scheme is given by

$$I^{\mathrm{XOR}} = \min\left\{I_{\mathsf{A}}^{\mathrm{BRC}}, I_{\mathsf{B}}^{\mathrm{BRC}}\right\}, \tag{4.38}$$

i.e., $R_{\mathsf{A}} \leq I^{\mathrm{XOR}}$ and $R_{\mathsf{B}} \leq I^{\mathrm{XOR}}$, with

$$I_{\mathsf{A}}^{\mathrm{BRC}} = \frac{1}{2}\log\left(\mathbf{I}_{N_{\mathsf{R}}} + \frac{P_{\mathsf{R}}}{N_{\mathsf{R}}\sigma_{\mathsf{B}}^2}\mathbf{H}_{\mathsf{B}}\boldsymbol{\Omega}\mathbf{H}_{\mathsf{B}}^H\right) \tag{4.39}$$

$$I_{\mathsf{B}}^{\mathrm{BRC}} = \frac{1}{2}\log\left(\mathbf{I}_{N_{\mathsf{R}}} + \frac{P_{\mathsf{R}}}{N_{\mathsf{R}}\sigma_{\mathsf{A}}^2}\mathbf{H}_{\mathsf{A}}\boldsymbol{\Omega}\mathbf{H}_{\mathsf{A}}^H\right). \tag{4.40}$$

The choice of the transmit signal covariance matrix determines the achievable rates in the BRC phase using the network coding scheme.

I^{XOR} in (4.38) is the minimum of the mutual information between the relay and both stations subject to the transmit power constraint of the relay. The maximum value of the information rates, i.e., $I^{\mathrm{XOR}\star}$ can be obtained by solving

$$\begin{array}{ll} \text{maximize} & \min\left\{I_{\mathsf{A}}^{\mathrm{BRC}}, I_{\mathsf{B}}^{\mathrm{BRC}}\right\} \\ \text{subject to} & \mathrm{tr}(\boldsymbol{\Omega}) \leq N_{\mathsf{R}}, \quad \boldsymbol{\Omega} \succeq 0 \\ \text{variable} & \boldsymbol{\Omega} \end{array} \tag{4.41}$$

where $\succeq$ denotes the positive semidefinite generalized inequality. This problem can be converted into the following convex optimization problem by introducing a *slack variable* τ:

$$\begin{array}{ll} \text{maximize} & \tau \\ \text{subject to} & I_{\mathsf{A}}^{\mathrm{BRC}} \geq \tau; \quad I_{\mathsf{B}}^{\mathrm{BRC}} \geq \tau \\ & \mathrm{tr}(\boldsymbol{\Omega}) \leq N_{\mathsf{R}}, \quad \boldsymbol{\Omega} \succeq 0 \\ \text{variable} & \boldsymbol{\Omega}. \end{array} \tag{4.42}$$

This optimization problem is equivalent to the problem to determine the capacity of a Gaussian broadcast channel with common information [123], which is also denoted as *multicast* scenario. It can be seen that different to the superposition coding scheme we have to find one covariance matrix for two links, instead of one matrix for each link. The optimization problem (4.42) can be solved by *semidefinite programming* techniques [218, 237].

Since both receiving stations have to decode the data contained in the symbol s_{R}, traditional network coding scheme let the relay transmit at a data rate that can be supported by both links, which sacrifices the stronger link. Compared to the superposition coding scheme, an advantage of the network coding scheme is that it does not suffer from the loss in transmit power when the same amount of information bits are contained in the symbol.

Maximizing the Sum Rate: For the given channel realization, the set of achievable rate pair for the two-way DF relaying system using the network coding scheme must satisfy both the MAC phase and the BRC phase constraints. However, once the maximum achievable rate for the BRC phase, i.e., $I^{\mathrm{XOR}\star}$ is calculated, those constraint becomes linear, and the maximum sum rate can be obtained as

$$\begin{aligned} R_{\text{sum}} &= R_{\mathsf{A}} + R_{\mathsf{B}} \\ &= \min\left\{\min(I_{\mathsf{A}}^{\mathrm{MAC}}, I^{\mathrm{XOR}\star}) + \min(I_{\mathsf{B}}^{\mathrm{MAC}}, I^{\mathrm{XOR}\star}), I_{\text{sum}}^{\mathrm{MAC}}\right\}. \end{aligned} \tag{4.43}$$

4.4.3 A Unified View of Superposition Coding and Network Coding Schemes

For two-way DF relaying schemes, the SPC scheme operates on the symbol level and the network coding scheme operates on the bit level. However, they have one thing in common, i.e., the relay introduces a binary operation to combine two elements each chosen from a set. Based on the result after the binary operation and one of the elements, the other element can be uniquely determined. Thus we introduce the following concept of *quasigroup* to unify the data combining schemes at the relay for two-way relaying schemes. A quasigroup is an algebraic structure resembling a group in the sense that "division" is always possible. Quasigroups differ from groups mainly in that they need not be associative.

Definition 4.1. *A quasigroup $\langle Q, *\rangle$ is a set Q with a binary operation $*$ (that is, a magma), such that for each a and b in Q, there exist unique elements x and y in Q such that:*

$$\begin{aligned} a * x &= b; \\ y * a &= b. \end{aligned}$$

The unique solutions to these equations are written $x = a \setminus b$ and $y = b \,/\, a$. "$\setminus$" and "/" denote, respectively, the defined binary operations of left and right division (sometimes called parastrophe).

For the SPC scheme, the set Q is chosen as complex symbol space $\mathbb{C}$ or $\mathbb{C}^{N_{\mathsf{R}}}$ and the magma operation is the linear sum "+". For the network coding scheme, Q is chosen as binary set $\{0, 1\}$ and the magma operation is the XOR operation "$\oplus$".

Actually, any combining scheme that obeys the definition of the quasigroup can be applied to the two-way relaying protocol. However, the SPC scheme and the network coding scheme are the two most widely used ones up to now.

4.4.4 Capacity Region of Bidirectional Broadcast Channel

In the MAC phase of a two-way DF relaying system, we have a classical multiple access channel. The optimal coding strategies for the multiple access channel and its capacity region have been characterized in [9,148], which has been shown in (4.15)–(4.17). In the BRC phase of a two-way DF relaying system, the relay combine the information decoded from the two stations and send the combined information back to the two stations, where the receiver of each station decodes the unknown information based on the received signal and its known information transmitted to its partner. Such a channel model is different from the broadcast channel in information theory. We use the definition of [170] and call it *bidirectional broadcast channel*. The capacity region and its optimal coding strategies have been proposed in [170,261,264]. Here we summarize the definitions for the Gaussian MIMO bidirectional broadcast channel and its capacity region.

The Gaussian MIMO bidirectional broadcast channel consists of the transmitting relay station and two receiving stations A and B, where the signal model is defined in (4.33)–(4.36). At any time instant $t = 1, 2, \cdots$, the transmitter sends the transmit symbol $\mathbf{s}_{\mathsf{R}}(t) \in \mathcal{S}_{\mathsf{R}} \subset \mathbb{C}^{N_{\mathsf{R}} \times 1}$. Here $\mathcal{S}_{\mathsf{R}}$ denotes the transmit symbol set. The transmit signal sequence must satisfy the relay transmit power constraint, i.e., for a transmit signal sequence of length N we have $\frac{1}{N}\sum_{t=1}^{N} \mathbf{s}_{\mathsf{R}}(t)^H \mathbf{s}_{\mathsf{R}}(t) \leq P_{\mathsf{R}}$. Equivalently, the transmit signal normalized with respect to the power constraint is defined as $\tilde{\mathbf{s}}_{\mathsf{R}} = \sqrt{\frac{N_{\mathsf{R}}}{P_{\mathsf{R}}}}\mathbf{s}_{\mathsf{R}}$. The transmit signal covariance matrix is defined as $\boldsymbol{\Omega} = \mathrm{E}\{\tilde{\mathbf{s}}_{\mathsf{R}}\tilde{\mathbf{s}}_{\mathsf{R}}^H\}$. In order to satisfy the transmit power constraint, we require $\mathrm{tr}(\boldsymbol{\Omega}) \leq N_{\mathsf{R}}$.

We denote w_{A} and w_{B} as the independent information messages at the stations A and B, respectively, which are also known at the relay station. The transmit data symbol for time instant 1 to N at the relay encoder is chosen according to the messages $w_{\mathsf{A}} \in \mathcal{W}_{\mathsf{A}}$ and $w_{\mathsf{B}} \in \mathcal{W}_{\mathsf{B}}$ jointly, i.e.,

$$\mathbf{s}_{\mathsf{R}}^N : \mathcal{W}_{\mathsf{A}} \times \mathcal{W}_{\mathsf{B}} \mapsto \mathcal{S}_{\mathsf{R}}^N. \tag{4.44}$$

$\mathcal{W}_{\mathsf{A}}$ and $\mathcal{W}_{\mathsf{B}}$ denote the message sets for the information from Station A and B, respectively.

The relay wants to sent the message w_{A} to station B and the message w_{B} to station A simultaneously. Each receiver $i \in \{\mathsf{A}, \mathsf{B}\}$ decodes based on $\mathbf{y}_i(1), \cdots, \mathbf{y}_i(N)$ and the message of its own $w_i \in \mathcal{W}_i$. That is, the decoders at stations A and B work as follows

$$g_{\mathsf{A}} : \mathcal{Y}_{\mathsf{A}}^N \times \mathcal{W}_{\mathsf{A}} \mapsto \mathcal{W}_{\mathsf{B}}, \tag{4.45}$$

$$g_{\mathsf{B}} : \mathcal{Y}_{\mathsf{B}}^N \times \mathcal{W}_{\mathsf{B}} \mapsto \mathcal{W}_{\mathsf{A}}. \tag{4.46}$$

For the given message set $v = [w_{\mathsf{A}}, w_{\mathsf{B}}]$, the receiver of station A is in error if $g(\mathbf{y}_{\mathsf{A}}^N, w_{\mathsf{A}}) \neq w_{\mathsf{B}}$. The probability of error event at the receiver of station A is defined as

$$\lambda_{\mathsf{A}}(v) = \mathrm{P}[g(\mathbf{y}_{\mathsf{A}}^N, w_{\mathsf{A}}) \neq w_{\mathsf{B}} \mid \mathbf{s}_{\mathsf{R}}^N(v) \text{ sent}] \tag{4.47}$$

Accordingly, we denote the probability that the receiver of station B is in error by

$$\lambda_{\mathsf{B}}(v) = \mathrm{P}[g(\mathbf{y}_{\mathsf{B}}^N, w_{\mathsf{B}}) \neq w_{\mathsf{A}} \mid \mathbf{s}_{\mathsf{R}}^N(v) \text{ sent}] \tag{4.48}$$

Now, we able to introduce the notation for the average probability of error for the station i, where $i \in \{\mathsf{A}, \mathsf{B}\}$

$$\mu_i^{(N)} = \frac{1}{|\mathcal{V}|} \sum_{v \in \mathcal{V}} \lambda_i(v) \tag{4.49}$$

where $\mathcal{V}$ is defined as the overall message set $\mathcal{V} = \mathcal{W}_{\mathsf{A}} \times \mathcal{W}_{\mathsf{B}}$.

A rate pair $R_{\mathsf{A}}, R_{\mathsf{B}}$ is *achievable* for the Gaussian MIMO bidirectional broadcast channel with the transmit power constraint P_{R} if for any $\delta > 0$ there exists an $N(\delta) \in \mathbb{N}$ and a sequence of codes $\mathbf{s}_{\mathsf{R}}^N$ satisfying the power constraint such that for all $N \geq N(\delta)$ we have $\frac{\log|\mathcal{W}_{\mathsf{A}}|}{N} \geq R_{\mathsf{B}} - \delta$ and $\frac{\log|\mathcal{W}_{\mathsf{B}}|}{N} \geq R_{\mathsf{A}} - \delta$ while $\mu_{\mathsf{A}}^{(N)}, \mu_{\mathsf{B}}^{(N)} \to 0$ as $N \to \infty$.

The capacity region describes the transmission rate pairs $R_{\mathsf{A}}, R_{\mathsf{B}}$ that are simultaneously achievable by terminals A and B under certain channel conditions, i.e., $C = \{[R_{\mathsf{A}}, R_{\mathsf{B}}] \in \mathbb{R}_+^2 : [R_{\mathsf{A}}, R_{\mathsf{B}}] \text{ is achievable}\}$. Its capacity region is described by the following theorem

Theorem 4.4.1 ([262]). *For given covariance matrix $\mathbf{\Omega} \succeq 0$ with $\mathrm{tr}(\mathbf{\Omega}) \leq N_R$ satisfying the power constraint, the rate pairs R_A, R_B satisfying the following inequalities are achievable*

$$R_A \leq I_A^{\mathrm{BRC}}(\mathbf{\Omega}) = \frac{1}{2} \log \left(\mathbf{I}_{N_B} + \frac{P_R}{N_B \sigma_B^2} \mathbf{H}_B \mathbf{\Omega} \mathbf{H}_B^H \right), \tag{4.50}$$

$$R_B \leq I_B^{\mathrm{BRC}}(\mathbf{\Omega}) = \frac{1}{2} \log \left(\mathbf{I}_{N_A} + \frac{P_R}{N_A \sigma_A^2} \mathbf{H}_A \mathbf{\Omega} \mathbf{H}_A^H \right). \tag{4.51}$$

Here the pre-log factor $1/2$ is due to two channel uses (MAC and BRC phases) in the two-way DF relaying protocol. The capacity region of the Gaussian MIMO bidirectional broadcast channel is given by

$$C = \bigcup_{\mathrm{tr}(\boldsymbol{\Omega}) \le P_R, \boldsymbol{\Omega} \succeq 0} \mathrm{conv}[I_A^{\mathrm{BRC}}(\boldsymbol{\Omega}), I_B^{\mathrm{BRC}}(\boldsymbol{\Omega})] \tag{4.52}$$

The technique to prove the capacity region is random binning (random coding approaches). Note Theorem 4.4.1 has an interesting implication that if both terminals have perfect information about the messages that is intended for the other side, they can decode the messages transmitted by the relay as if the other side does not exist.

Characterizing the Capacity Region: The capacity region of the BRC phase bidirectional broadcast channel is convex. One method to characterize the capacity region by its boundary is to calculate the weighted sum rate optimal rate pairs. That is, for the given weighting coefficient $0 < w < 1$, the aim is to find the optimal covariance matrix that maximizes the weighted sum rate:

$$\begin{aligned} &\text{maximize} && w I_{\mathsf{A}}^{\mathrm{BRC}}(\boldsymbol{\Omega}) + (1-w) I_{\mathsf{B}}^{\mathrm{BRC}}(\boldsymbol{\Omega}) \\ &\text{subject to} && \mathrm{tr}(\boldsymbol{\Omega}) \le N_{\mathsf{R}}, \quad \boldsymbol{\Omega} \succeq 0 \\ &\text{variable} && \boldsymbol{\Omega} \end{aligned} \tag{4.53}$$

After the weighting factor w is taken all the values from 0 to 1, we can characterize all the boundary points for the capacity region.

Another method to determine the boundary of the BRC phase bidirectional broadcast channel capacity region is to solve the following convex optimization problem:

$$\begin{aligned} &\text{maximize} && I_{\mathsf{A}}^{\mathrm{BRC}}(\boldsymbol{\Omega}) \\ &\text{subject to} && I_{\mathsf{B}}^{\mathrm{BRC}}(\boldsymbol{\Omega}) \ge q \\ & && \mathrm{tr}(\boldsymbol{\Omega}) \le N_{\mathsf{R}}, \quad \boldsymbol{\Omega} \succeq 0 \\ &\text{variable} && \boldsymbol{\Omega} \end{aligned} \tag{4.54}$$

where q takes the value from 0 to $\max_{\mathrm{tr}(\boldsymbol{\Omega}) \le N_{\mathsf{R}}, \boldsymbol{\Omega} \succeq 0} I_{\mathsf{B}}^{\mathrm{BRC}}(\boldsymbol{\Omega})$

Maximizing the Sum Rate: The overall achievable rate region of the MIMO two-way DF relaying system must satisfy both the MAC phase capacity region described by (4.15)–(4.17) and the BRC phase capacity region (4.50)–(4.51). Since we assume the stations A and B do not have the channel knowledge on $\mathbf{G}_{\mathsf{A}}$ and $\mathbf{G}_{\mathsf{B}}$ in the MAC and equal power is allocated to the data streams in $\mathbf{x}_{\mathsf{A}}$ and $\mathbf{x}_{\mathsf{B}}$, $I_{\mathsf{A}}^{\mathrm{MAC}}$, $I_{\mathsf{B}}^{\mathrm{MAC}}$ and $I_{\mathsf{sum}}^{\mathrm{MAC}}$ in (4.15)–(4.17) are constants for given channels $\mathbf{G}_{\mathsf{A}}$ and $\mathbf{G}_{\mathsf{B}}$. However, we assume the relay has the channel knowledge of

$\mathbf{H}_\mathsf{A}$ and $\mathbf{H}_\mathsf{B}$ in the BRC phase so that the transmit covariance matrix $\boldsymbol{\Omega}$ can be optimized. Furthermore, we can choose the operational point in the overall achievable rate region to maximize the sum rate. In this case, the points that achieves the maximum sum rate only depends on the transmit signal covariance matrix $\boldsymbol{\Omega}$.

First of all, the point $(I_\mathsf{A}^{\mathrm{BRC}}(\boldsymbol{\Omega}^\star), I_\mathsf{B}^{\mathrm{BRC}}(\boldsymbol{\Omega}^\star))$ that maximizes the BRC phase capacity region can be calculated by solving the following convex optimization problem:

$$\begin{array}{ll} \text{maximize} & I_\mathsf{A}^{\mathrm{BRC}}(\boldsymbol{\Omega}) + I_\mathsf{B}^{\mathrm{BRC}}(\boldsymbol{\Omega}) \\ \text{subject to} & \mathrm{tr}(\boldsymbol{\Omega}) \leq N_\mathsf{R}, \quad \boldsymbol{\Omega} \succeq 0 \\ \text{variable} & \boldsymbol{\Omega} \end{array} \tag{4.55}$$

The optimum transmit covariance matrix that solves the problem (4.55) is denoted as $\boldsymbol{\Omega}^\star$. By comparing $(I_\mathsf{A}^{\mathrm{BRC}}(\boldsymbol{\Omega}^\star), I_\mathsf{B}^{\mathrm{BRC}}(\boldsymbol{\Omega}^\star))$ with $I_\mathsf{A}^{\mathrm{MAC}}$ and $I_\mathsf{B}^{\mathrm{BRC}}$ in (4.15) and (4.16), we can obtain the following four cases as shown in Fig. 4.5:

1. $I_\mathsf{A}^{\mathrm{MAC}} \geq I_\mathsf{A}^{\mathrm{BRC}}(\boldsymbol{\Omega}^\star)$ and $I_\mathsf{B}^{\mathrm{MAC}} \geq I_\mathsf{B}^{\mathrm{BRC}}(\boldsymbol{\Omega}^\star)$: In this case, the maximum achievable sum rate is $\min\{I_{\mathrm{sum}}^{\mathrm{MAC}}, I_\mathsf{A}^{\mathrm{BRC}}(\boldsymbol{\Omega}^\star) + I_\mathsf{B}^{\mathrm{BRC}}(\boldsymbol{\Omega}^\star)\}$.

   If $I_{\mathrm{sum}}^{\mathrm{MAC}} \geq I_\mathsf{A}^{\mathrm{BRC}}(\boldsymbol{\Omega}^\star) + I_\mathsf{B}^{\mathrm{BRC}}(\boldsymbol{\Omega}^\star)$, the point $(I_\mathsf{A}^{\mathrm{BRC}}(\boldsymbol{\Omega}^\star), I_\mathsf{B}^{\mathrm{BRC}}(\boldsymbol{\Omega}^\star))$ is also located within the MAC phase capacity region. Since the overall achievable rate region is the intersection of the MAC phase and BRC phase capacity regions, it can be no larger than either of them. Thus the point $(I_\mathsf{A}^{\mathrm{BRC}}(\boldsymbol{\Omega}^\star), I_\mathsf{B}^{\mathrm{BRC}}(\boldsymbol{\Omega}^\star))$ achieves the maximum sum rate in the overall achievable rate region and them maximum sum rate is $(I_\mathsf{A}^{\mathrm{BRC}}(\boldsymbol{\Omega}^\star) + I_\mathsf{B}^{\mathrm{BRC}}(\boldsymbol{\Omega}^\star))$;

   If $I_{\mathrm{sum}}^{\mathrm{MAC}} \leq I_\mathsf{A}^{\mathrm{BRC}}(\boldsymbol{\Omega}^\star) + I_\mathsf{B}^{\mathrm{BRC}}(\boldsymbol{\Omega}^\star)$, at least one point within the region $R_\mathsf{A} \leq I_\mathsf{A}^{\mathrm{BRC}}(\boldsymbol{\Omega}^\star)$ and $R_\mathsf{B} \leq I_\mathsf{B}^{\mathrm{BRC}}(\boldsymbol{\Omega}^\star)$ achieves the sum rate $I_{\mathrm{sum}}^{\mathrm{MAC}}$. This is because the region $R_\mathsf{A} \leq I_\mathsf{A}^{\mathrm{BRC}}(\boldsymbol{\Omega}^\star)$ and $R_\mathsf{B} \leq I_\mathsf{B}^{\mathrm{BRC}}(\boldsymbol{\Omega}^\star)$ is continuous. The sum rate of the point $(0, 0)$ is smaller than $I_{\mathrm{sum}}^{\mathrm{MAC}}$ while the sum rate of the point $(I_\mathsf{A}^{\mathrm{BRC}}(\boldsymbol{\Omega}^\star), I_\mathsf{B}^{\mathrm{BRC}}(\boldsymbol{\Omega}^\star))$ is larger. So there exist one point in the region that achieves the sum rate $I_{\mathrm{sum}}^{\mathrm{MAC}}$. Since $I_{\mathrm{sum}}^{\mathrm{MAC}}$ is the maximum sum rate achievable in the MAC phase capacity region, it is thus also the maximum sum rate in the overall achievable rate region.

2. $I_\mathsf{A}^{\mathrm{MAC}} \leq I_\mathsf{A}^{\mathrm{BRC}}(\boldsymbol{\Omega}^\star)$ and $I_\mathsf{B}^{\mathrm{MAC}} \leq I_\mathsf{B}^{\mathrm{BRC}}(\boldsymbol{\Omega}^\star)$: In this case, the maximum achievable sum rate is $I_{\mathrm{sum}}^{\mathrm{MAC}}$. This is because the MAC phase capacity region is inside the BRC phase capacity region in this case.

3. $I_\mathsf{A}^{\mathrm{MAC}} \geq I_\mathsf{A}^{\mathrm{BRC}}(\boldsymbol{\Omega}^\star)$ and $I_\mathsf{B}^{\mathrm{MAC}} \leq I_\mathsf{B}^{\mathrm{BRC}}(\boldsymbol{\Omega}^\star)$: In this case, the point $(I_\mathsf{A}^{\mathrm{BRC}}(\boldsymbol{\Omega}^\star), I_\mathsf{B}^{\mathrm{BRC}}(\boldsymbol{\Omega}^\star))$ is outside the MAC phase capacity region and we have to solve

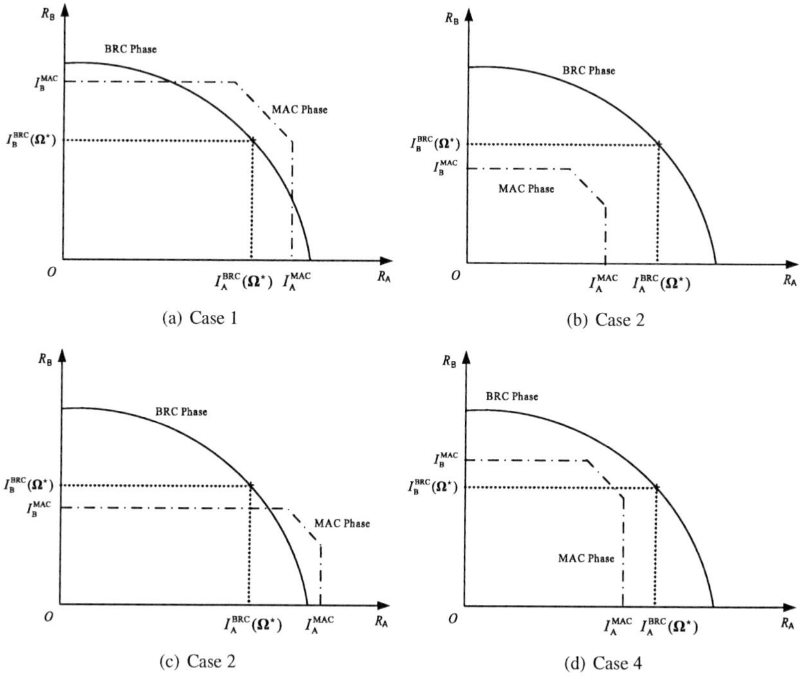

(a) Case 1

(b) Case 2

(c) Case 2

(d) Case 4

Fig. 4.5: Four cases depicting the capacity regions of the MAC and BRC phases.

the following convex optimization problem

$$\begin{aligned} \text{maximize} \quad & I_{\mathsf{A}}^{\mathrm{BRC}}(\boldsymbol{\Omega}) \\ \text{subject to} \quad & I_{\mathsf{B}}^{\mathrm{BRC}}(\boldsymbol{\Omega}) \geq I_{\mathsf{B}}^{\mathrm{MAC}} \\ & \mathrm{tr}(\boldsymbol{\Omega}) \leq N_{\mathsf{R}}, \quad \boldsymbol{\Omega} \succeq 0 \\ \text{variable} \quad & \boldsymbol{\Omega} \end{aligned} \tag{4.56}$$

The optimum value of the objective function in (4.56) is denoted as $p^{\star}$. In this case, the maximum sum rate is $\min\{I_{\text{sum}}^{\mathrm{MAC}}, p^{\star} + I_{\mathsf{B}}^{\mathrm{MAC}}\}$ following similar discussion as in Case 1.

4. $I_{\mathsf{A}}^{\mathrm{MAC}} \leq I_{\mathsf{A}}^{\mathrm{BRC}}(\boldsymbol{\Omega}^{\star})$ and $I_{\mathsf{B}}^{\mathrm{MAC}} \geq I_{\mathsf{B}}^{\mathrm{BRC}}(\boldsymbol{\Omega}^{\star})$: This case is similar to Case 3. We have to first solve the following convex optimization problem

$$\begin{aligned} \text{maximize} \quad & I_{\mathsf{B}}^{\mathrm{BRC}}(\boldsymbol{\Omega}) \\ \text{subject to} \quad & I_{\mathsf{A}}^{\mathrm{BRC}}(\boldsymbol{\Omega}) \geq I_{\mathsf{A}}^{\mathrm{MAC}} \\ & \mathrm{tr}(\boldsymbol{\Omega}) \leq N_{\mathsf{R}}, \quad \boldsymbol{\Omega} \succeq 0 \\ \text{variable} \quad & \boldsymbol{\Omega} \end{aligned} \tag{4.57}$$

The optimum value of the objective function in (4.57) is denoted as $p_2^{\star}$. In this case, the maximum sum rate is $\min\{I_{\text{sum}}^{\mathrm{MAC}}, p_2^{\star} + I_{\mathsf{A}}^{\mathrm{MAC}}\}$ following similar discussion as in Case 3.

The algorithm for calculating the maximum sum rate considering the MAC phase and the BRC phase capacity region can be described by Algorithm 2

Algorithm 2 Maximizing R_{sum} for two-way DF relaying systems considering the MAC and BRC phase capacity region

Find the point $(I_{\mathsf{A}}^{\mathrm{BRC}}(\boldsymbol{\Omega}^{\star}), I_{\mathsf{B}}^{\mathrm{BRC}}(\boldsymbol{\Omega}^{\star}))$ that solves the convex optimization problem (4.55)
if $I_{\mathsf{A}}^{\mathrm{MAC}} \geq I_{\mathsf{A}}^{\mathrm{BRC}}(\boldsymbol{\Omega}^{\star})$ and $I_{\mathsf{B}}^{\mathrm{MAC}} \geq I_{\mathsf{B}}^{\mathrm{BRC}}(\boldsymbol{\Omega}^{\star})$ **then**
  $R_{\text{sum}} = \min\{I_{\text{sum}}^{\mathrm{MAC}}, I_{\mathsf{A}}^{\mathrm{BRC}}(\boldsymbol{\Omega}^{\star}) + I_{\mathsf{B}}^{\mathrm{BRC}}(\boldsymbol{\Omega}^{\star})\}$;
else if $I_{\mathsf{A}}^{\mathrm{MAC}} \leq I_{\mathsf{A}}^{\mathrm{BRC}}(\boldsymbol{\Omega}^{\star})$ and $I_{\mathsf{B}}^{\mathrm{MAC}} \leq I_{\mathsf{B}}^{\mathrm{BRC}}(\boldsymbol{\Omega}^{\star})$ **then**
  $R_{\text{sum}} = I_{\text{sum}}^{\mathrm{MAC}}$;
else if $I_{\mathsf{A}}^{\mathrm{MAC}} \geq I_{\mathsf{A}}^{\mathrm{BRC}}(\boldsymbol{\Omega}^{\star})$ and $I_{\mathsf{B}}^{\mathrm{MAC}} \leq I_{\mathsf{B}}^{\mathrm{BRC}}(\boldsymbol{\Omega}^{\star})$ **then**
  The optimum value of the objective function in (4.56) is denoted as $p^{\star}$;
  $R_{\text{sum}} = \min\{I_{\text{sum}}^{\mathrm{MAC}}, p^{\star} + I_{\mathsf{B}}^{\mathrm{MAC}}\}$;
else
  The optimum value of the objective function in (4.57) is denoted as $p_2^{\star}$;
  $R_{\text{sum}} = \min\{I_{\text{sum}}^{\mathrm{MAC}}, p_2^{\star} + I_{\mathsf{A}}^{\mathrm{MAC}}\}$.
end if
return R_{sum}.

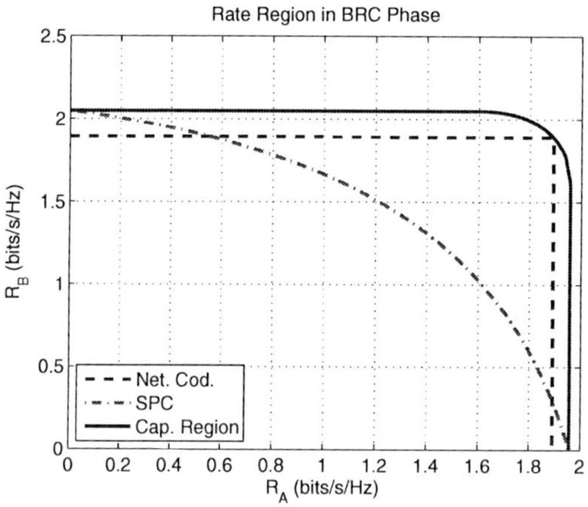

Fig. 4.6: Achievable rate region in the BRC phase of a MIMO two-way DF relaying system.

4.5 Performance Results

The achievable rate regions using the SPC scheme and the network coding scheme in the BRC phase of two-way DF relaying systems for a given channel realization are shown in Fig. 4.6, and they are compared with the capacity region according to Theorem 4.4.1. Here $N_\mathsf{A} = N_\mathsf{R} = N_\mathsf{B} = 4$, $P_\mathsf{R}/\sigma_\mathsf{A}^2 = P_\mathsf{R}/\sigma_\mathsf{B}^2 = 1$. The channel realization $\mathbf{H}_\mathsf{A}$ and $\mathbf{H}_\mathsf{B}$ are chosen as follows

$$\mathbf{H}_\mathsf{A} = \begin{bmatrix} -0.31+0.75i & -0.81+0.21i & 0.23-0.49i & -0.42-1.02i \\ -1.18+0.04i & 0.84-0.94i & 0.12+0.61i & 1.54+0.40i \\ 0.09-0.07i & 0.84+0.51i & -0.13+0.89i & -0.10-0.28i \\ 0.20-0.59i & -0.03+1.15i & 0.51-1.13i & 0.08+0.49i \end{bmatrix} \tag{4.58}$$

$$\mathbf{H}_\mathsf{B} = \begin{bmatrix} 0.58-1.53i & 0.84+0.36i & -1.13+0.27i & -0.57+0.00i \\ 0.50-0.04i & -0.85+1.20i & 0.18-0.71i & 0.37-0.22i \\ 0.91-0.71i & -0.01+0.42i & -0.75-0.01i & 0.16+0.77i \\ 0.47+0.43i & -0.11-0.46i & 1.00-0.03i & -0.65-1.33i \end{bmatrix}. \tag{4.59}$$

It can be seen from Fig. 4.6 that the maximum achievable sum rate of the SPC scheme in lower than that of the network coding scheme. This is because the SPC scheme modulates the two sets of decoded data bits separately and each set of data consumes part of the transmit power, whereas the network coding scheme combines the two sets of data into one and does

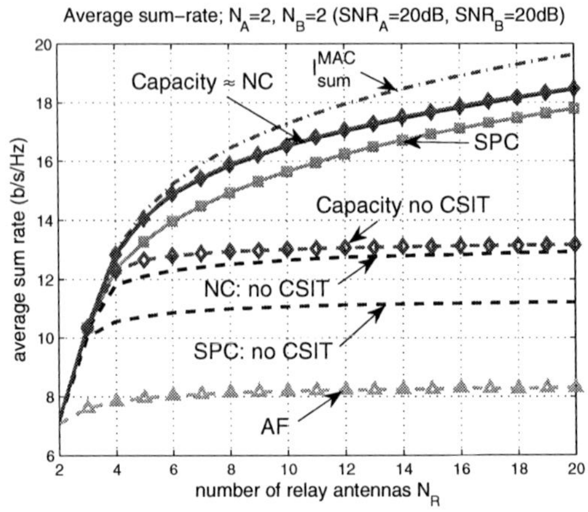

Fig. 4.7: Average sum rates with and without CSIT with respect to the number of relay antennas N_{R}. The considered schemes are: BRC phase capacity region according to Theorem 4.4.1, network coding scheme, SPC scheme and AF. The AF relay linear processing matrix is calculated according to (4.5).

not split the power. However, the SPC scheme can easily transmit different data rates to the two stations in the BRC phase. The network coding scheme achieves the point that is the intersection of the bisector of the first quadrant and the capacity region boundary. In such a symmetric channel condition, i.e., the channel to both stations are equally strong, the network coding scheme achieves most of the internal area of the capacity region.

Fig. 4.7 shows the average sum rates with and without CSIT for two-way relaying systems with respect to the number of relay antennas N_{R}. Here the sum rates consider the achievable rates after both the MAC phase and the BRC phase. We consider frequency-flat block fading channels between all stations; the elements of the channel matrices $\mathbf{G}_{\mathsf{A}}$, $\mathbf{G}_{\mathsf{B}}$, $\mathbf{H}_{\mathsf{A}}$ and $\mathbf{H}_{\mathsf{B}}$ are i.i.d. $\mathcal{CN}(0,1)$ random variables. We assume the channels to be constant over the two phases of the two-way relaying scheme. The noise vectors at the relay and the stations are circular symmetric complex Gaussian noise vectors with zero mean. Furthermore, we define $\mathrm{SNR}_{\mathsf{A}} = P_{\mathsf{A}}/\sigma^2 = P_{\mathsf{R}}/\sigma_{\mathsf{A}}^2$ and $\mathrm{SNR}_{\mathsf{B}} = P_{\mathsf{B}}/\sigma^2 = P_{\mathsf{R}}/\sigma_{\mathsf{B}}^2$. In Fig. 4.7, the number of antennas at the stations A and B is fixed to $N_{\mathsf{A}} = N_{\mathsf{B}} = 2$ and the SNR for the channel between the station A, B and the relay is $\mathrm{SNR}_{\mathsf{A}} = \mathrm{SNR}_{\mathsf{B}} = 20\mathrm{dB}$. When $N_{\mathsf{R}} = 2$, i.e., all the three stations have two antennas, there is no significant difference in the sum rates for all the considered schemes, neither with nor without CSIT. This is mainly due to the fact that the sum rate for

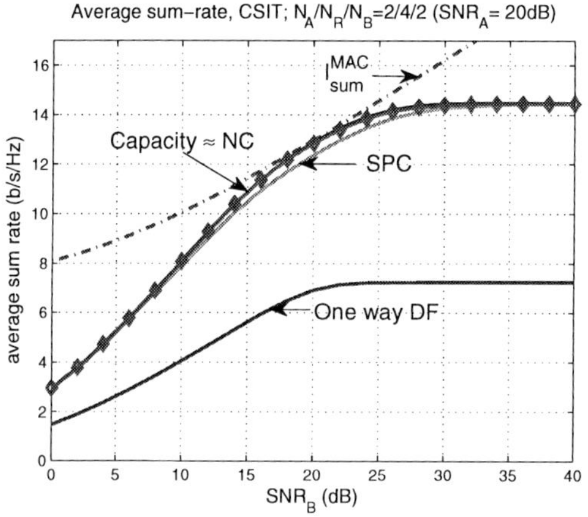

Fig. 4.8: Average sum rates with CSIT: two-way DF relaying protocol using the network coding scheme and the SPC scheme compared to one-way DF relaying protocol.

the $N_{\mathsf{A}} = N_{\mathsf{B}} = N_{\mathsf{R}} = 2$ case is limited by the MAC phase constraint $I_{\text{sum}}^{\text{MAC}}$ (4.17) compared to the curve of $I_{\text{sum}}^{\text{MAC}}$. For $N_{\mathsf{R}} = 4$ antennas at the relay the SPC scheme without CSIT shows a significantly lower sum rate than the other schemes. This is due to the fact that in this case $\beta = 1/2$ is chosen, i.e. power $P_{\mathsf{R}}/2$ is used for s_{A} and $P_{\mathsf{R}}/2$ for s_{B}. Since the network coding scheme combines the two decoded information bits and transmits the combined bits using the total power P_{R}, the network coding scheme has 3dB power gain compared to the SPC scheme when no CSIT is available. The maximum achievable sum rate using random coding according to Theorem 4.4.1 is approximately the same as the network coding case when CSIT is available. When no CSIT is available, the gain in the maximum sum rate using random coding according to Theorem 4.4.1 compared to the the network coding scheme is marginal. When no CSIT is available, the maximum achievable sum rates for all the considered schemes show almost no increase with the number of the relay antennas N_{R} when N_{R} is large, i.e., $N_{\mathsf{R}} \geq 6$. However, when CSIT is available at the relay, the maximum achievable sum rates increase linearly with the number of relay antennas N_{R} when N_{R} is large, i.e., $N_{\mathsf{R}} \geq 6$. As a result, the difference between the maximum achievable sum rates with and without CSIT cases increases with the increase of the number of relay antennas N_{R}. Another observation of in Fig. 4.7 is that the maximum achievable sum rate of two-way AF relaying scheme is significantly lower than that of the DF relaying scheme, when the relay linear processing matrix is calculated according to (4.5).

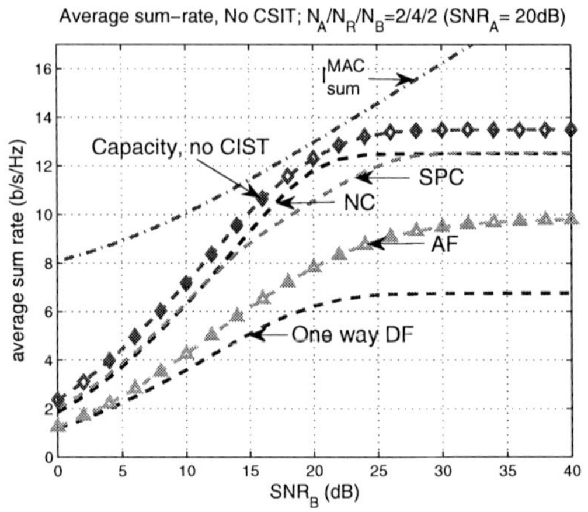

Fig. 4.9: Average sum rates without CSIT: two-way DF relaying protocol using the network coding scheme and the SPC scheme compared to one-way DF relaying protocol and two-way AF relaying protocol (relay linear processing matrix is calculated according to (4.5)).

In Fig. 4.8 and Fig. 4.9, we study the impact of unbalanced link quality, i.e. $\mathrm{SNR}_\mathsf{A} = 20\mathrm{dB}$ and $0\mathrm{dB} \leq \mathrm{SNR}_\mathsf{B} \leq 40\mathrm{dB}$. The average sum-rates are given for a scenario with $N_\mathsf{A} = N_\mathsf{B} = 2$ antennas at the nodes A and B and $N_\mathsf{R} = 4$ antennas at the relay. The elements of the channel matrices $\mathbf{G}_\mathsf{A}$, $\mathbf{G}_\mathsf{B}$, $\mathbf{H}_\mathsf{A}$ and $\mathbf{H}_\mathsf{B}$ are i.i.d. $\mathcal{CN}(0,1)$ random variables, and the channels remain constant over their corresponding phases of the two-way relaying scheme. Fig. 4.8 shows the average sum rate when CSIT is available at the relay. We consider the random coding scheme that achieves the capacity region in the BRC phase, the network coding scheme and the SPC scheme for the two-way DF relaying protocol, as well as the one-way DF relaying protocol. The sum-rate of the one-way DF relaying scheme is calculated by averaging over four hops: from node A over the relay to node B and vice versa. When the random coding scheme that achieves the capacity region in the BRC phase is applied, it achieves approximately the same average sum rate as the network coding scheme, and shows marginal advantage over the SPC scheme for the two-way DF relaying protocol. When there is no CSIT available as shown in Fig. 4.9, the difference between the three schemes increases. Without CSIT, the network coding scheme shows an advantage for $13\mathrm{dB} \leq \mathrm{SNR}_2 \leq 27\mathrm{dB}$ compared to the SPC scheme. Generally, all two-way relaying schemes with CSIT or without almost double the average sum rate of the one-way DF relaying scheme in those scenario.

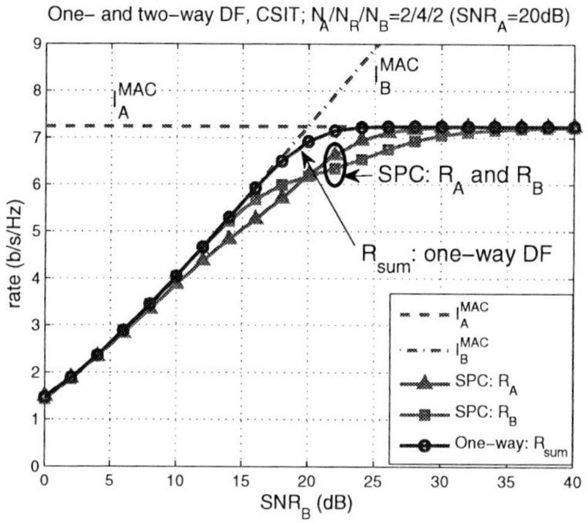

Fig. 4.10: Average user rates for two-way DF relaying protocol using the SPC scheme compared to the sum rate for one-way DF relaying.

In Fig. 4.9, it is also shown that the two-way DF relaying protocol outperforms the two-way AF relaying protocol in the average sum rate when CSIT is not available at the relay.

Fig. 4.10 shows for the same scenario as Fig. 4.8 where a comparison of the user rates in the SPC scheme and the average sum rate of one-way DF relaying is presented. In addition, $I_{\mathrm{A}}^{\mathrm{MAC}}$ and $I_{\mathrm{B}}^{\mathrm{MAC}}$ defined in (4.15) and (4.16) are shown, too. In the considered scenario the user rates differ only slightly; the second interesting observation is that both SPC user rates are not far from the sum rate of the one-way DF protocol.

4.6 Chapter Summary and Conclusions

In this chapter, we extended the two-way AF and DF relaying protocol [193] to multiple antennas at all stations, i.e., MIMO two-way AF and DF relaying systems. We presented the signal models for the MIMO two-way AF and DF relaying protocols and assume further the knowledge of transmit CSI at the DF relay. In the TDD transmission mode, the relay has to estimate the MIMO channels for decoding in the MAC anyway. So, in the BRC phase this knowledge can be used for precoding if the bursts are short enough compared to the coherence time of the MIMO channels. In the FDD transmission mode, the relay may require feedback of channel knowledge from the two receiving stations. Equal time (or frequency) resources were allocated to the MAC and BRC phases. We compared the two practical approaches, i.e., the SPC and the network coding schemes, for combining the information at the relay in the BRC phase of two-way DF relaying protocol. Furthermore, we also presented the method of characterizing the capacity region in the MIMO bidirectional broadcast channel and calculating its maximum sum rate. We showed that two-way relaying achieves a quite substantial improvement in spectral efficiency compared to conventional relaying with and without transmit CSI at the relay. We showed that the difference in sum-rate compared to the case where no CSIT is used, increases with increasing ratio between number of relay antennas and number of node antennas. We further showed that the network coding scheme achieves nearly the optimal sum rate when CSIT is available at the relay. When CSIT is not available at the relay, the two-way DF relaying protocol always significantly outperforms the two-way AF relaying protocol.

Chapter 5

Optimum Time-Division in Two-Way Relaying Systems

The capacity region of a general two-way relay channel is still an open problem up to now. The optimal relaying strategy is therefore also unknown. However, specific relaying strategies, such as amplify-and-forward (AF), compress-and-forward (CF) and decode-and-forward (DF), have been proposed for the two-way relay channel. Among them, the two-way DF relaying scheme is of particular interest due to its practical applicability in real-world systems. We consider two-phase two-way relaying protocols, i.e., we divide the two-way communication into the multiple access (MAC) phase and the broadcast (BRC) phase in accordance with other papers. Under such a assumption, the rate region of the two-way DF relaying system depends on the time-division (TD) between the MAC and BRC phases, i.e., the temporal or spectral resources allocated to the two phases. In Chapter 4, we only considered equal TD and no power scaling. In this chapter, we discuss the achievable rate regions in two-way DF relaying systems with optimum TD strategies between the MAC phase and the BRC phase and provide practical algorithms for characterizing them, which is especially complex for multiple-input multiple-output (MIMO) systems. Both peak power constraint and average power constraint are considered. We show that by optimizing the TD strategies, the achievable rate regions of the two-way DF relaying system can be significantly enhanced. Using those methods, we also compare the ergodic sum rates of the system and the average achievable rates of one user given the minimum quality-of-service (QoS) requirement of the other in different scenarios.

5.1 Introduction

After the proposal of the two-way DF relaying protocol [193], people are looking for the optimal transmission schemes that maximize the achievable rate region. The MAC phase in the two-way DF relaying protocol represents a conventional multiple access channel. Its capacity region and optimal coding schemes have been summarized in [78]. The encoding schemes at the relay in the BRC phase has aroused much research interest in recent years. Practical encoding schemes, such as the *superposition coding* (SPC) scheme [193] and the *networking coding* scheme [260], have been proposed (see Chapter 4). The difference between the two schemes lies in combining the data at the relay on the symbol level or on the bit level. From the information theoretic perspective, the BRC phase in the two-way DF relaying protocol represents a two-user broadcast scenario where both receiving stations have perfect side information about the messages intended for its partner. This channel model is called *bidirectional broadcast channel* and its optimal coding strategies have recently been characterized in [262, 264].

Even though the optimal coding schemes and the rate regions for the MAC phase and the BRC phase in the two-way DF relaying system can be separately characterized according to those in the multiple access channel and the broadcast channel, the achievable rate region of the two-way DF relaying system also depends on the "coupling" between the two phases. This is because the MAC and BRC phases have to be separated either in time (time-division duplex, TDD) or in frequency (frequency-division duplex, FDD) due to the half-duplex constraint of relaying systems. The total time or frequency resources can not be spent solely on the MAC or the BRC phase. We call this coupling the time-division (TD) between the MAC and the BRC phases. Note that the name "time-division" does not imply it is only valid for TDD relaying systems. The same discussions can be applied to the FDD systems as well. Many papers, such as [239], allocate equal time or frequency resources to the MAC and BRC phases. The author of [169] considered the problem of finding the optimum TD strategies between the two phases when the SPC scheme is applied in the BRC phase. The authors of [263] considered the optimal TD strategies between the MAC and BRC phases for two-way DF relaying system with multiple antennas at the relay and single antenna at user stations. They characterized the overall achievable rate regions using the optimum coding strategies for the MAC and BRC phases separately, and compared it with the rate regions when the SPC scheme and the network coding scheme are applied in the BRC phase. In [171], the authors presented the method to characterized the overall achievable rate regions using the optimum coding strategies for the MAC and BRC phases for the same system model as in [263], i.e., multiple antennas are equipped at the relay and each user station only has a single antenna. They showed that the achievable rate region using optimal TD strategies is convex and no additional time-sharing within the rate region is necessary. In [168], the optimum TD strategy for single antenna two-way DF relaying system, i.e., each station including the relay is equipped with a single antenna, was discussed in detail. However, how to determine the optimum TD strategies for such a system with multiple antennas, so that the achievable rate region is maximized, has not yet been solved. In this case, both the TD strategies and the transmit signal covariance matrices have to be optimized if all the stations have channel knowledge.

In this chapter, we require the relay to fully decode the data from the two transmitting stations in the MAC phase and propose algorithms for calculating the achievable rate regions with optimum TD strategies between the MAC and BRC phases for the following two cases: One is under the peak power constraint and the other is under the average power constraint. Here peak power constraint means that the transmit power at each station cannot exceed certain constraints at any given time. In this case, TD strategies only affect the pre-log factor in the rate expression, and the rate regions of the MAC phase or the BRC phase are scaled by the TD factor. Average power constraint means that the transmit power at the stations can be varied if the transmit power averaged over the whole time period does not exceed certain constraints. In this case, both the pre-log factor and the transmit power inside the rate expressions are affected by the TD strategies. For both power constraints, we present optimization methods to characterize their achievable rate regions considering both the MAC phase and the BRC phase rate constraints. The increase of the achievable rate regions compared to the equal TD case is shown. Using the proposed methods, we also compare the average achievable rates of one user station given the minimum quality-of-service (QoS) requirement of the other under different scenarios. To the best of our knowledge, the proposed method achieves the largest rate region for the MIMO two-way DF relaying protocol up to now.

Our Contributions: The contributions of this chapter are summarized as follows:

- We characterize the optimal TD strategies in MIMO two-way DF relaying systems using convex optimization methods. The optimal transmit covariance matrices at the MIMO user stations and the relay station under the optimal TD strategies are determined as well.
- The ergodic sum rate and the average achievable rates of one user station given the minimum quality-of-service (QoS) requirement of the other are presented. Insights are provided for the results under different system setups.

This chapter is organized as follows: The system model and the capacity regions of the MAC and BRC phases are shown in Section 5.2. An algorithm to find the optimum TD

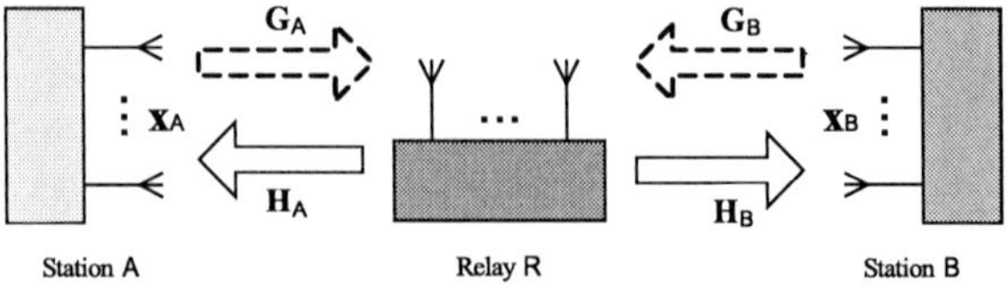

Fig. 5.1: MIMO two-way DF relaying system. The dashed arrows and solid arrows represent the transmissions in the MAC and BRC phases, respectively.

strategy under the peak power constraint is presented in Section 5.3. Section 5.4 presents an algorithm to find the optimum TD strategy under the average power constraint. Comprehensive simulation results are presented in Section 5.5, where the achievable rate regions and the ergodic sum rates of two-way relaying systems with optimum TD strategies are compared to those with equal TD strategy. Furthermore, we also show the average achievable rate of one user given the QoS requirement of the other. After that, conclusions are drawn in Section 5.6.

5.2 System Model

We consider a two-way DF relaying system as shown in Fig. 5.1, where the number of antennas at station A, the relay and station B are denoted as N_A, N_R and N_B, respectively. Each station operates in time division duplex (TDD) mode, i.e., it transmits and receives data consecutively in time. $\mathbf{G}_k \in \mathbb{C}^{N_\mathsf{R} \times N_k}$ and $\mathbf{H}_k \in \mathbb{C}^{N_k \times N_\mathsf{R}}$, where $k \in \{\mathsf{A}, \mathsf{B}\}$, denote the channel matrices between station k and the relay in the MAC and BRC phases, respectively. All the channels are frequency-flat and remain constant during its corresponding transmission phase. Both the transmitters and receivers have their corresponding channel knowledge. We define the *time-division (TD) factor* α as the portion of the total transmission time assigned to the MAC phase as shown in Fig. 5.2, where $0 \leq \alpha \leq 1$. The BRC phase occupies $1 - \alpha$ portion of the total transmission time. The transmit power constraints at station A, the relay and station B are P_A, P_R and P_B, respectively. We distinguish between the following two cases: One is *peak power constraint* where the actual transmit power in the MAC and BRC phases cannot exceed P_k, $k \in \{\mathsf{A}, \mathsf{B}, \mathsf{R}\}$; the other is *average power constraint* where the transmitter k can vary its actual transmit power in the MAC and BRC phases according to α if the transmit power averaged over the whole relaying process does not exceed P_k, $k \in \{\mathsf{A}, \mathsf{B}, \mathsf{R}\}$. We use P_k^* to denote the actual transmit power in the MAC or BRC phase for transmitter k. R_A denotes the information rate of the data to be transmitted from station A to B, and R_B

denotes the information rate of the data to be transmitted from station B to A.

In the MAC phase, station A and B transmit their data simultaneously to the relay, and the relay decodes the received data. The received signal $\mathbf{y} \in \mathbb{C}^{N_\mathsf{R}\times 1}$ at the relay in the MAC phase is

$$\begin{aligned}\mathbf{y} &= \mathbf{G}_\mathsf{A}\mathbf{x}_\mathsf{A} + \mathbf{G}_\mathsf{B}\mathbf{x}_\mathsf{B} + \mathbf{n} &(5.1)\\ &= \sqrt{\frac{P_\mathsf{A}^*}{N_\mathsf{A}}}\mathbf{G}_\mathsf{A}\tilde{\mathbf{x}}_\mathsf{A} + \sqrt{\frac{P_\mathsf{B}^*}{N_\mathsf{B}}}\mathbf{G}_\mathsf{B}\tilde{\mathbf{x}}_\mathsf{B} + \mathbf{n}, &(5.2)\end{aligned}$$

where $P_k^* = P_k$ under the peak power constraint and $P_k^* = P_k/\alpha$ under the average power constraint for $k \in \{\mathsf{A},\mathsf{B}\}$. $\tilde{\mathbf{x}}_\mathsf{A} \in \mathbb{C}^{N_\mathsf{A}\times 1}$ and $\tilde{\mathbf{x}}_\mathsf{B} \in \mathbb{C}^{N_\mathsf{B}\times 1}$ denotes the normalized transmit signal vectors at station A and B in the MAC phase, respectively. The transmit signal covariance matrices are defined as $\boldsymbol{\Omega}_\mathsf{A} = \mathrm{E}(\tilde{\mathbf{x}}_\mathsf{A}\tilde{\mathbf{x}}_\mathsf{A}^H)$ and $\boldsymbol{\Omega}_\mathsf{B} = \mathrm{E}(\tilde{\mathbf{x}}_\mathsf{B}\tilde{\mathbf{x}}_\mathsf{B}^H)$. In order to satisfy the power constraint, we have $\mathrm{tr}(\boldsymbol{\Omega}_\mathsf{A}) \le N_\mathsf{A}$ and $\mathrm{tr}(\boldsymbol{\Omega}_\mathsf{B}) \le N_\mathsf{B}$. $\mathbf{n} \sim \mathcal{CN}(0, \sigma^2\mathbf{I}_{N_\mathsf{R}})$ is the additive white Gaussian noise (AWGN) at the relay in the MAC phase.

Given the TD factor α, the MAC phase capacity region $\mathcal{C}_{\mathrm{MAC}}(\alpha) = (R_\mathsf{A}, R_\mathsf{B})$ can be characterized as [78]

$$\begin{aligned} R_\mathsf{A} &\le \alpha \log_2 \det\left(\mathbf{I}_{N_\mathsf{R}} + \frac{P_\mathsf{A}^*}{N_\mathsf{A}\sigma^2}\mathbf{G}_\mathsf{A}\boldsymbol{\Omega}_\mathsf{A}\mathbf{G}_\mathsf{A}^H\right), &(5.3)\\ R_\mathsf{B} &\le \alpha \log_2 \det\left(\mathbf{I}_{N_\mathsf{R}} + \frac{P_\mathsf{B}^*}{N_\mathsf{B}\sigma^2}\mathbf{G}_\mathsf{B}\boldsymbol{\Omega}_\mathsf{B}\mathbf{G}_\mathsf{B}^H\right), &(5.4)\\ \sum_{k\in\{\mathsf{A},\mathsf{B}\}} R_k &\le \alpha \log_2 \det\left(\mathbf{I}_{N_\mathsf{R}} + \sum_{k\in\{\mathsf{A},\mathsf{B}\}} \frac{P_k^*}{N_k\sigma^2}\mathbf{G}_k\boldsymbol{\Omega}_k\mathbf{G}_k^H\right), &(5.5)\end{aligned}$$

where $\mathrm{tr}(\boldsymbol{\Omega}_k) \le N_k$ and $\boldsymbol{\Omega}_k \succeq 0$ for $k \in \{\mathsf{A},\mathsf{B}\}$. For simplicity reasons, we denote the special case $\mathcal{C}_{\mathrm{MAC}}(\alpha = 1)$ as $\mathcal{C}_{\mathrm{MAC}}(1)$.

In the BRC phase, the relay combines the decoded data into the data symbol vector $\mathbf{x} \in \mathbb{C}^{N_\mathsf{R}\times 1}$ and sends it to the two stations. The received signal vectors at station A and B are $\mathbf{y}_\mathsf{A} \in \mathbb{C}^{N_\mathsf{A}\times 1}$ and $\mathbf{y}_\mathsf{B} \in \mathbb{C}^{N_\mathsf{B}\times 1}$, respectively. We have

$$\begin{aligned}\mathbf{y}_\mathsf{A} &= \mathbf{H}_\mathsf{A}\mathbf{x} + \mathbf{n}_\mathsf{A} &(5.6)\\ &= \sqrt{\frac{P_\mathsf{R}^*}{N_\mathsf{R}}}\mathbf{H}_\mathsf{A}\tilde{\mathbf{x}} + \mathbf{n}_\mathsf{A}, &(5.7)\\ \mathbf{y}_\mathsf{B} &= \mathbf{H}_\mathsf{B}\mathbf{x} + \mathbf{n}_\mathsf{B} &(5.8)\\ &= \sqrt{\frac{P_\mathsf{R}^*}{N_\mathsf{R}}}\mathbf{H}_\mathsf{B}\tilde{\mathbf{x}} + \mathbf{n}_\mathsf{B}, &(5.9)\end{aligned}$$

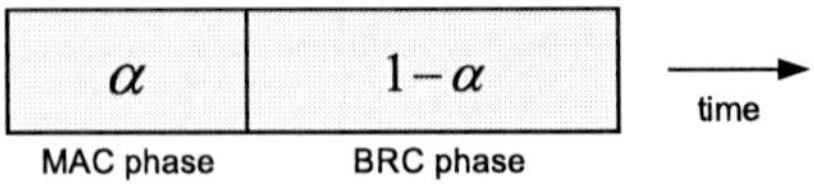

Fig. 5.2: Time-division between the MAC and BRC phases

where $P_{\mathsf{R}}^{*} = P_{\mathsf{R}}$ under the peak power constraint and $P_{\mathsf{R}}^{*} = P_{\mathsf{R}}/(1-\alpha)$ under the average power constraint. $\mathbf{n}_{\mathsf{A}} \sim \mathcal{CN}(0, \sigma_{\mathsf{A}}^2 \mathbf{I}_{N_{\mathsf{A}}})$ and $\mathbf{n}_{\mathsf{B}} \sim \mathcal{CN}(0, \sigma_{\mathsf{B}}^2 \mathbf{I}_{N_{\mathsf{B}}})$ are the AWGN at the receivers of station A and B, respectively. $\tilde{\mathbf{x}}$ denotes the normalized transmit signal vector at the relay, and its covariance matrix is defined as $\mathbf{\Omega} = \mathrm{E}(\tilde{\mathbf{x}}\tilde{\mathbf{x}}^H)$. Furthermore, we require $\mathrm{tr}(\mathbf{\Omega}) \leq N_{\mathsf{R}}$ in order to satisfy the power constraint at the relay. Assuming both stations A and B have perfect side information about the messages intended for the other station, the capacity region of the BRC phase $\mathcal{C}_{\mathrm{BRC}}(\alpha) = (R_{\mathsf{A}}, R_{\mathsf{B}})$ can be characterized as [262]

$$R_{\mathsf{A}} \leq (1-\alpha)\log_2 \det\left(\mathbf{I}_{N_{\mathsf{B}}} + \frac{P_{\mathsf{R}}^{*}}{N_{\mathsf{R}}\sigma_{\mathsf{B}}^2}\mathbf{H}_{\mathsf{B}}\mathbf{\Omega}\mathbf{H}_{\mathsf{B}}^H\right), \tag{5.10}$$

$$R_{\mathsf{B}} \leq (1-\alpha)\log_2 \det\left(\mathbf{I}_{N_{\mathsf{A}}} + \frac{P_{\mathsf{R}}^{*}}{N_{\mathsf{R}}\sigma_{\mathsf{A}}^2}\mathbf{H}_{\mathsf{A}}\mathbf{\Omega}\mathbf{H}_{\mathsf{A}}^H\right), \tag{5.11}$$

where $\mathrm{tr}(\mathbf{\Omega}) \leq N_{\mathsf{R}}$ and $\mathbf{\Omega} \succeq 0$. For simplicity reasons, we denote the special case $\mathcal{C}_{\mathrm{BRC}}(\alpha = 0)$ as $\mathcal{C}_{\mathrm{BRC}}(0)$.

$\mathcal{C}_{\mathrm{MAC}}(\alpha)\bigcap\mathcal{C}_{\mathrm{BRC}}(\alpha)$ depicts the achievable rate region of the two-way relaying system for the given TD factor α. Considering all possible TD factors, we have the following achievable rate region of the two-way relaying system

$$\mathcal{R}_{\mathrm{OPT}} = \bigcup_{0\leq\alpha\leq 1}\left(\mathcal{C}_{\mathrm{MAC}}(\alpha)\bigcap\mathcal{C}_{\mathrm{BRC}}(\alpha)\right). \tag{5.12}$$

Here we restrict ourselves to the case that the transmit signal covariance matrices $\mathbf{\Omega}_{\mathsf{A}}$, $\mathbf{\Omega}_{\mathsf{B}}$ and $\mathbf{\Omega}$ remain unchanged in their corresponding transmission phase. Furthermore, we have the following conjecture:

Conjecture 5.2.1. *$\mathcal{R}_{\mathrm{OPT}}$ is always convex for Gaussian multiple-input multiple-output (MIMO) two-way relaying channels.*

It has been proved in [168] that $\mathcal{R}_{\mathrm{OPT}}$ is always convex for two-way relaying systems with only single-antenna stations and relay under both peak and average power constraints. Recently, the authors of [171] proved that $\mathcal{R}_{\mathrm{OPT}}$ is convex when A and B each has single antenna and R has multiple antennas. Whether $\mathcal{R}_{\mathrm{OPT}}$ is always convex in general MIMO

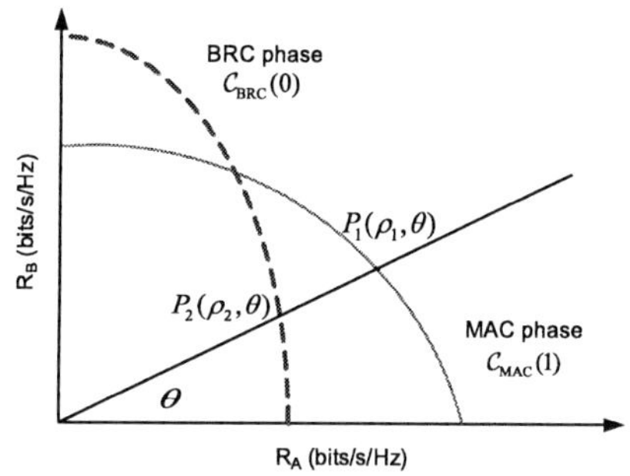

Fig. 5.3: The MAC and BRC capacity regions $\mathcal{C}_{\text{MAC}}(1)$ and $\mathcal{C}_{\text{BRC}}(0)$ of a two-way relaying system. The ray θ intersects with $\mathcal{C}_{\text{MAC}}(1)$ and $\mathcal{C}_{\text{BRC}}(0)$ on the point P_1 and P_2, respectively. The two points are indicated in polar coordinates.

two-way relaying systems is still an open problem to the best of our knowledge. If $\mathcal{R}_{\text{OPT}}$ is not always convex, *time-sharing* between two TD strategies is further required to achieve certain boundary points of $\mathcal{C}_{\text{OPT}} = \text{conv}\mathcal{R}_{\text{OPT}}$, where conv denotes the convex hull. However, this time-sharing strategy requires the transmit signal covariance matrices Ω_{A}, Ω_{B} and Ω to be changed in different time-sharing phases, which is not considered here.

When the rate region $\mathcal{R}_{\text{OPT}}$ is convex, its boundary points can be determined by maximizing the weighted sum rate as in Appendix 5.7.1. However, this method is subject to certain implementation constraints, and it is hard to be solved by standard second-order cone and semidefinite programming solvers [230].

In the following, we propose two implementable algorithms to determine the boundary points of the achievable rate region $\mathcal{R}_{\text{OPT}}$ regardless of its convexity. They find both the optimum values of α and the signal covariance matrices for each boundary point of $\mathcal{R}_{\text{OPT}}$. Section 5.3 considers the problem under the peak power constraint, and Section 5.4 discusses the problem under the average power constraint.

5.3 Optimum Time-Division Under Peak Power Constraint

Under the peak power constraint, we have $P_k^* = P_k$ for $k \in \{\text{A}, \text{R}, \text{B}\}$ in (5.3)–(5.5) and (5.10)–(5.11). Fig. 5.3 shows the capacity regions $\mathcal{C}_{\text{MAC}}(1)$ and $\mathcal{C}_{\text{BRC}}(0)$ of a two-way relaying system. For a given TD factor α, the boundary points of $\mathcal{C}_{\text{MAC}}(1)$ and $\mathcal{C}_{\text{BRC}}(0)$ are scaled respectively by the factors of α and $1 - \alpha$ along the line through the origin to get

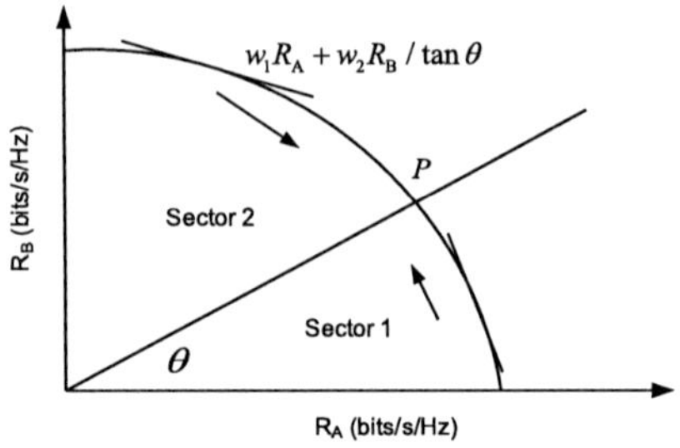

Fig. 5.4: Finding the point P where the ray θ intersects with the capacity region $\mathcal{C}$'s boundary

the boundary points of $\mathcal{C}_{\text{MAC}}(\alpha)$ and $\mathcal{C}_{\text{BRC}}(\alpha)$. We can write $\mathcal{C}_{\text{MAC}}(\alpha) = \alpha\mathcal{C}_{\text{MAC}}(1)$ and $\mathcal{C}_{\text{BRC}}(\alpha) = (1-\alpha)\mathcal{C}_{\text{BRC}}(0)$. It is more convenient for us to use polar coordinates to represent the rate pair $(R_{\mathsf{A}}, R_{\mathsf{B}})$ in our discussions of this section.

The algorithm to characterize the boundary points of the achievable rate region $\mathcal{R}_{\text{OPT}}$ under the peak power constraint consists of the following two steps: Firstly, for a given angle θ, we determine the two intersection points $P_1(\rho_1, \theta)$ and $P_2(\rho_2, \theta)$ on the boundaries of $\mathcal{C}_{\text{MAC}}(1)$ and $\mathcal{C}_{\text{BRC}}(0)$ as shown in Fig. 5.3, where they are indicated by polar coordinates. Secondly, we calculate the optimum TD factor $\alpha^\star$ between P_1 and P_2, and get the boundary point on $\mathcal{R}_{\text{OPT}}$ for the given angle θ. The first step will be discussed in Section 5.3.1, and the second step will be discussed in Section 5.3.2. Without causing confusions, we use θ to denote both the angle and the ray that forms the angle with the R_{A}-axis.

5.3.1 Boundary Point on Capacity Regions for Given θ

The capacity regions $\mathcal{C}_{\text{MAC}}(1)$ and $\mathcal{C}_{\text{BRC}}(0)$ are both convex. Efficient algorithms for calculating the maximum weighted sum rate $\sum_k \mu_k R_k$, $k \in \{\mathsf{A}, \mathsf{B}\}$, for the MAC phase $\mathcal{C}_{\text{MAC}}(1)$ are available in e.g., [149, 242]. Here μ_k, $k \in \{\mathsf{A}, \mathsf{B}\}$ are non-negative weighting constants. Similar algorithms can be applied to calculate the maximum weighted sum rate for the BRC phase $\mathcal{C}_{\text{BRC}}(0)$ (see Appendix 5.7.2). Let $\mathcal{C}$ represent the capacity region $\mathcal{C}_{\text{MAC}}(1)$ or $\mathcal{C}_{\text{BRC}}(0)$. For a given angle $0 \le \theta \le \pi/2$ as shown in Fig. 5.4, we denote the intersection point of the ray θ with the boundary of the capacity region $\mathcal{C}$ as P. Since $\mathcal{C}$ is convex, there is only one intersection point. Furthermore, the following lemma can be utilized to determine the intersection point P.

Lemma 5.3.1 ([138]). *Let w denote the weighting factor, where $0 \le w \le 1$. The intersection* *point $P = (R^\star_{\mathsf{A}}, R^\star_{\mathsf{B}})$ of the ray θ with the boundary of the capacity region $\mathcal{C}$ satisfies*

$$R^\star_A = \frac{R^\star_B}{\tan\theta} \tag{5.13}$$

$$= \max_{(R_A,R_B)\in\mathcal{C}} \min\left(R_A, \frac{R_B}{\tan\theta}\right) \tag{5.14}$$

$$= \max_{(R_A,R_B)\in\mathcal{C}} \min_{0\le w\le 1} \left[wR_A + (1-w)\frac{R_B}{\tan\theta}\right] \tag{5.15}$$

$$= \min_{0\le w\le 1} \max_{(R_A,R_B)\in\mathcal{C}} \left[wR_A + (1-w)\frac{R_B}{\tan\theta}\right]. \tag{5.16}$$

Furthermore, we have

$$wR^\star_A + (1-w)\frac{R^\star_B}{\tan\theta} = R^\star_A = \frac{R^\star_B}{\tan\theta}, \quad \forall w. \tag{5.17}$$

Proof. Equation (5.13) is because the point $P = (R^\star_{\mathsf{A}}, R^\star_{\mathsf{B}})$ is on the ray θ. Equation (5.14) is due to the fact that the ray θ divides the first quadrant into two sectors. In Sector 1 that is below the ray θ, $R_{\mathsf{B}}/R_{\mathsf{A}} \le \tan\theta$ and thus $R_{\mathsf{B}}/\tan\theta \le R_{\mathsf{A}}$. In Sector 1, $\min(R_{\mathsf{A}}, R_{\mathsf{B}}/\tan\theta) = R_{\mathsf{B}}/\tan\theta$ and (5.14) is equivalent to maximizing R_{B}, which is achieved by the intersection point $P = (R^\star_{\mathsf{A}}, R^\star_{\mathsf{B}})$. The same argument also applies to Sector 2. Equation (5.15) follows from the fact that $wx + (1-w)y \ge (w + 1 - w)\min(x,y) = \min(x,y)$, $\forall x, y \ge 0$ and $0 \le w \le 1$. The equality is achieved when

$$w = \begin{cases} 1 & \text{if } x \le y \\ 0 & \text{if } x > y. \end{cases}$$

Thus $\min_{0\le w\le 1}[wx + (1-w)y] = \min(x,y)$. Equation (5.16) follows from Fan's Minimax Theorem [63] (see Appendix 5.7.3). This is because the capacity region $\mathcal{C}$ and the set $\mathcal{W} = \{w|0 \le w \le 1\}$ are both convex and compact. Moreover, the function $f(w, R_{\mathsf{A}}, R_{\mathsf{B}}) = wR_{\mathsf{A}} + (1-w)R_{\mathsf{B}}/\tan\theta$ is continuous and linear on the set $\mathcal{W}$ and $\mathcal{C}$. So the strong max-min property holds. □

The algorithm to determine the intersection point of the given ray θ with the boundary of the capacity region $\mathcal{C}$ is presented in Algorithm 3. Here we utilize the bisection method [23] to determine the optimum weighting factor w.

Algorithm 3 Calculating the intersection point $P = (R_A^\star, R_B^\star)$ for the given ray θ

Initialize $w^{\min} = 0$, $w^{\max} = 1$ and $w = (w^{\max} + w^{\min})/2$.
repeat
  Determine the rate pair $(R_A, R_B) \in \mathcal{C}$ that maximizes the weighted sum rate using the algorithms described in [149, 242], i.e., determine

$$(R_A(w), R_B(w)) = \arg\max_{(R_A, R_B)\in\mathcal{C}} wR_A + (1-w)\frac{R_B}{\tan\theta}.$$

  if $R_A(w) < R_B(w)/\tan\theta$ **then**
    $w^{\min} = w$; $w^{\max}$ unchanged; $w = (w^{\max} + w^{\min})/2$;
  else if $R_A(w) > R_B(w)/\tan\theta$ **then**
    $w^{\max} = w$; $w^{\min}$ unchanged; $w = (w^{\max} + w^{\min})/2$;
  end if
until $|R_A - R_B/\tan\theta| < \epsilon$ or $|w^{\max} - w^{\min}| < \zeta$
return $R_A^\star = wR_A(w) + (1-w)R_B(w)/\tan\theta$, $R_B^\star = R_A^\star \tan\theta$.

5.3.2 Calculating the Optimum Time-Division Factor

For a given angle θ, Algorithm 3 can determine the intersection points $P_1 = (R_{A1}^\star, R_{B1}^\star)$ and $P_2 = (R_{A2}^\star, R_{B2}^\star)$ of the ray θ with the capacity region boundaries $\mathcal{C}_{\text{MAC}}(1)$ and $\mathcal{C}_{\text{BRC}}(0)$. In order to calculate the optimum TD factor between P_1 and P_2, we first convert them into polar coordinate representations $P_1(\rho_1, \theta)$ and $P_2(\rho_2, \theta)$ as shown in Fig. 5.3. The following lemma can be utilized to determine the optimum TD factor $\alpha^\star$ between P_1 and P_2:

Lemma 5.3.2. *We assume the ray θ intersects with the boundaries of $\mathcal{C}_{\text{MAC}}(1)$ and $\mathcal{C}_{\text{BRC}}(0)$ respectively on $P_1(\rho_1, \theta)$ and $P_2(\rho_2, \theta)$. The point $(\rho^\star, \theta)$ in polar coordinates is on the boundary of $\mathcal{R}_{\text{OPT}}$ with optimum TD factor $\alpha^\star$, where*

$$\rho^\star = \frac{\rho_1\rho_2}{\rho_1 + \rho_2} \tag{5.18}$$

$$\alpha^\star = \frac{\rho_2}{\rho_1 + \rho_2}. \tag{5.19}$$

The transmit covariance matrices of the MAC and BRC phases for that point are the corresponding ones obtained for the points P_1 and P_2, respectively.

Proof. For a given TD factor α, the point $Q(\rho(\alpha), \theta)$ represented in polar coordinates is on the boundary of the rate region $\mathcal{C}_{\text{MAC}}(\alpha) \bigcap \mathcal{C}_{\text{BRC}}(\alpha)$, where $\rho(\alpha) = \min(\alpha\rho_1, (1-\alpha)\rho_2)$. $\rho(\alpha)$ is maximized when $\alpha^\star\rho_1 = (1-\alpha^\star)\rho_2$, i.e., when $\alpha^\star = \rho_2/(\rho_1 + \rho_2)$. The same result is obtained for two-way relaying systems with single antenna [168]. When we choose that

TD factor $\alpha^\star$, the corresponding $\rho(\alpha^\star) = \rho_1\rho_2/(\rho_1+\rho_2)$, which is the maximum value that $\rho(\alpha)$ can achieve for $0 \le \alpha \le 1$. □

In summary, we have Algorithm 4 to calculate the boundary point of $\mathcal{R}_{\mathrm{OPT}}$ for the given angle θ. Evaluating every boundary point of $\mathcal{R}_{\mathrm{OPT}}$ for $0 \le \theta \le \pi/2$ yields the whole boundary of the achievable rate region $\mathcal{R}_{\mathrm{OPT}}$ under the peak power constraint.

Algorithm 4 Calculating the boundary point of $\mathcal{R}_{\mathrm{OPT}}$ for the given angle θ

1: For given θ, Use Algorithm 3 to determine the intersection points of the ray θ with $\mathcal{C}_{\mathrm{MAC}}(1)$ and $\mathcal{C}_{\mathrm{BRC}}(0)$;
2: Use (5.18) and (5.19) to determine the boundary point in polar coordinate $(\rho^\star, \theta)$.

5.4 Optimum Time-Division Under Average Power Constraint

Under the average power constraint, $\mathcal{C}_{\mathrm{MAC}}(\alpha)$ and $\mathcal{C}_{\mathrm{BRC}}(\alpha)$ are not simply the scaled versions of $\mathcal{C}_{\mathrm{MAC}}(1)$ and $\mathcal{C}_{\mathrm{BRC}}(0)$. In order to find the optimum TD strategies under the average power constraint, we first prove the following lemma:

Lemma 5.4.1. *For any given covariance matrices Ω_{A}, Ω_{B} and Ω, the right-hand sides (RHS) of* (5.3)–(5.5) *are monotonically increasing functions of α, and the RHS of* (5.10)–(5.11) *are monotonically decreasing functions of α.*

Proof. For any given transmit signal covariance matrix Ω_{A}, the RHS of (5.3) under the average power constraint can be written as

$$\alpha \log_2 \det\left(\mathbf{I}_{N_{\mathsf{R}}} + \frac{P_{\mathsf{A}}}{\alpha N_{\mathsf{A}}\sigma^2}\mathrm{G}_{\mathsf{A}}\Omega_{\mathsf{A}}\mathrm{G}_{\mathsf{A}}^H\right) = \sum_{i=1}^{N_{\mathsf{R}}} \alpha \log_2(1+\frac{\lambda_i}{\alpha}), \tag{5.20}$$

where $\lambda_i \ge 0$, $i = 1, \cdots, N_{\mathsf{R}}$, denote the eigenvalues of $\frac{P_{\mathsf{A}}}{N_{\mathsf{A}}\sigma^2}\mathrm{G}_{\mathsf{A}}\Omega_{\mathsf{A}}\mathrm{G}_{\mathsf{A}}^H$. $\forall \lambda_i \ge 0$, $f(\alpha) = \alpha\log_2(1+\lambda_i/\alpha)$ is a monotonically increasing function of α when $0 \le \alpha \le 1$. Thus (5.20) is a monotonically increasing function of α. The same discussion also applies to (5.4)–(5.5). Since the RHS of (5.10)–(5.11) are monotonically increasing functions of $1-\alpha$, they are monotonically decreasing functions of α. □

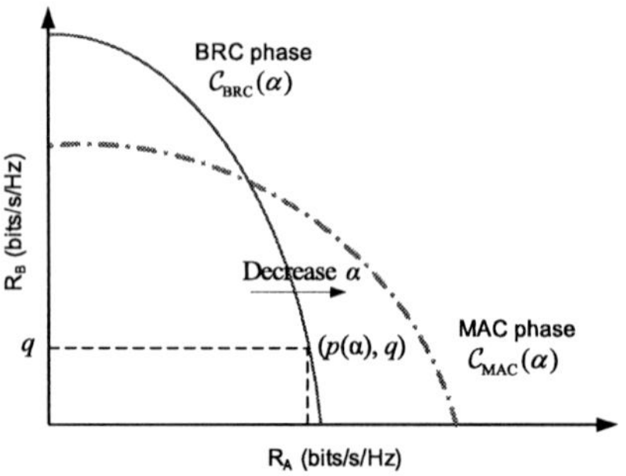

Fig. 5.5: For a given value q and TD factor α, calculate the value $p(\alpha)$ such that $(p(\alpha), q)$ is on the boundary of $\mathcal{C}_{\mathrm{MAC}}(\alpha)$. Decreasing α expands $\mathcal{C}_{\mathrm{BRC}}(\alpha)$ and shrinks $\mathcal{C}_{\mathrm{MAC}}(\alpha)$.

This lemma implies that the region $\mathcal{C}_{\mathrm{MAC}}(\alpha)$ swells, while the region $\mathcal{C}_{\mathrm{BRC}}(\alpha)$ diminishes as α increases. In this section, we propose an algorithm to determine the boundary points of $\mathcal{R}_{\mathrm{OPT}}$ under the average power constraint. We first define

$$q_{\max} = \max_{\mathrm{tr}(\mathbf{\Omega}) \leq N_{\mathsf{R}}, \mathbf{\Omega} \succeq 0} \log_2 \det \left(\mathbf{I}_{N_{\mathsf{A}}} + \frac{P_{\mathsf{R}}}{N_{\mathsf{R}} \sigma_{\mathsf{A}}^2} \mathbf{H}_{\mathsf{A}} \mathbf{\Omega} \mathbf{H}_{\mathsf{A}}^H \right),$$

where $q_{\max}$ is the maximum possible value of R_{B} in the BRC phase. For a given value q, where $0 \leq q \leq q_{\max}$, the proposed algorithm finds the optimum value $p^\star$ and $\alpha^\star$ such that $(p^\star, q)$ is on the boundary of $\mathcal{R}_{\mathrm{OPT}}$ under the average power constraint, and the TD factor for that point is $\alpha^\star$, i.e., $(p^\star, q) \in \mathcal{C}_{\mathrm{MAC}}(\alpha^\star) \bigcap \mathcal{C}_{\mathrm{BRC}}(\alpha^\star)$. The idea of the algorithm is shown in Fig. 5.5. For each value of the TD factor α, we can determine the point $(p(\alpha), q)$ on the boundary of $\mathcal{C}_{\mathrm{BRC}}(\alpha)$ for the given value q, and $(p(\alpha), q) \in \mathcal{C}_{\mathrm{BRC}}(\alpha)$ is satisfied. $p^\star$ is the largest *achievable* value of $p(\alpha)$, i.e.,

$$p^\star = \max_{0 \leq \alpha \leq 1} \{p(\alpha) | (p(\alpha), q) \in \mathcal{C}_{\mathrm{MAC}}(\alpha)\}. \tag{5.21}$$

For the given value of α, we can determine whether $(p(\alpha), q) \in \mathcal{C}_{\mathrm{MAC}}(\alpha)$. If $(p(\alpha), q) \in \mathcal{C}_{\mathrm{MAC}}(\alpha)$, a smaller TD factor β, where $\beta \leq \alpha$, can be chosen with increased $p(\beta)$ while still keeping the point $(p(\beta), q)$ to be inside $\mathcal{C}_{\mathrm{MAC}}(\beta)$. Otherwise, the present value of α is too small and should be increased. The fact that decreasing the TD factor α expands $\mathcal{C}_{\mathrm{BRC}}(\alpha)$ and shrinks $\mathcal{C}_{\mathrm{MAC}}(\alpha)$ is due to Lemma 5.4.1.

The details of the algorithm work as follows: For a given value q, where $0 \leq q \leq q_{\max}$, we first choose an initial value of α, where $0 \leq \alpha \leq 1$. For the given value q and the TD factor α, the point $(p(\alpha), q)$ on the boundary of $\mathcal{C}_{\text{BRC}}(\alpha)$ under the average power constraint can be calculated by solving the following convex optimization problem:

$$\begin{array}{ll} \text{maximize} & (1-\alpha)\log_2\det\left(\mathbf{I}_{N_\mathsf{B}} + \frac{P_\mathsf{R}^*}{N_\mathsf{R}\sigma_\mathsf{B}^2}\mathbf{H}_\mathsf{B}\boldsymbol{\Omega}\mathbf{H}_\mathsf{B}^H\right) \\ \text{subject to} & (1-\alpha)\log_2\det\left(\mathbf{I}_{N_\mathsf{A}} + \frac{P_\mathsf{R}^*}{N_\mathsf{R}\sigma_\mathsf{A}^2}\mathbf{H}_\mathsf{A}\boldsymbol{\Omega}\mathbf{H}_\mathsf{A}^H\right) \geq q \\ & \text{tr}(\boldsymbol{\Omega}) \leq N_\mathsf{R}, \quad \boldsymbol{\Omega} \succeq 0 \\ \text{variable} & \boldsymbol{\Omega} \end{array} \tag{5.22}$$

where $P_\mathsf{R}^* = P_\mathsf{R}/(1-\alpha)$. The maximum value of the objective function in (5.22) is $p(\alpha)$. Secondly, we can check whether $(p(\alpha), q) \in \mathcal{C}_{\text{MAC}}(\alpha)$ with the TD factor α under the average power constraint. That is, we solve the following convex feasibility problem:

$$\begin{array}{ll} \text{find} & \boldsymbol{\Omega}_\mathsf{A}, \boldsymbol{\Omega}_\mathsf{B} \\ \text{subject to} & \alpha\log_2\det\left(\mathbf{I}_{N_\mathsf{R}} + \frac{P_\mathsf{A}^*}{N_\mathsf{A}\sigma^2}\mathbf{G}_\mathsf{A}\boldsymbol{\Omega}_\mathsf{A}\mathbf{G}_\mathsf{A}^H\right) \geq p(\alpha) \\ & \alpha\log_2\det\left(\mathbf{I}_{N_\mathsf{R}} + \frac{P_\mathsf{B}^*}{N_\mathsf{B}\sigma^2}\mathbf{G}_\mathsf{B}\boldsymbol{\Omega}_\mathsf{B}\mathbf{G}_\mathsf{B}^H\right) \geq q \\ & \alpha\log_2\det\left(\mathbf{I}_{N_\mathsf{R}} + \sum\limits_{k\in\{\mathsf{A},\mathsf{B}\}} \frac{P_k^*}{N_k\sigma^2}\mathbf{G}_k\boldsymbol{\Omega}_k\mathbf{G}_k^H\right) \geq p(\alpha) + q \\ & \text{tr}(\boldsymbol{\Omega}_\mathsf{A}) \leq N_\mathsf{A}, \quad \boldsymbol{\Omega}_\mathsf{A} \succeq 0 \\ & \text{tr}(\boldsymbol{\Omega}_\mathsf{B}) \leq N_\mathsf{B}, \quad \boldsymbol{\Omega}_\mathsf{B} \succeq 0. \end{array} \tag{5.23}$$

where $P_\mathsf{A}^* = P_\mathsf{A}/\alpha$ and $P_\mathsf{B}^* = P_\mathsf{B}/\alpha$. By solving the problem (5.23), we get a *feasibility certificate* to show whether suitable matrices $\boldsymbol{\Omega}_\mathsf{A}$ and $\boldsymbol{\Omega}_\mathsf{B}$ can be found satisfying the constraints. If suitable covariance matrices $\boldsymbol{\Omega}_\mathsf{A}$ and $\boldsymbol{\Omega}_\mathsf{B}$ can be found satisfying the constraints of (5.23), then $(p(\alpha), q) \in \mathcal{C}_{\text{MAC}}(\alpha)$. This indicates that $\alpha \geq \alpha^\star$ and the present choice of TD factor should be decreased; if no suitable covariance matrices $\boldsymbol{\Omega}_\mathsf{A}$ and $\boldsymbol{\Omega}_\mathsf{B}$ can be found, then $(p(\alpha), q) \notin \mathcal{C}_{\text{MAC}}(\alpha)$, which indicates that $\alpha < \alpha^\star$ and the present choice of TD factor should be increased. Here we utilized Lemma 5.4.1. This feasibility certificate can be considered as a subgradient [23] for finding the optimum factor $\alpha^\star$. This process repeats until α converges. This algorithm is summarized in Algorithm 5.

We get the corresponding $\boldsymbol{\Omega}$, $\boldsymbol{\Omega}_\mathsf{A}$ and $\boldsymbol{\Omega}_\mathsf{B}$ when we solve (5.22) and (5.23). Evaluation for every point of $0 \leq q \leq q_{\max}$ yields the boundary of the whole achievable rate region $\mathcal{R}_{\text{OPT}}$ under the average power constraint.

Algorithm 5 Bisection method to determine the boundary point $(p^\star, q)$ on $\mathcal{R}_{\mathrm{OPT}}$ and the optimum TD factor $\alpha^\star$ for given QoS requirement $R_{\mathsf{B}} = q$

Initialize $\alpha^{\min} = 0$, $\alpha^{\max} = 1$ and $\alpha = (\alpha^{\min} + \alpha^{\max})/2$.
while $\alpha_{\max} - \alpha_{\min} > \epsilon$ **do**
  Solve $(p(\alpha), q)$ for the convex optimization problem (5.22);
  Check the feasibility of $(p(\alpha), q)$ in the feasibility check problem (5.23);
  if $(p(\alpha), q)$ is feasible **then**
    $\alpha^{\max} = \alpha$; $\alpha^{\min}$ unchanged; $\alpha = (\alpha^{\max} + \alpha^{\min})/2$
  else
    $\alpha^{\min} = \alpha$; $\alpha^{\max}$ unchanged; $\alpha = (\alpha^{\max} + \alpha^{\min})/2$
  end if
end while
return $p^\star = p(\alpha)$, $\alpha^\star = \alpha$.

5.5 Simulation Results

Using the methods proposed in Section 5.3 and Section 5.4, we can characterize the achievable rate regions with optimum TD strategies for given channels. In a real mobile system, the channel knowledge may not always be available to the transmitters. So we also show the rate regions when there is no channel state information at the transmitters (CSIT). In such a case, the transmit covariance matrices are chosen as $\mathbf{\Omega}_{\mathsf{A}} = \mathbf{I}_{N_{\mathsf{A}}}$, $\mathbf{\Omega}_{\mathsf{B}} = \mathbf{I}_{N_{\mathsf{B}}}$ and $\mathbf{\Omega} = \mathbf{I}_{N_{\mathsf{R}}}$. As stated in Appendix 5.7.1, the rate region under the peak power constraint can be characterized using linear programming for no CSIT case, while its rate region under the average power constraint is calculated using Algorithm 5 in the simulations.

5.5.1 Achievable Rate Region $\mathcal{R}_{\mathrm{OPT}}$ for Static Channels

The achievable rate regions $\mathcal{R}_{\mathrm{OPT}}$ with optimum TD strategies for static channels with full CSIT and no CSIT are shown in Fig. 5.6 and Fig. 5.7, respectively. Both peak and average power constraints are considered. The channel matrices are randomly generated as follows

$$\mathbf{G}_{\mathsf{A}} = \begin{bmatrix} -1.30 - 0.45i & -0.60 + 0.13i \\ -1.88 + 0.33i & 0.34 + 0.66i \end{bmatrix} \tag{5.24}$$

$$\mathbf{G}_{\mathsf{B}} = \begin{bmatrix} -0.05 - 0.45i & -0.55 + 0.19i \\ 0.60 - 0.99i & 1.13 - 0.01i \end{bmatrix} \tag{5.25}$$

$$\mathbf{H}_{\mathsf{A}} = \mathbf{G}_{\mathsf{A}}^T, \quad \mathbf{H}_{\mathsf{B}} = \mathbf{G}_{\mathsf{B}}^T. \tag{5.26}$$

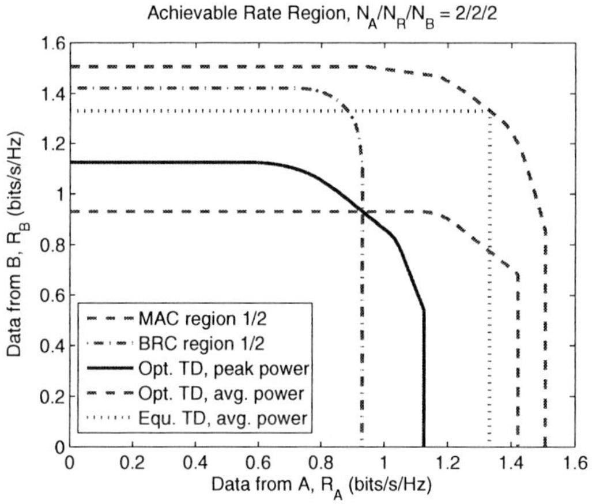

Fig. 5.6: Achievable rate regions under peak and average power constraints. "MAC region 1/2" and "BRC region 1/2" correspond to $1/2\mathcal{C}_{\mathrm{MAC}}(1)$ and $1/2\mathcal{C}_{\mathrm{BRC}}(0)$, respectively. Their intersection region is the achievable rate region with equal TD under the peak power constraint.

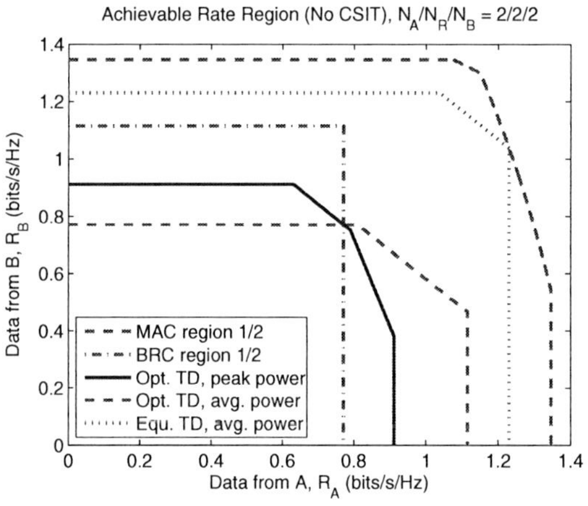

Fig. 5.7: Achievable rate regions under peak and average power constraints for no CSIT case. "MAC region 1/2" and "BRC region 1/2" correspond to $1/2\mathcal{C}_{\mathrm{MAC}}(1)$ and $1/2\mathcal{C}_{\mathrm{BRC}}(0)$, respectively. Their intersection region is the achievable rate region with equal TD under the peak power constraint.

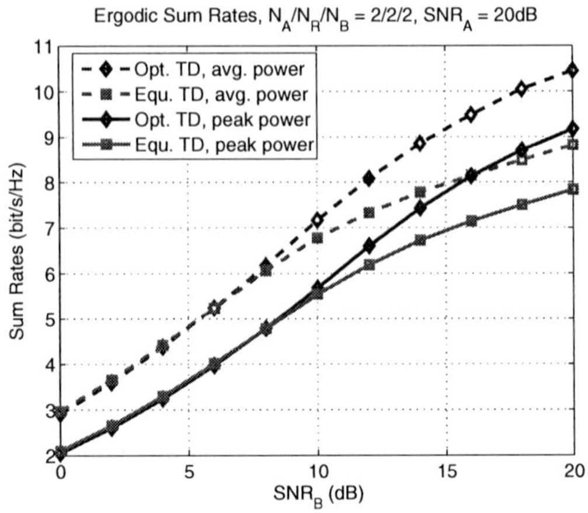

Fig. 5.8: Ergodic sum rate under peak and average power constraints, $N_{\mathsf{A}}/N_{\mathsf{R}}/N_{\mathsf{B}} = 2/2/2$, $\mathrm{SNR}_{\mathsf{A}} = 20\mathrm{dB}$

The channels remain constant during their corresponding transmission phase. Furthermore, we have $P_k/\sigma^2 = 1$, where $k \in \{\mathsf{A}, \mathsf{B}, \mathsf{R}\}$. Given $R_{\mathsf{B}} = 0.5$ bits/s/Hz, the maximum achievable rate of R_{A} is shown in Table 5.1. The gain of the achievable rate R_{A} in optimum TD case is significant compared to equal TD case. For average power constraint, the actual transmit power P_k^* is larger than P_k. Due to power scaling, the achievable rate region under the average power constraint is also larger than that under the peak power constraint.

Table 5.1: Maximum Achievable Rate of R_{A} for Given QoS Requirement $R_{\mathsf{B}} = 0.5$ bits/s/Hz

	Equal TD	Optimum TD	Increase
Peak Power Constraint (CSIT)	0.93 bits/s/Hz	1.12 bits/s/Hz	20.43%
Average Power Constraint (CSIT)	1.33 bits/s/Hz	1.51 bits/s/Hz	13.53%
Peak Power Constraint (no CSIT)	0.77 bits/s/Hz	0.87 bits/s/Hz	13.19%
Average Power Constraint (no CSIT)	1.23 bits/s/Hz	1.35 bits/s/Hz	9.43%

5.5.2 Ergodic Sum Rate in Rayleigh Fading Channels

Fig. 5.8–Fig. 5.11 show the ergodic sum rate of a two-way relaying system in Rayleigh fading channels for different antenna configurations with and without CSIT. Each entry in $\mathbf{G}_{\mathsf{A}}$, $\mathbf{G}_{\mathsf{B}}$, $\mathbf{H}_{\mathsf{A}}$ and $\mathbf{H}_{\mathsf{B}}$ are $\mathcal{CN}(0,1)$ random variables. We have $\mathrm{SNR}_{\mathsf{A}} = P_{\mathsf{A}}/\sigma^2 = P_{\mathsf{R}}/\sigma_{\mathsf{A}}^2 = 20\mathrm{dB}$,

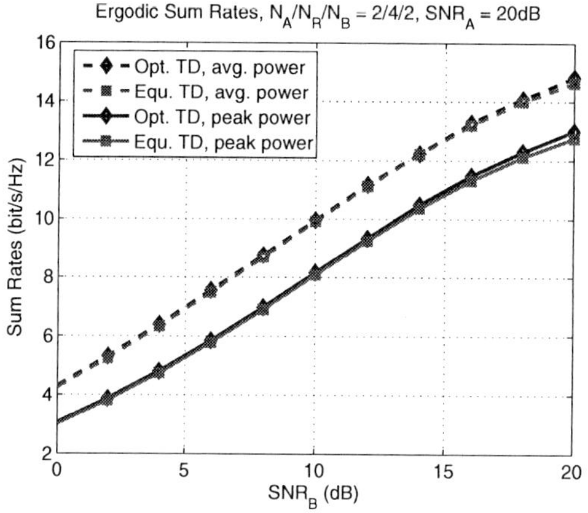

Fig. 5.9: Ergodic sum rate under peak and average power constraints, $N_A/N_R/N_B = 2/4/2$, $\mathrm{SNR}_A = 20\mathrm{dB}$

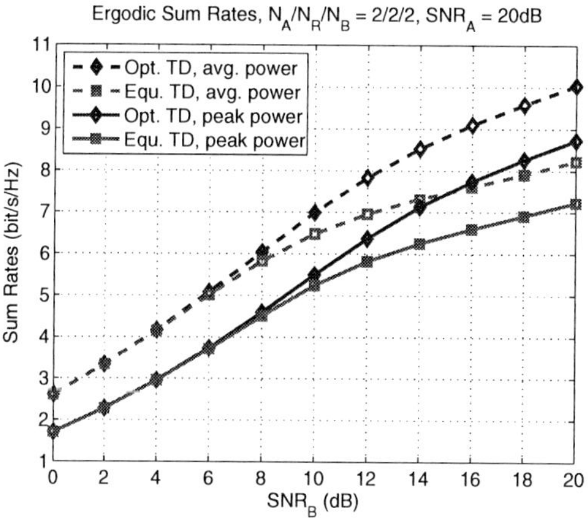

Fig. 5.10: Ergodic sum rate under peak and average power constraints (no CSIT case), $N_A/N_R/N_B = 2/2/2$, $\mathrm{SNR}_A = 20\mathrm{dB}$

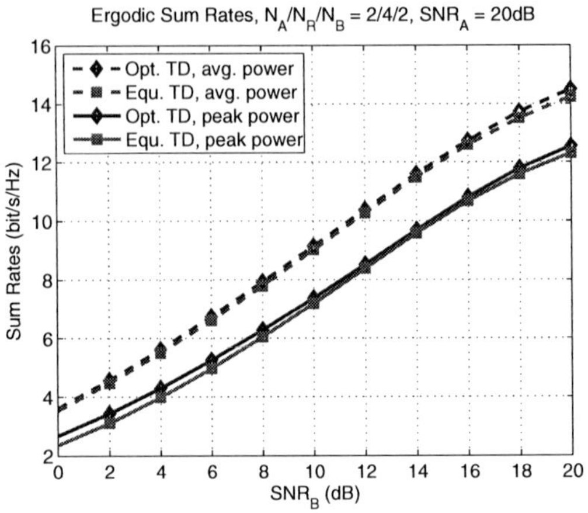

Fig. 5.11: Ergodic sum rate under peak and average power constraints (no CSIT case), $N_{\mathsf{A}}/N_{\mathsf{R}}/N_{\mathsf{B}} = 2/4/2$, $\mathrm{SNR}_{\mathsf{A}} = 20\mathrm{dB}$

which means that the distance between station A and the relay is fixed. We also define $\mathrm{SNR}_{\mathsf{B}} = P_{\mathsf{B}}/\sigma^2 = P_{\mathsf{R}}/\sigma_{\mathsf{B}}^2$, which is shown as the x-axis in the figure. In Fig. 5.8, the number of antennas at station A, the relay and station B are $N_{\mathsf{A}} = N_{\mathsf{R}} = N_{\mathsf{B}} = 2$. Fig. 5.8 shows that the optimum TD strategies do not gain in the ergodic sum rate when $\mathrm{SNR}_{\mathsf{B}}$ is below 8dB. Since $\mathrm{SNR}_{\mathsf{A}}$ is high, the weak link, i.e., the link between the relay and station B, determines the achievable rate region when $\mathrm{SNR}_{\mathsf{B}}$ is low. By choosing different TD factors, we can increase either R_{A} or R_{B}, but at the price of the other. In this case, optimum TD strategies do not increase the sum rate of the system. However, when $\mathrm{SNR}_{\mathsf{B}} = 20\mathrm{dB}$, optimum TD strategies increase the sum rate by 1.2 bits/s/Hz and 1.5 bits/s/Hz compared to equal TD case under the peak power constraint and average power constraint, respectively. This is because the relay has only two antennas. When $\mathrm{SNR}_{\mathsf{B}}$ is high, the MAC phase becomes the bottleneck of the system. By increasing the duration of the MAC phase, more data from station A and B can be decoded at the relay. Those data can still be retransmitted back to the two stations in the BRC phase even though the duration of the BRC phase is shorter now. By choosing the optimum TD strategies, the sum rate can be increased. On the other hand, optimum TD strategies do not increase the sum rate much when the number of antennas is increased to $N_{\mathsf{R}} = 4$ as shown in Fig. 5.9. In this case, the MAC phase is not a constraint of the system any more. By increasing the duration of the MAC phase, more data from station A and B can be decoded at the relay. However, those data cannot be transmitted to the two

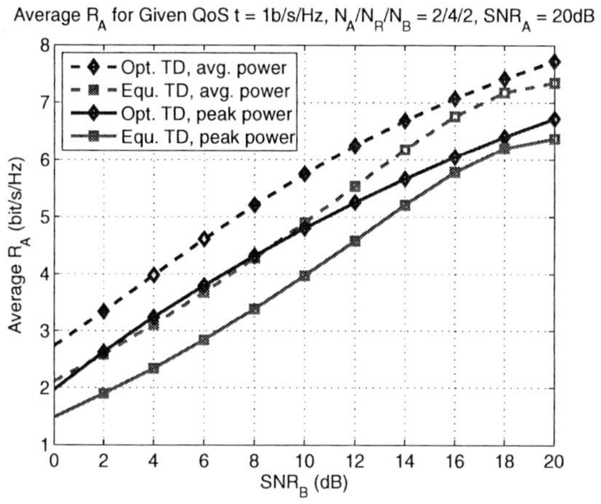

Fig. 5.12: Average achievable rate of R_A for given QoS requirement $R_\mathsf{B} \geq$ 1bit/s/Hz, $N_\mathsf{A}/N_\mathsf{R}/N_\mathsf{B} = 2/4/2$, $\mathrm{SNR}_\mathsf{A} = 20$dB

stations in the BRC phase due to its shorter duration. By comparing Fig. 5.9 and Fig. 5.8, we can observe that by increasing the number of antennas N_R at the relay, the sum rates have been improved by about 1 bits/s/Hz both under peak and average power constraints. Similar observations can be found for no CSIT case in Fig. 5.10 and Fig. 5.11.

5.5.3 Average Achievable Rate R_A for Given QoS Requirement in R_B

The maximum achievable ergodic sum rate does not consider the QoS for individual users, and is only one out of many figures of merit for communication systems. In some systems, especially cellular communication networks, the data traffic is asymmetric: the data rate transmitted by one user (e.g., the uplink data rate) is limited, while the data rate transmitted by the other user (e.g., the downlink data rate) should be increased as large as possible. This motivates us to consider the following problem: Given the setup of a two-way relaying system and the QoS requirement in R_B, what is the maximum average achievable rate R_A if optimum TD strategy is applied. We apply Algorithm 5 to answer this problem.

We consider the antenna configuration that $N_\mathsf{A} = N_\mathsf{B} = 2$ and $N_\mathsf{R} = 4$, where it has been shown in Section 5.5.2 that optimum TD strategy does not improve the ergodic sum rate much. The setup is similar to Section 5.5.2, where $\mathrm{SNR}_\mathsf{A} = P_\mathsf{A}/\sigma^2 = P_\mathsf{R}/\sigma_\mathsf{A}^2 = 20$dB and $\mathrm{SNR}_\mathsf{B} = P_\mathsf{B}/\sigma^2 = P_\mathsf{R}/\sigma_\mathsf{B}^2$. We consider a Ricean fading channel, where the K-factor

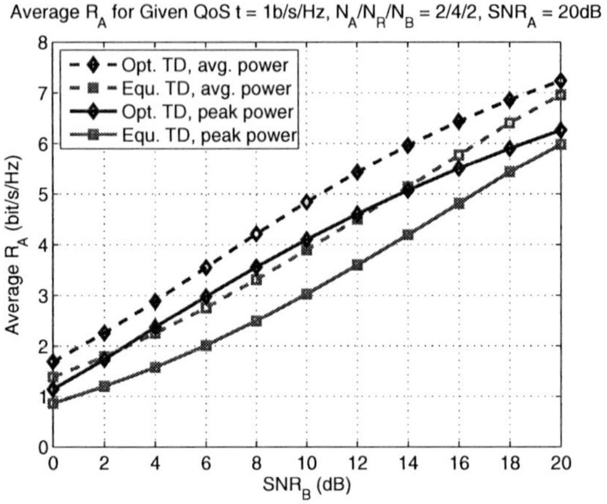

Fig. 5.13: Average achievable rate of R_{A} for given QoS requirement $R_{\mathsf{B}} \geq 1$bit/s/Hz (for no CSIT case), $N_{\mathsf{A}}/N_{\mathsf{R}}/N_{\mathsf{B}} = 2/4/2$, $\mathrm{SNR}_{\mathsf{A}} = 20$dB

between A and R is $K_{\mathsf{A}} = 5$ (i.e., 7dB) and the K-factor between B and R is $K_{\mathsf{B}} = 1$ (i.e., 0dB). We choose a Ricean fading channel model since we do not want the channel to be frequently in outage. The QoS requirement for the data rate transmitted from Station B is $R_{\mathsf{B}} \geq 1$bit/s/Hz.

Fig. 5.12 and Fig. 5.13 show the average achievable rate R_{A} for given QoS requirement $R_{\mathsf{B}} \geq 1$bit/s/Hz for full CSIT and no CSIT cases, respectively. When R_{B} is in outage for the given channel, R_{A} is set to be 0. We can observe that the optimum TD strategy increases the average achievable rate of R_{A} by about 1bit/s/Hz under peak or average power constraint when $\mathrm{SNR}_{\mathsf{B}} = 10$dB, no matter whether CSIT is available. Combined with the results obtained in Section 5.5.1, we can see that using the optimum TD strategies can increase the achievable rates of one user given the QoS requirement of its partner not only in static channels but also in fading channels. In Fig. 5.12 and Fig. 5.13, we can observe that curve of the average rate R_{A} with peak power constraint is almost parallel to that with average power constraint in both the optimum TD strategies and the equal TD strategy. This is true no matter whether the CSIT is available or not. Another observation is that the gain using optimum TD strategies compared to the equal TD strategy decreases when $\mathrm{SNR}_{\mathsf{B}}$ is high at around 18dB.

5.6 Chapter Summary and Conclusions

We proposed two methods for characterizing the achievable rate regions with optimum TD strategies for two-way DF relaying systems with multiple antennas. Both peak power constraint and average power constraint were considered. Simulation results showed that the achievable rate region can be further increased by choosing optimum TD strategies. At high SNR, the optimum TD strategies improve the ergodic sum rate when the MAC phase is the bottleneck of the system, e.g., when $N_\mathsf{A} + N_\mathsf{B} > N_\mathsf{R}$. The gain in ergodic sum rate by using the optimum TD strategies is small when $N_\mathsf{A} + N_\mathsf{B} \leq N_\mathsf{R}$. However, even when $N_\mathsf{A} + N_\mathsf{B} \leq N_\mathsf{R}$, the average achievable rate of one user given the QoS requirement of the other is increased a lot by using optimum TD strategies under peak power and average power constraint.

5.7 Appendices

5.7.1 Maximum Weighted Sum Rate Calculation for $\mathcal{R}_{\text{OPT}}$

When the rate region $\mathcal{R}_{\text{OPT}}$ is convex, its boundary points can be determined by solving the following optimization problem:

$$
\begin{aligned}
\text{maximize} \quad & \mu R_\mathsf{A} + (1-\mu) R_\mathsf{B} \\
\text{subject to} \quad & (R_\mathsf{A}, R_\mathsf{B}) \in \mathcal{C}_{\text{MAC}}(\alpha) \bigcap \mathcal{C}_{\text{BRC}}(\alpha) \\
& 0 \leq \alpha \leq 1 \\
\text{variables} \quad & R_\mathsf{A}, R_\mathsf{B}, \alpha, \boldsymbol{\Omega}_\mathsf{A}, \boldsymbol{\Omega}_\mathsf{B}, \boldsymbol{\Omega},
\end{aligned}
\tag{5.27}
$$

where μ is a weighting constant and $0 \leq \mu \leq 1$.

For a fixed weighting constant μ, the problem (5.27) can be solved by decomposition methods [23]: Firstly, we observe that for a given TD factor α, where $0 \leq \alpha \leq 1$, the following is a convex optimization problem:

$$
\begin{aligned}
\text{maximize} \quad & \mu R_\mathsf{A} + (1-\mu) R_\mathsf{B} \\
\text{subject to} \quad & (R_\mathsf{A}, R_\mathsf{B}) \in \mathcal{C}_{\text{MAC}}(\alpha) \bigcap \mathcal{C}_{\text{BRC}}(\alpha) \\
& \alpha \text{ is given} \\
\text{variables} \quad & R_\mathsf{A}, R_\mathsf{B}, \boldsymbol{\Omega}_\mathsf{A}, \boldsymbol{\Omega}_\mathsf{B}, \boldsymbol{\Omega}.
\end{aligned}
\tag{5.28}
$$

(5.28) is a convex optimization problem because $\mathcal{C}_{\text{MAC}}(\alpha)$ and $\mathcal{C}_{\text{BRC}}(\alpha)$ are both convex regions for the fixed value α. Their intersection is thus also convex. For the fixed α, we denote the optimum objective value of (5.28) as $r(\alpha) = \mu R_{\mathsf{A}}^{\star}(\alpha) + (1-\mu)R_{\mathsf{B}}^{\star}(\alpha)$, and its value can be determined by convex optimization methods, e.g., interior-point methods [33]. That is, $r(\alpha)$ can be considered as a function of the TD factor α. Secondly, we solve the following problem:

$$\begin{array}{ll} \text{maximize} & r(\alpha) = \mu R_{\mathsf{A}}^{\star}(\alpha) + (1-\mu)R_{\mathsf{B}}^{\star}(\alpha) \\ \text{subject to} & 0 \le \alpha \le 1 \\ \text{variables} & \alpha. \end{array} \tag{5.29}$$

Since there is only one variable α in (5.29), it can be solved by using the *bisection* method [23]. The solution $\alpha^{\star}$ and the optimum objective value in (5.29) are also respectively the optimum TD factor and the optimum objective value of (5.27) for the given weighting constant μ. In this way, the original problem (5.27) is decomposed into two subproblems (5.28) and (5.29). However, this decomposition method suffers from difficulties in implementation because it is hard for (5.28) to be converted to a second-order cone programming problem or a semidefinite programming problem and solved by standard solvers [230].

There is an exception when the transmitters do not have channel state information (CSIT) under the peak power constraint. In such a case, the transmit covariance matrices are chosen as $\mathbf{\Omega}_{\mathsf{A}} = \mathbf{I}_{N_{\mathsf{A}}}$, $\mathbf{\Omega}_{\mathsf{B}} = \mathbf{I}_{N_{\mathsf{B}}}$ and $\mathbf{\Omega} = \mathbf{I}_{N_{\mathsf{R}}}$. Thus the problem (5.27) degenerates into a linear programming problem with variable α and can be solved by standard linear programming tools [69].

5.7.2 Maximum Weighted Sum Rate Calculation for $\mathcal{C}_{\mathrm{MAC}}(1)$ and $\mathcal{C}_{\mathrm{BRC}}(0)$

We briefly summarizes the method for calculating the maximum weighted sum rate for the MAC phase and BRC phase channels used in Section 5.3.1. The discussions follow that of [78, 242].

5.7.2.1 MAC Phase Channel

The MAC phase capacity region $\mathcal{C}_{\mathrm{MAC}}(1)$ is convex. Given two weighting factors $\mu_{\mathsf{A}}, \mu_{\mathsf{B}} \ge 0$ and assuming $\mu_{\mathsf{A}} \ge \mu_{\mathsf{B}}$ without loss of generality, the problem of maximizing the function

$\mu_{\mathsf{A}} R_{\mathsf{A}} + \mu_{\mathsf{B}} R_{\mathsf{B}}$ over all rate vectors in $(R_{\mathsf{A}}, R_{\mathsf{B}}) \in \mathcal{C}_{\mathrm{MAC}}(1)$ is equivalent to [78]

$$\begin{aligned} \text{maximize} \quad & \mu_{\mathsf{B}} \log_2 \det \left(\mathbf{I}_{N_{\mathsf{R}}} + \textstyle\sum_{k \in \{\mathsf{A},\mathsf{B}\}} \frac{P_k}{N_k \sigma^2} \mathbf{G}_k \boldsymbol{\Omega}_k \mathbf{G}_k^H \right) \\ & + (\mu_{\mathsf{A}} - \mu_{\mathsf{B}}) \log_2 \det \left(\mathbf{I}_{N_{\mathsf{R}}} + \tfrac{P_{\mathsf{A}}}{N_{\mathsf{A}} \sigma^2} \mathbf{G}_{\mathsf{A}} \boldsymbol{\Omega}_{\mathsf{A}} \mathbf{G}_{\mathsf{A}}^H \right) \\ \text{subject to} \quad & \mathrm{tr}(\boldsymbol{\Omega}_k) \le N_k, \quad \boldsymbol{\Omega}_k \succeq 0, \quad \text{for } k \in \{\mathsf{A},\mathsf{B}\} \\ \text{variables} \quad & \boldsymbol{\Omega}_{\mathsf{A}}, \boldsymbol{\Omega}_{\mathsf{B}}. \end{aligned} \tag{5.30}$$

The covariance matrices that maximize the objective function in (5.30) are also the corresponding optimal covariances for the maximum weighted sum rate problem. Furthermore, (5.30) is a convex optimization problem and can be solved using convex optimization tools numerically or using the following algorithm [242].

Denote the objective function of (5.30) as $f(\boldsymbol{\Omega}_{\mathsf{A}}, \boldsymbol{\Omega}_{\mathsf{B}})$. The gradient of the $f(\boldsymbol{\Omega}_{\mathsf{A}}, \boldsymbol{\Omega}_{\mathsf{B}})$ with respect to the covariance matrix $\boldsymbol{\Omega}_{\mathsf{A}}$ and $\boldsymbol{\Omega}_{\mathsf{B}}$ is respectively given by

$$\begin{aligned} \nabla_{\mathsf{A}} f(\boldsymbol{\Omega}_{\mathsf{A}}, \boldsymbol{\Omega}_{\mathsf{B}}) = & \frac{\mu_{\mathsf{B}}}{\ln 2} \left[\mathbf{G}_{\mathsf{A}}^H \left(\mathbf{I}_{N_{\mathsf{R}}} + \sum_{k \in \{\mathsf{A},\mathsf{B}\}} \frac{P_k}{N_k \sigma^2} \mathbf{G}_k \boldsymbol{\Omega}_k \mathbf{G}_k^H \right)^{-1} \mathbf{G}_{\mathsf{A}} \right] \\ & + \frac{(\mu_{\mathsf{A}} - \mu_{\mathsf{B}})}{\ln 2} \left[\mathbf{G}_{\mathsf{A}}^H \left(\mathbf{I}_{N_{\mathsf{R}}} + \frac{P_{\mathsf{A}}}{N_{\mathsf{A}} \sigma^2} \mathbf{G}_{\mathsf{A}} \boldsymbol{\Omega}_{\mathsf{A}} \mathbf{G}_{\mathsf{A}}^H \right)^{-1} \mathbf{G}_{\mathsf{A}} \right]; \end{aligned} \tag{5.31}$$

$$\nabla_{\mathsf{B}} f(\boldsymbol{\Omega}_{\mathsf{A}}, \boldsymbol{\Omega}_{\mathsf{B}}) = \frac{\mu_{\mathsf{B}}}{\ln 2} \left[\mathbf{G}_{\mathsf{B}}^H \left(\mathbf{I}_{N_{\mathsf{R}}} + \sum_{k \in \{\mathsf{A},\mathsf{B}\}} \frac{P_k}{N_k \sigma^2} \mathbf{G}_k \boldsymbol{\Omega}_k \mathbf{G}_k^H \right)^{-1} \mathbf{G}_{\mathsf{B}} \right]. \tag{5.32}$$

The algorithm proceeds iteratively as follows: Given the nth iterate $\boldsymbol{\Omega}_k(n), k \in \{\mathsf{A},\mathsf{B}\}$, determine the principle eigenvectors $\mathbf{v}_k$ (of unit norm) and the corresponding principle eigenvalues λ_k of the gradients $\nabla_k f(\boldsymbol{\Omega}_{\mathsf{A}}, \boldsymbol{\Omega}_{\mathsf{B}})$ for $k \in \{\mathsf{A},\mathsf{B}\}$. Let $j^* = \arg\max_{k \in \{\mathsf{A},\mathsf{B}\}} \lambda_k$.

The $(n+1)$st iterate $\boldsymbol{\Omega}_k(n+1), k \in \{\mathsf{A},\mathsf{B}\}$ are updated as follows:

$$\boldsymbol{\Omega}_{j^*}(n+1) = t^* \boldsymbol{\Omega}_{j^*}(n) + (1 - t^*) N_{j^*} \mathbf{v}_{j^*} \mathbf{v}_{j^*}^H \tag{5.33}$$

$$\boldsymbol{\Omega}_i(n+1) = t^* \boldsymbol{\Omega}_i(n), \quad \text{for } i \neq j^* \tag{5.34}$$

where t^* is the solution to the following one-dimensional optimization that can be solved through bisection

$$t^* = \begin{cases} \arg\max_{0 \le t \le 1} f\left(t\boldsymbol{\Omega}_{\mathsf{A}}(n) + (1-t) N_{\mathsf{A}} \mathbf{v}_{\mathsf{A}} \mathbf{v}_{\mathsf{A}}^H, t\boldsymbol{\Omega}_{\mathsf{B}}(n) \right), & \text{when } j^* = \mathsf{A}, \\ \arg\max_{0 \le t \le 1} f\left(t\boldsymbol{\Omega}_{\mathsf{A}}(n), t\boldsymbol{\Omega}_{\mathsf{B}}(n) + (1-t) N_{\mathsf{B}} \mathbf{v}_{\mathsf{B}} \mathbf{v}_{\mathsf{B}}^H \right), & \text{when } j^* = \mathsf{B}. \end{cases} \tag{5.35}$$

As $n \to \infty$, the covariance matrices converge to the optimum covariance matrices. When

$\mu_{\mathsf{A}} < \mu_{\mathsf{B}}$, we only need to swap the subscripts in (5.30), and the same algorithm applies.

5.7.2.2 BRC Phase Channel

The BRC phase capacity region $\mathcal{C}_{\mathrm{BRC}}(0)$ is also convex. Given two weighting factors $\mu_{\mathsf{A}}, \mu_{\mathsf{B}} \geq 0$, the problem of maximizing the function $\mu_{\mathsf{A}} R_{\mathsf{A}} + \mu_{\mathsf{B}} R_{\mathsf{B}}$ over all rate vectors in $(R_{\mathsf{A}}, R_{\mathsf{B}}) \in \mathcal{C}_{\mathrm{BRC}}(0)$ is equivalent to

$$\begin{array}{ll} \text{maximize} & \mu_{\mathsf{A}} \log_2 \det\left(\mathbf{I}_{N_{\mathsf{B}}} + \frac{P_{\mathsf{R}}}{N_{\mathsf{R}}\sigma_{\mathsf{B}}^2} \mathbf{H}_{\mathsf{B}} \boldsymbol{\Omega} \mathbf{H}_{\mathsf{B}}^H\right) + \mu_{\mathsf{B}} \log_2 \det\left(\mathbf{I}_{N_{\mathsf{A}}} + \frac{P_{\mathsf{R}}}{N_{\mathsf{R}}\sigma_{\mathsf{A}}^2} \mathbf{H}_{\mathsf{A}} \boldsymbol{\Omega} \mathbf{H}_{\mathsf{A}}^H\right) \\ \text{subject to} & \mathrm{tr}(\boldsymbol{\Omega}) \leq N_{\mathsf{R}}, \quad \boldsymbol{\Omega} \succeq 0 \\ \text{variables} & \boldsymbol{\Omega}. \end{array} \tag{5.36}$$

The covariance matrices that maximize the objective function in (5.36) are also the corresponding optimal covariances for the maximum weighted sum rate problem. Furthermore, (5.36) is a convex optimization problem and can be solved using convex optimization tools numerically or using the following algorithm.

We denote the objective function of (5.36) as $g(\boldsymbol{\Omega})$. Its gradient with respect to the covariance matrix Ω is then given by

$$\nabla g(\boldsymbol{\Omega}) = \mu_{\mathsf{A}} \mathbf{H}_{\mathsf{B}}^H \left(\mathbf{I}_{N_{\mathsf{B}}} + \frac{P_{\mathsf{R}}}{N_{\mathsf{R}}\sigma_{\mathsf{B}}^2} \mathbf{H}_{\mathsf{B}} \boldsymbol{\Omega} \mathbf{H}_{\mathsf{B}}^H\right)^{-1} \mathbf{H}_{\mathsf{B}} + \mu_{\mathsf{B}} \mathbf{H}_{\mathsf{A}}^H \left(\mathbf{I}_{N_{\mathsf{A}}} + \frac{P_{\mathsf{R}}^*}{N_{\mathsf{R}}\sigma_{\mathsf{A}}^2} \mathbf{H}_{\mathsf{A}} \boldsymbol{\Omega} \mathbf{H}_{\mathsf{A}}^H\right)^{-1} \mathbf{H}_{\mathsf{A}}. \tag{5.37}$$

Given the nth iterate $\boldsymbol{\Omega}(n)$, determine the principle eigenvectors $\mathbf{v}$ (of unit norm) and the corresponding principle eigenvalues λ of the gradients $\nabla g(\boldsymbol{\Omega})$. The $(n+1)$st iterate $\boldsymbol{\Omega}(n+1)$ is updated as follows:

$$\boldsymbol{\Omega}(n+1) = t^* \boldsymbol{\Omega}(n) + (1 - t^*) N_{\mathsf{R}} \mathbf{v}\mathbf{v}^H \tag{5.38}$$

where t^* is the solution to the following one-dimensional optimization that can be solved through bisection

$$t^* = \arg\max_{0 \leq t \leq 1} g\left(t\boldsymbol{\Omega}(n) + (1-t) N_{\mathsf{R}} \mathbf{v}\mathbf{v}^H\right). \tag{5.39}$$

As $n \to \infty$, the covariance matrices converge to the optimum covariance matrices.

5.7.3 Minimax Theorem

Theorem 5.7.1 ([63, 189]). *Let A and B be nonempty sets and $f : A \times B \mapsto \mathbb{R}$ a function on $A \times B$. The primal problem associated with the function f is given by*

$$v(P) = \inf_{b \in B} \sup_{a \in A} f(a, b) \tag{5.40}$$

while the dual problem has the form

$$v(D) = \sup_{a \in A} \inf_{b \in B} f(a, b). \tag{5.41}$$

We have $v(P) = v(D)$, whenever both A and B are convex, A is compact and f is concave and upper-semicontinuous in a and convex in B.

Chapter 6

Self-Interference-Aided Channel Estimation in Two-Way Relaying Systems

The two-way decode-and-forward (DF) relaying protocol and its applications have been discussed in Chapter 2 and Chapter 4. The core idea of the two-way DF relaying protocol is the *self-interference* (SI) cancellation in the broadcast (BRC) phase. That is, the known data information transmitted from the relay does not interfere with the decoding at the receivers in the BRC phase since it can be canceled in the decoding process. However, conventional two-way relaying protocols only *cancel* instead of *utilize* the SI. In this chapter and Chapter 8, we propose practical transmission schemes that exploit the SI to provide additional benefits in two-way DF relaying protocols. In this chapter, we propose a novel approach that utilizes the SI for channel estimation in the BRC phase of MIMO two-way DF relaying system when the superposition coding (SPC) scheme is applied. In this case, the SI contains the known data symbols at the receivers, which can play a similar role as pilot sequences for the purpose of channel estimation. Firstly, we propose the SI-aided joint maximum likelihood (ML) channel estimation and data detection scheme. We show that its channel estimation performance can approach the Cramér-Rao lower bound when the signal-to-noise ratio at the receiver is high. Then we propose an SI-aided linear channel estimator in iterative receivers. Its channel estimate is achieved with lower complexity compared to the joint ML scheme. The proposed SI-aided channel estimation can be applied together with pilot sequence to improve the channel estimation accuracy. In order to make fair comparisons, we consider two types of systems in this chapter: one purely relies on SI to estimate the channel, and the other purely relies on the pilot sequences to estimate the channel. In the purely SI-aided channel estimator, higher bandwidth efficiency can be achieved since pilot sequences are no longer transmitted. We show the conditions under which the SI-aided linear channel estimator is able to outperform the pilot-aided channel estimator. Taking into account the channel estimation errors, we propose an optimized power allocation scheme to provide fairness for the different stations. Under the same average power constraint and transmitting the same amount of data symbols, simulations show that the performance of the proposed SI-aided iterative receiver can significantly outperform that of the iterative receiver which solely relies on the pilots for channel estimation in realistic scenarios.

6.1 Introduction

The *two-way relaying protocol* [193] is applicable to systems with bidirectional information flow and recovers a significant portion of the half-duplex loss in traditional relaying systems. A large number of transmission scenarios, such as the base station communicates with a mobile user via a dedicated relay in a cellular system, or two mobile clients transmit data to each other via the access point in a wireless local area network (WLAN), is suitable for the application of the two-way relaying protocol. Compared to traditional relaying protocols that require four phases (in time or frequency) to achieve bidirectional communication between the two stations, the two-way relaying protocol only needs two phases to exchange the data, namely, the multiple access (MAC) phase and the broadcast (BRC) phase (see Chapter 2).

The two-way decode-and-forward (DF) relaying protocol is particularly interesting due to its practical applicability in real-world communication systems. The idea of the two-way DF relaying protocol is to combine the decoded data information at the relay, and send back the combined data information to the two user stations in the BRC phase. The back-propagated data information, called the *self-interference* (SI), can be canceled at the receiver if the channel from the relay to the receiver is known. Up to now, there are two major types of practical schemes proposed for combining the data at the relay, i.e., the *superposition coding* (SPC) scheme and the *network coding* scheme. The SPC scheme combines the decoded data information on the symbol level, whereas the network coding scheme combines the decoded data on the bit level. The comparison of the two schemes can be found in, e.g., [88, 128, 151, 193, 260] (see Chapter 4). The SI cancellation is the core idea of the two-way relaying protocol. That is, the known data information transmitted from the relay does not interfere with the decoding at the receivers in the BRC phase since it can be canceled in the decoding process. However, conventional two-way relaying protocols only *cancel* instead of *utilize* the SI. In this chapter and Chapter 8, we propose schemes that exploit the SI to provide additional benefits in two-way DF relaying protocols. In this chapter, we propose a novel approach that utilizes the SI for channel estimation in the BRC phase of the multiple-input multiple-output (MIMO) two-way DF relaying system when the superposition coding (SPC) scheme is applied. In this case, the SI contains the known data symbols at the receivers, which can play a similar role as pilot sequences for the purpose of channel estimation.

Future wireless communication systems are envisioned to be equipped with multiple antennas because multiple-input multiple-output (MIMO) technology can provide significant increase in channel capacity [224] and great enhancement in link reliability [221]. To obtain that advantage, channel knowledge is required at least at the receiver side. However, accurate channel estimation in MIMO systems is generally difficult. Conventional pilot-aided channel estimation schemes send orthogonal pilot sequences on different transmit antennas to estimate the channel, which wastes system resources, especially when the number of transmit antennas is large. The major algorithms for time-multiplexed pilot-aided channel estimation, i.e., the least-square (LS) algorithm and the minimum mean square error (MMSE) algorithm, are summarized in [26], where the authors also investigated the optimal choice of pilot signals. A pilot-embedding method was proposed in [103, 109] and further investigated in [100, 126, 166, 283], where low-power orthogonal pilot sequences are transmitted concurrently with the data. Such schemes trade transmit power for higher spectrum efficiency. Decision-directed iterative channel estimation can be found in, e.g., [56, 108]. All those channel estimation schemes rely on an initial estimate of the MIMO channel based on the transmission of pilot sequences.

Blind and semi-blind channel estimation are discussed in, e.g., [11, 29, 164]. The purpose is to identify the channel solely based on some known properties of the channel or data symbols. For example, the use of cyclostationary statistics to accomplish blind MIMO channel estimation has first been proposed for frequency-flat fading channels in [199], and was extended to MIMO-OFDM systems in [29], which can recover the transmitted symbol streams up to a phase rotation. A blind channel estimation approach combining different modulation schemes on adjacent subcarriers was proposed in [164] for OFDM systems. Compared to the statistically blind channel estimation approaches in [199], which have a slow convergence rate, the proposed scheme in [164] achieves fast convergence rate exploiting the specific symbol constellation of the system based on the maximum likelihood (ML) principle. A semi-blind channel estimation and data detection scheme based on the ML principle exploiting the discrete symbol constellation was proposed in [11]. The advantage of blind channel estimation is that the resources occupied by pilot sequences can be released. However, the blind channel estimation methods suffer from the ambiguity problem, i.e., the channel and data cannot be uniquely identified without transmitting additional pilots for conventional modulation schemes. That problem is inherent in blind channel estimation schemes themselves. In order to uniquely determine the channel and data, some pilots are still required to be transmitted.

Channel estimation in relaying systems is a relatively new topic. The authors of [71] investigated the channel estimation strategies and optimal pilot sequence design for conventional unidirectional AF relaying systems with one source, one destination and multiple relays. Both the LS and the MMSE algorithms were considered. The authors investigated the linear precoding matrix design at the relays to minimize the channel estimation errors subject to the individual power constraints at each relays. The same setup is considered for DF relaying systems in [72], where the authors investigated two types of channel estimation algorithms: MMSE and ML estimation algorithms. Pilot symbol spacing in unidirectional AF relaying systems is address in [182]. Channel estimator design in a two-way AF relaying system with one source, one destination and one relay is considered in [73]. All those works only consider relaying systems with single antenna on each station.

In this chapter, we consider channel estimation in a two-way DF relaying system with multiple antennas. We propose a novel approach that exploits the known data symbols in the SI to estimate the channel for the BRC phase. We call it *SI-aided channel estimation.* The SI plays a similar role in the proposed scheme as the superimposed pilots in [103]. However, pilots contain no information and waste power, while the proposed approach does not suffer from the power penalty since SI is inherent in the considered scenario. Furthermore, the data contained in the SI is random and does not have the special structure as pilots in [103]. Specific problems, such as the power allocation to provide fairness for the two receivers, arise in two-way DF relaying systems. The channel estimation schemes in the MAC phase is left aside in this paper, since the channel knowledge at the relay in the MAC phase can be obtained from existing multiuser channel estimation schemes, e.g., in [240], or from the feedback of the stations. The SI-aided channel estimation approach can be used together with pilot-aided channel estimation to improve the accuracy of the estimated channel. In order to make fair comparisons, we consider two types of systems in this chapter: one purely relies on SI to estimate the channel, and the other purely relies on the pilot sequences to estimate the channel.

Our Contributions: The contributions of this chapter are summarized as follows:

- We propose the SI-aided joint maximum likelihood (ML) channel estimation and data detection scheme when no *a priori* knowledge of the channel is available. The problem of unique identifiability of the channel and data symbols is addressed. We observe that the SI implicitly helps to uniquely determine the channel, and channel estimates with high precision can be obtained without using pilots. Furthermore, we show that its channel estimation performance can approach the Cramér-Rao lower bound for joint channel and data symbol estimation when the signal-to-noise ratio (SNR) at the receiver is high.

- We propose an SI-aided linear channel estimator. We provide its theoretical mean square error (MSE) performance analysis, and show the conditions under which the SI-aided linear channel estimation outperforms the pure pilot-aided channel estimation. The advantage of the proposed scheme is more conspicuous when the coherence interval in block-fading channels or the observation frame length in time-varying channels is large. In order to provide fairness for the receivers at different stations in the BRC phase, we propose an optimized power allocation at the relay taking into account the channel estimation errors.

- We show how SI-aided linear channel estimator is integrated in commonly used receivers, e.g., the iterative receiver structure for channel estimation and data detection. Considering SI, we show that only small modifications to existing receiver structures are required.

- Our proposed SI-aided channel estimation scheme can be used alone or be combined with pilot-aided channel estimation. In the purely SI-aided channel estimator, higher bandwidth efficiency in the BRC phase can be achieved since pilot sequences are no longer transmitted. Simulation results verify our conclusions and show that the performance of the SI-aided iterative receiver can greatly outperform that of the pilot-aided iterative receiver under the same average power constraint and transmitting the same amount of data. The proposed scheme is particularly interesting for multi-carrier systems because SI can track the channel in all subcarriers, and there is no need of doing interpolations in time and frequency as pilot-aided schemes.

- To the best of our knowledge, this is the first scheme that exploits the SI for channel estimation in two-way relaying systems. Prior works focus on canceling interference when it contains known data [193]. This work goes one step further and shows that SI can be *utilized*. Since interferences with known data is very common in cooperative communication systems, we believe more application areas can be found for them.

The following of the chapter is organized as follows: The system model is shown in Section 6.2. The SI-aided joint ML channel estimation and data detection scheme is proposed in Section 6.3. Afterwards, we consider receiver structures with separate channel estimation and data detection. In Section 6.4, we focus on the SI-aided linear channel estimator

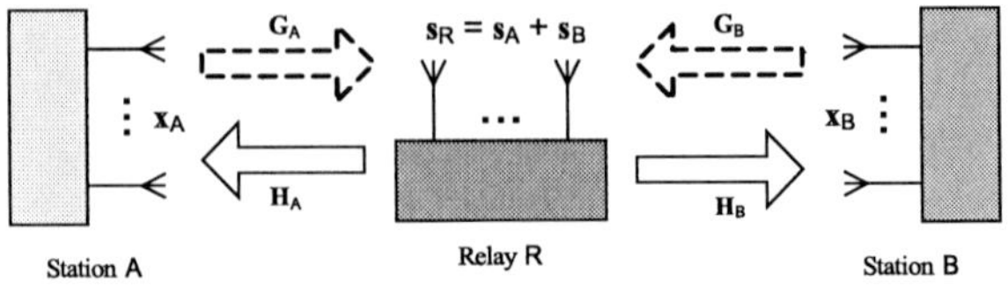

Fig. 6.1: MIMO two-way DF relaying system, where the dashed arrows represent the transmission in the MAC phase, and the solid arrows represent the transmission in the BRC phase.

and provide the performance analysis. The estimated channel can serve as the initial channel estimate for the iterative receiver structure in Section 6.5, where the subsequent channel estimation and MIMO symbol demapping are discussed in detail. Comprehensive simulation results are presented in Section 6.6, where we compare the proposed schemes with the pilot-aided schemes in block-fading and time-varying channels. Conclusions are drawn in Section 6.7.

6.2 System Model

We consider a relaying system where two wireless stations A and B exchange data via a half-duplex relay as shown in Fig. 6.1. The number of antennas at station A, the relay and station B are denoted as N_{A}, N_{R} and N_{B}, respectively.

The data of stations A and B are exchanged in two or three phases when the two-way relaying protocol is applied. Their difference only lies in the MAC phase and not in the BRC phase. The MAC and BRC phases can be separated in time (time-division duplex, TDD) or in frequency (frequency-division duplex, FDD). Considering the two-phase protocol, at one time slot in the MAC phase, stations A and B transmit their information-bearing data symbol vectors $\mathbf{x}_{\mathsf{A}} \in \mathbb{C}^{N_{\mathsf{A}} \times 1}$ and $\mathbf{x}_{\mathsf{B}} \in \mathbb{C}^{N_{\mathsf{B}} \times 1}$ simultaneously to the relay. The received signal $\mathbf{y}_{\mathsf{R}}$ at the relay can be expressed as

$$\mathbf{y}_{\mathsf{R}} = \mathbf{G}_{\mathsf{A}}\mathbf{x}_{\mathsf{A}} + \mathbf{G}_{\mathsf{B}}\mathbf{x}_{\mathsf{B}} + \mathbf{n}_{\mathsf{R}} \tag{6.1}$$

where $\mathbf{G}_{\mathsf{A}} \in \mathbb{C}^{N_{\mathsf{R}} \times N_{\mathsf{A}}}$ and $\mathbf{G}_{\mathsf{B}} \in \mathbb{C}^{N_{\mathsf{R}} \times N_{\mathsf{B}}}$ are the channel matrices from stations A and B to the relay, respectively. The additive noise vector at the relay is $\mathbf{n}_{\mathsf{R}} \sim \mathcal{CN}(0, \sigma^2 \mathbf{I}_{N_{\mathsf{R}}})$. The relay decodes the received signals and extracts the data bits contained in $\mathbf{x}_{\mathsf{A}}$ and $\mathbf{x}_{\mathsf{B}}$. In this multiuser detection scenario, the receiver structures for channel estimation and decoding can

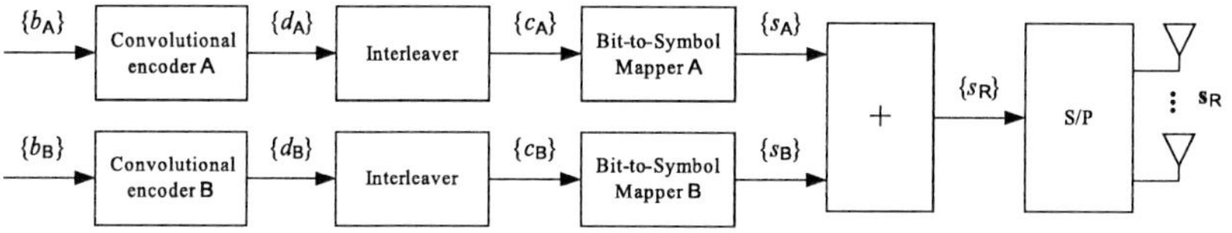

Fig. 6.2: Re-encoding and retransmission at the relay in the BRC phase.

be found in, e.g., [120, 207, 240]. In order to guarantee the decoded data at the relay to be correct, error-detecting codes, e.g., cyclic redundancy check (CRC) codes, can be applied in the data. In the following, we focus on the BRC phase and assume the data transmitted from stations A and B in the MAC phase have already been decoded by the relay.

In the BRC phase, the decoded data from stations A and B are re-encoded and retransmitted. Fig. 6.2 shows the transmitter structure with spatial multiplexing. The bit sequences $\{b_A\}$ and $\{b_B\}$ respectively denote the decoded data bits from stations A and B, where each element $b_A, b_B \in \{0,1\}$, and their length is $L_{b,A}$ and $L_{b,B}$, respectively. After being processed by the convolutional encoders with coding rate r_A and r_B, the output coded bit sequences are respectively denoted as $\{d_A\}$ and $\{d_B\}$, where each of their elements $d_A, d_B \in \{0,1\}$. The two sequences are bitwise interleaved to form the code sequences $\{c_A\}$ and $\{c_B\}$. Then the bit-interleaved codewords are respectively partitioned into groups of m_A and m_B bits. Each bit group of $\{c_A\}$ and $\{c_B\}$ is mapped to a complex symbol within the M_A-ary and M_B-ary quadrature amplitude modulation (QAM) or phase-shift keying (PSK) symbol alphabets, where $M_A = 2^{m_A}$ and $M_B = 2^{m_B}$. In order to guarantee the length of sequence $\{s_A\}$ and $\{s_B\}$ to be equal, we require $L_{b,A}/(r_A m_A) = L_{b,B}/(r_B m_B)$. The two symbol sequences $\{s_A\}$ and $\{s_B\}$ are added together element-wise to form the symbol sequence $\{s_R\}$, i.e., $\{s_R\} = \{s_A + s_B\}$. The resulting complex symbol sequence is further grouped into blocks of size N_R and each block forms one transmit symbol vector $\mathbf{s}_R$ at the relay, namely $\mathbf{s}_R = [s_{R,1}, \cdots, s_{R,N_R}]^T = [s_{A,1}, \cdots, s_{A,N_R}]^T + [s_{B,1}, \cdots, s_{B,N_R}]^T = \mathbf{s}_A + \mathbf{s}_B$, where $\mathbf{s}_R, \mathbf{s}_A, \mathbf{s}_B \in \mathbb{C}^{N_R \times 1}$. Furthermore, we have $\mathrm{E}(\mathbf{s}_A\mathbf{s}_A^H) = P_A/N_R\mathbf{I}_{N_R}$, $\mathrm{E}(\mathbf{s}_B\mathbf{s}_B^H) = P_B/N_R\mathbf{I}_{N_R}$ and $P_A + P_B = P_R$, where P_R is the transmit power constraint at the relay in the BRC phase. Stations A and B both know the modulation schemes and the power allocation of $\mathbf{s}_A$ and $\mathbf{s}_B$.

At a given time slot k in the BRC phase, the signals received at stations A and B are

$$\mathbf{y}_{A,k} = \mathbf{H}_{A,k}\mathbf{s}_{R,k} + \mathbf{n}_{A,k} = \underbrace{\mathbf{H}_{A,k}\mathbf{s}_{A,k}}_{\text{SI for station A}} + \mathbf{H}_{A,k}\mathbf{s}_{B,k} + \mathbf{n}_{A,k}, \tag{6.2}$$

$$\mathbf{y}_{B,k} = \mathbf{H}_{B,k}\mathbf{s}_{R,k} + \mathbf{n}_{B,k} = \mathbf{H}_{B,k}\mathbf{s}_{A,k} + \underbrace{\mathbf{H}_{B,k}\mathbf{s}_{B,k}}_{\text{SI for station B}} + \mathbf{n}_{B,k}, \tag{6.3}$$

where $\mathbf{H}_{\mathsf{A},k} \in \mathbb{C}^{N_\mathsf{A} \times N_\mathsf{R}}$ and $\mathbf{H}_{\mathsf{B},k} \in \mathbb{C}^{N_\mathsf{B} \times N_\mathsf{R}}$ denote the channel from the relay to stations A and B, respectively. $\mathbf{n}_{\mathsf{A},k} \sim \mathcal{CN}(0, \sigma_\mathsf{A}^2 \mathbf{I}_{N_\mathsf{A}})$ and $\mathbf{n}_{\mathsf{B},k} \sim \mathcal{CN}(0, \sigma_\mathsf{B}^2 \mathbf{I}_{N_\mathsf{B}})$ are the additive noise vectors. Since $\mathbf{s}_{\mathsf{A},k}$ (resp. $\mathbf{s}_{\mathsf{B},k}$) is modulated from the data of station A (resp. station B), the received signal part containing the known data is called *self-interference* (SI) for its receiver in (6.2) and (6.3).

If the channel knowledge from the relay to its receiver is available, the SI is "harmless" since it can be canceled at the receiver without actually degrading the system performance [193]. On the other hand, we observe that the SI also contains the channel information and can be utilized for channel estimation at the receivers. Before we discuss the details of the SI-aided channel estimation schemes, we present the channel models in Section 6.2.1 and Section 6.2.2. The discussion is based on a flat fading channel, which can be generalized to orthogonal frequency division multiplexing (OFDM) transmission systems straightforwardly. We only consider the channel $\mathbf{H}_\mathsf{B}$ and the receiver of station B in the following sections. The same discussions also apply to $\mathbf{H}_\mathsf{A}$ and the receiver of station A.

6.2.1 Block-Fading Channel Model

The block-fading channel model has been introduced in [25] to model the slowly varying fading in a low-mobility environment. This model is particularly relevant in wireless communication situations involving slow time-frequency hopping (e.g., Global System for Mobile Communications (GSM), Enhanced Data GSM Environment (EDGE)) or multicarrier modulation systems using OFDM. In the block-fading channel model, the channel remains constant for the *coherence interval* of L time slots, where $L \geq N_\mathsf{R}$. The channels in different coherence intervals are independent. We denote the transmitted symbol vectors from the relay during one coherence interval as $\mathbf{s}_{\mathsf{R},1}, \cdots, \mathbf{s}_{\mathsf{R},L} \in \mathbb{C}^{N_\mathsf{R} \times 1}$. We omit the time index of the channel in the coherence interval and denote it as $\mathbf{H}_\mathsf{B}$. The corresponding $N_\mathsf{B} \times L$ received signal matrix $\mathbf{Y}_\mathsf{B} = [\mathbf{y}_{\mathsf{B},1}, \cdots, \mathbf{y}_{\mathsf{B},L}]$ at station B can be expressed as

$$\mathbf{Y}_\mathsf{B} = \mathbf{H}_\mathsf{B}\mathbf{S}_\mathsf{R} + \mathbf{N}_\mathsf{B} \tag{6.4}$$

$$= \mathbf{H}_\mathsf{B}\mathbf{S}_\mathsf{B} + \underbrace{\mathbf{H}_\mathsf{B}\mathbf{S}_\mathsf{A} + \mathbf{N}_\mathsf{B}}_{\mathbf{V}} \tag{6.5}$$

where $\mathbf{S}_\mathsf{R} = [\mathbf{s}_{\mathsf{R},1}, \cdots, \mathbf{s}_{\mathsf{R},L}]$, $\mathbf{S}_\mathsf{A} = [\mathbf{s}_{\mathsf{A},1}, \cdots, \mathbf{s}_{\mathsf{A},L}]$ and $\mathbf{S}_\mathsf{B} = [\mathbf{s}_{\mathsf{B},1}, \cdots, \mathbf{s}_{\mathsf{B},L}]$ are the $N_\mathsf{R} \times L$ transmitted symbol matrices, and $\mathbf{N}_\mathsf{B} = [\mathbf{n}_{\mathsf{B},1}, \cdots, \mathbf{n}_{\mathsf{B},L}]$ is the $N_\mathsf{B} \times L$ matrix of additive noise. The received signals except the SI is denoted as $\mathbf{V}$ in (6.5).

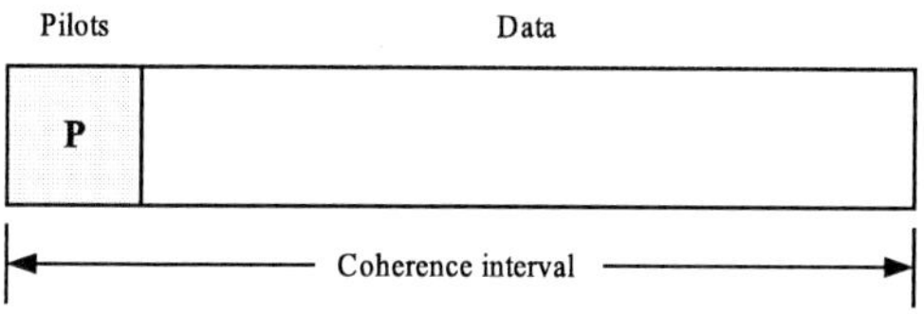

Fig. 6.3: Pilot-aided channel estimation in block-fading channel

The traditional way of estimating the block-fading channel is by transmitting pilot sequences as shown in Fig. 6.3. As the beginning of each coherence interval, the pilot sequences that occupy L_p time slots are transmitted from the relay, where $L_p \geq N_R$. The pilot sequence matrix $\mathbf{P} \in \mathbb{C}^{N_R \times L_p}$ is known to the receivers. The corresponding received signal matrix at Station B can be expressed as

$$\mathbf{Y}_B^{(p)} = \mathbf{H}_B \mathbf{P} + \mathbf{N}_B^{(p)} \tag{6.6}$$

The LS method estimates the channel matrix $\mathbf{H}_B$ as

$$\hat{\mathbf{H}}_B = \mathbf{Y}_B^{(p)} \mathbf{P}^{\dagger} \tag{6.7}$$

where $\mathbf{P}^{\dagger} = \mathbf{P}^H (\mathbf{P}\mathbf{P}^H)^{-1}$ is the pseudoinverse of $\mathbf{P}$. Subject to the transmitting training power constraint:

$$||\mathbf{P}||_F^2 = L_p P_R. \tag{6.8}$$

It has been shown in [26] that the optimal training matrix satisfies the following equation

$$\mathbf{P}\mathbf{P}^H = \frac{L_p P_R}{N_R} \mathbf{I}_{N_R}. \tag{6.9}$$

Therefore, any training matrix with orthogonal rows of the same norm $L_p P_R / N_R$ is optimal for the LS channel estimation.

The linear MMSE channel estimate of the channel $\mathbf{H}_B$ based on the received signal $\mathbf{Y}_B^{(p)}$ can be expressed as

$$\hat{\mathbf{H}}_B = \mathbf{Y}_B^{(p)} (\mathbf{P}^H \mathbf{R}_\mathbf{H} \mathbf{P} + \sigma_B^2 N_B \mathbf{I}_{L_p})^{-1} \mathbf{P}^H \mathbf{R}_\mathbf{H} \tag{6.10}$$

where $\mathbf{R}_\mathbf{H} = \mathrm{E}\left(\mathbf{H}_B^H \mathbf{H}_B\right)$. The optimal training matrix for MMSE channel estimation depends on the channel correlation matrix $\mathbf{R}_\mathbf{H}$ and its property has been characterized in [26]. However, when the channel is uncorrelated, i.e., when $\mathbf{R}_\mathbf{H}$ is a scaled identity matrix, orthogonal training matrices are optimal in minimizing the MSE of the channel estimation error.

6.2.2 Time-Varying Channel Model

In mobile communications, the channel often changes fast due to Doppler effects. The channel gains in the time-varying channel model are assumed to change in each time slot but are correlated with each other according to some statistical factors. We consider the transmission of a frame of K symbol vectors: $\mathbf{s}_{\mathsf{R},1}, \cdots, \mathbf{s}_{\mathsf{R},K} \in \mathbb{C}^{N_\mathsf{R} \times 1}$. The $K \times N_\mathsf{R}K$ data matrix $\boldsymbol{\mathcal{S}}_\mathsf{R}$ is defined as

$$\boldsymbol{\mathcal{S}}_\mathsf{R} = \begin{pmatrix} \mathbf{s}_{\mathsf{R},1}^T & \mathbf{0}^T & \cdots & \mathbf{0}^T \\ \mathbf{0}^T & \mathbf{s}_{\mathsf{R},2}^T & \cdots & \mathbf{0}^T \\ \vdots & \vdots & \ddots & \vdots \\ \mathbf{0}^T & \mathbf{0}^T & \cdots & \mathbf{s}_{\mathsf{R},K}^T \end{pmatrix} \tag{6.11}$$

$$= \begin{pmatrix} \mathbf{s}_{\mathsf{A},1}^T & \mathbf{0}^T & \cdots & \mathbf{0}^T \\ \mathbf{0}^T & \mathbf{s}_{\mathsf{A},2}^T & \cdots & \mathbf{0}^T \\ \vdots & \vdots & \ddots & \vdots \\ \mathbf{0}^T & \mathbf{0}^T & \cdots & \mathbf{s}_{\mathsf{A},K}^T \end{pmatrix} + \begin{pmatrix} \mathbf{s}_{\mathsf{B},1}^T & \mathbf{0}^T & \cdots & \mathbf{0}^T \\ \mathbf{0}^T & \mathbf{s}_{\mathsf{B},2}^T & \cdots & \mathbf{0}^T \\ \vdots & \vdots & \ddots & \vdots \\ \mathbf{0}^T & \mathbf{0}^T & \cdots & \mathbf{s}_{\mathsf{B},K}^T \end{pmatrix} \tag{6.12}$$

$$= \boldsymbol{\mathcal{S}}_\mathsf{A} + \boldsymbol{\mathcal{S}}_\mathsf{B}. \tag{6.13}$$

where $\mathbf{0}$ represents an $N_\mathsf{R} \times 1$ all-zero vector. We respectively denote the received signal matrix and the noise matrix at station B as $\boldsymbol{\mathcal{Y}}_\mathsf{B} = [\mathbf{y}_{\mathsf{B},1}, \cdots, \mathbf{y}_{\mathsf{B},K}]^T \in \mathbb{C}^{K \times N_\mathsf{B}}$ and $\boldsymbol{\mathcal{N}}_\mathsf{B} = [\mathbf{n}_{\mathsf{B},1}, \cdots, \mathbf{n}_{\mathsf{B},K}]^T \in \mathbb{C}^{K \times N_\mathsf{B}}$. Furthermore, we stack the channel matrices in those K time slots and define $\boldsymbol{\mathcal{H}}_\mathsf{B} = [\mathbf{H}_{\mathsf{B},1}, \cdots, \mathbf{H}_{\mathsf{B},K}]^T \in \mathbb{C}^{N_\mathsf{R}K \times N_\mathsf{B}}$. Therefore, we can write the system model as

$$\boldsymbol{\mathcal{Y}}_\mathsf{B} = \boldsymbol{\mathcal{S}}_\mathsf{R}\boldsymbol{\mathcal{H}}_\mathsf{B} + \boldsymbol{\mathcal{N}}_\mathsf{B} \tag{6.14}$$

$$= (\boldsymbol{\mathcal{S}}_\mathsf{A} + \boldsymbol{\mathcal{S}}_\mathsf{B})\boldsymbol{\mathcal{H}}_\mathsf{B} + \boldsymbol{\mathcal{N}}_\mathsf{B}. \tag{6.15}$$

Using pilot sequences to estimate the time-varying channel is shown in Fig. 6.4, where the pilot sequences are placed with each other for a certain distance and the estimated channel in those time slots are interpolated to get the estimate of the channel state information (CSI) at the data positions [59].

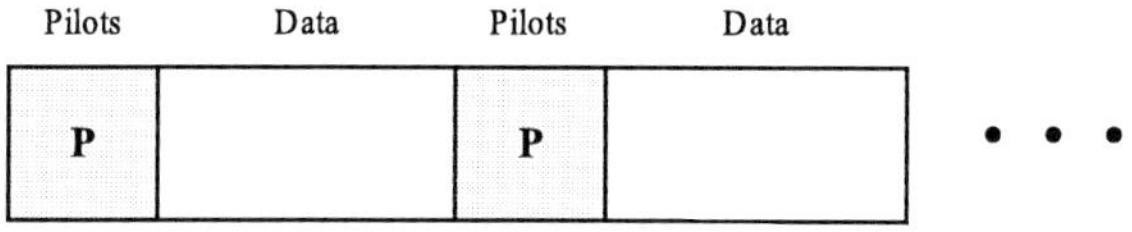

Fig. 6.4: Pilot-aided channel estimation in time-varying channel

6.3 SI-Aided Joint ML Channel Estimation and Data Detection

In this section, we propose the SI-aided joint ML channel estimation and data detection scheme that determines the channel and data symbols together. Such scheme assumes that the receiver has no *a priori* knowledge about the channel. The channel estimates and the detected data are obtained solely based on the received signals with the help of the known data symbols contained in the SI. No pilots are transmitted, and this scheme exploits the property that the transmitted symbols belong to finite alphabets.

6.3.1 Block-Fading Channel

Since the noise term $\mathbf{N}_{\mathsf{B}}$ in (6.5) is Gaussian, the probability density function (PDF) of the received signal $\mathbf{Y}_{\mathsf{B}}$ conditioned on the channel $\mathbf{H}_{\mathsf{B}}$ and the symbol matrix $\mathbf{S}_{\mathsf{A}}$ can be written as [181]

$$p(\mathbf{Y}_{\mathsf{B}} \mid \mathbf{H}_{\mathsf{B}}, \mathbf{S}_{\mathsf{A}}) = \frac{1}{(\pi\sigma_{\mathsf{B}}^2)^{N_{\mathsf{B}}L}} \exp\left(-\frac{1}{\sigma_{\mathsf{B}}^2}\|\mathbf{Y}_{\mathsf{B}} - \mathbf{H}_{\mathsf{B}}(\mathbf{S}_{\mathsf{A}} + \mathbf{S}_{\mathsf{B}})\|_F^2\right). \quad (6.16)$$

The ML estimation of the channel $\mathbf{H}_{\mathsf{B}}$ and the symbol matrix $\mathbf{S}_{\mathsf{A}}$ can be obtained by maximizing $p(\mathbf{Y}_{\mathsf{B}} \mid \mathbf{H}_{\mathsf{B}}, \mathbf{S}_{\mathsf{A}})$ over all possible $\mathbf{S}_{\mathsf{A}}$ and $\mathbf{H}_{\mathsf{B}}$ jointly. That is, we want to find the channel matrix $\mathbf{H}_{\mathsf{B}}$ and the symbol matrix $\mathbf{S}_{\mathsf{A}}$ that optimally solve the following minimization problem

$$\min_{\mathbf{S}_{\mathsf{A}},\mathbf{H}_{\mathsf{B}}} \|\mathbf{Y}_{\mathsf{B}} - \mathbf{H}_{\mathsf{B}}(\mathbf{S}_{\mathsf{A}} + \mathbf{S}_{\mathsf{B}})\|_F^2. \quad (6.17)$$

Since $\mathbf{S}_{\mathsf{A}}$ is the modulated transmit symbol matrix, it belongs to a finite discrete symbol space depending on its modulation scheme and the coherence interval, while the channel matrix $\mathbf{H}_{\mathsf{B}}$ is chosen by Nature and is without constraint. The problem (6.17) is a least squares problem in $\mathbf{H}_{\mathsf{B}}$ and an *integer least-squares* problem in $\mathbf{S}_{\mathsf{A}} + \mathbf{S}_{\mathsf{B}}$ [33]. Since $\mathbf{H}_{\mathsf{B}}$ and $\mathbf{S}_{\mathsf{A}} + \mathbf{S}_{\mathsf{B}}$

are independent of each other, the problem (6.17) can be equivalently expressed as [8]

$$\min_{\mathbf{S}_\mathsf{A}} \left\{ \min_{\mathbf{H}_\mathsf{B}} \|\mathbf{Y}_\mathsf{B} - \mathbf{H}_\mathsf{B}(\mathbf{S}_\mathsf{A} + \mathbf{S}_\mathsf{B})\|_F^2 \right\}. \tag{6.18}$$

In order to solve the inner minimization, we reformulate the objective function as

$$\|\mathbf{Y}_\mathsf{B} - \mathbf{H}_\mathsf{B}(\mathbf{S}_\mathsf{A} + \mathbf{S}_\mathsf{B})\|_F^2 = \| \operatorname{vec}\left[\mathbf{Y}_\mathsf{B} - \mathbf{H}_\mathsf{B}(\mathbf{S}_\mathsf{A} + \mathbf{S}_\mathsf{B})\right] \|^2 \tag{6.19}$$

$$= \left\|\operatorname{vec}(\mathbf{Y}_\mathsf{B}) - \left((\mathbf{S}_\mathsf{A} + \mathbf{S}_\mathsf{B})^T \otimes \mathbf{I}_{N_\mathsf{B}}\right) \operatorname{vec}(\mathbf{H}_\mathsf{B})\right\|^2, \tag{6.20}$$

For each possible $\mathbf{S}_\mathsf{A}$, the corresponding $\hat{\mathbf{H}}_\mathsf{B}$ that minimizes (6.20) is given by

$$\operatorname{vec}(\hat{\mathbf{H}}_\mathsf{B}) = \left((\mathbf{S}_\mathsf{A} + \mathbf{S}_\mathsf{B})^T \otimes \mathbf{I}_{N_\mathsf{B}}\right)^\dagger \operatorname{vec}(\mathbf{Y}_\mathsf{B}). \tag{6.21}$$

By enumerating all possible $\mathbf{S}_\mathsf{A}$ and computing the value of the objective function in (6.18) for each $\mathbf{S}_\mathsf{A}$ and its corresponding $\hat{\mathbf{H}}_\mathsf{B}$ obtained by (6.21), we can find the pair $(\hat{\mathbf{S}}_\mathsf{A}, \hat{\mathbf{H}}_\mathsf{B})$ that jointly minimizes the objective function of (6.18). Since each entry of $\mathbf{S}_\mathsf{A}$ uses M_A-ary QAM or PSK modulation, the joint ML detector must search through all $M_\mathsf{A}^{N_\mathsf{R}L}$ possible $\mathbf{S}_\mathsf{A}$ candidates, which grows exponentially with the coherence interval L. In addition, each calculation of (6.21) involves matrix manipulation of size $N_\mathsf{B}L \times N_\mathsf{B}N_\mathsf{R}$. The computational complexity may be prohibitive when L is large.

6.3.1.1 Unique Identifiability

Compared to semi-blind estimation schemes, the proposed scheme does not transmit pilots but utilizes the SI in the received signal. In traditional blind channel estimation schemes, the channel and the transmitted data symbols cannot be uniquely identified for conventional modulation schemes [11, 29], and some pilots are still required to be transmitted in order to uniquely determine the channel. For example, suppose only $\mathbf{S}_\mathsf{A}$ is transmitted at the relay and each element of it is modulated using BPSK modulation. For its corresponding channel estimate $\hat{\mathbf{H}}_\mathsf{B}$, both $(\mathbf{S}_\mathsf{A}, \hat{\mathbf{H}}_\mathsf{B})$ and $(-\mathbf{S}_\mathsf{A}, -\hat{\mathbf{H}}_\mathsf{B})$ are possible candidate pairs. This is because

$$\hat{\mathbf{H}}_\mathsf{B}\mathbf{S}_\mathsf{A} = (-\hat{\mathbf{H}}_\mathsf{B})(-\mathbf{S}_\mathsf{A}) \tag{6.22}$$

and they produce the same value for the log-likelihood function $\|\mathbf{Y}_\mathsf{B} - \hat{\mathbf{H}}_\mathsf{B}\mathbf{S}_\mathsf{A}\|_F^2$. This ambiguity problem cannot be solved unless additional knowledge is available. Unlike those schemes, the known symbol $\mathbf{S}_\mathsf{B}$ contained in the SI implicitly helps to resolve the ambiguity problem, because the known data symbol $\mathbf{S}_\mathsf{B}$ in our proposed scheme offers the phase reference for the data symbols to be detected. Moreover, we have the following lemma:

Lemma 6.3.1. *The data symbol matrix* $\mathbf{S}_A$ *and the corresponding channel* $\mathbf{H}_B$ *are uniquely identifiable unless there exists an invertible matrix* $\mathbf{P} \in \mathbb{C}^{N_R \times N_R}$ *and another data symbol matrix candidate* $\breve{\mathbf{S}}_A$ *such that* $\mathbf{P}(\mathbf{S}_A + \mathbf{S}_B) = \breve{\mathbf{S}}_A + \mathbf{S}_B$*, i.e.,* $(\mathbf{P} - \mathbf{I}_{N_R})\mathbf{S}_B + \mathbf{P}\mathbf{S}_A = \breve{\mathbf{S}}_A$.

In some special cases, e.g., all the entries of $\mathbf{S}_\mathrm{A}$ contain the same data symbol and so do all the entries of $\mathbf{S}_\mathrm{B}$, the data symbol matrix $\mathbf{S}_\mathrm{A}$ is not uniquely identifiable. However, due to interleaving, the probability that such cases happen is very low when the coherence time is long enough.

6.3.1.2 Cramér-Rao Lower Bound for Joint Channel and Data Symbol Estimation With SI

We denote $\mathbf{s} = \mathrm{vec}(\mathbf{S}_\mathrm{A})$ and $\mathbf{h} = \mathrm{vec}(\mathbf{H}_\mathrm{B})$. Furthermore, we define $\boldsymbol{\theta} = [\mathbf{s}^T, \mathbf{h}^T]^T$ as the complex random parameter vector to be estimated. The Cramér-Rao lower bound on the channel estimate MSE for unbiased estimators that jointly estimate the channel and data symbols according to the system model (6.5) is summarized by the following lemma:

Lemma 6.3.2. *Without loss of generality, we assume the channel to be uncorrelated Rayleigh fading, i.e., each entry of* $\mathbf{H}_B$ *is an independent and identically distributed (i.i.d.)* $\mathcal{CN}(0,1)$ *random variable. For any unbiased joint channel and data symbol estimator with the channel output* $\hat{\mathbf{h}}$*, we have*

$$\frac{\mathrm{E}\left(\|\mathbf{h} - \hat{\mathbf{h}}\|^2\right)}{N_B N_R} \geq \left(1 + \frac{P_R L}{N_R \sigma_B^2}\right)^{-1}. \tag{6.23}$$

Proof. See Appendix 6.8.1. □

6.3.2 Time-Varying Channel

Following the same discussions on the block-fading channel, we can derive the joint ML channel estimation and data detection schemes for the time-varying channel. According to (6.15), the ML estimation of the channel $\boldsymbol{\mathcal{H}}_\mathrm{B}$ and the transmitted symbol $\boldsymbol{\mathcal{S}}_\mathrm{A}$ can be obtained by solving the following problem

$$\min_{\boldsymbol{\mathcal{S}}_\mathrm{A}, \boldsymbol{\mathcal{H}}_\mathrm{B}} \|\boldsymbol{\mathcal{Y}}_\mathrm{B} - (\boldsymbol{\mathcal{S}}_\mathrm{A} + \boldsymbol{\mathcal{S}}_\mathrm{B})\boldsymbol{\mathcal{H}}_\mathrm{B}\|_F^2. \tag{6.24}$$

The objective function of (6.24) can be written as

$$\|\mathcal{Y}_\mathsf{B} - \mathcal{S}_\mathsf{R}\mathcal{H}_\mathsf{B}\|_F^2 = \|\operatorname{vec}(\mathcal{Y}_\mathsf{B}) - (\mathbf{I}_{N_\mathsf{B}} \otimes (\mathcal{S}_\mathsf{A} + \mathcal{S}_\mathsf{B}))\operatorname{vec}(\mathcal{H}_\mathsf{B})\|^2. \tag{6.25}$$

The difference in the expression (6.25) and (6.20) is due to the different system modeling in (6.14) and (6.4). Furthermore, the size of matrix $\mathbf{I}_{N_\mathsf{B}} \otimes (\mathcal{S}_\mathsf{A} + \mathcal{S}_\mathsf{B})$ in (6.25) is $N_\mathsf{B}K \times N_\mathsf{B}N_\mathsf{R}K$. Since $N_\mathsf{B}K < N_\mathsf{B}N_\mathsf{R}K$, there are multiple solutions of $\mathcal{H}_\mathsf{B}$ that minimizes (6.25) for each given $\mathcal{S}_\mathsf{A}$. However, any of those solutions for the given $\mathcal{S}_\mathsf{A}$ produces the same value for the objective function in (6.24). So we only need to test the following solution

$$\operatorname{vec}(\hat{\mathcal{H}}_\mathsf{B}) = (\mathbf{I}_{N_\mathsf{B}} \otimes (\mathcal{S}_\mathsf{A} + \mathcal{S}_\mathsf{B}))^\dagger \operatorname{vec}(\mathcal{Y}_\mathsf{B}). \tag{6.26}$$

By enumerating all possible $\mathcal{S}_\mathsf{A}$, we can determine data symbol matrix $\mathcal{S}_\mathsf{A}$ that solves (6.24). The number of $\mathcal{S}_\mathsf{A}$ candidates is $M_\mathsf{A}^{N_\mathsf{R}K}$, which grows exponentially with the frame length K.

6.4 SI-Aided Linear Channel Estimator

If the coherence interval L or the frame length K is large, the computational complexity of the SI-aided joint ML detector proposed in Section 6.3 is prohibitive. In reality, practical receiver structures usually separate channel estimation and data detection. The most commonly used method to obtain channel knowledge is by transmitting pilots, which is called *pilot-aided channel estimation*. However, pilots consume system resources. In this section, we propose the SI-aided linear channel estimator that exploits SI for channel estimation without using pilots. Moreover, we provide its performance analysis, and compare it with that of the pilot-aided channel estimator. Taking into account the SI-aided channel estimation errors, we also propose an optimized power allocation of the data symbols at the relay to provide fairness for the different receiving stations in the BRC phase.

6.4.1 SI-Aided Linear Channel Estimation Algorithm

Linear channel estimation is widely adopted due to its simplicity. The least square (LS) and the linear minimum mean square error (LMMSE) algorithms are the two most commonly used linear channel estimation algorithms [26]. We only discuss the LMMSE channel estimation algorithm due to its better performance. For block-fading channels, the linear estimate of the channel based on the received signal matrix $\mathbf{Y}_\mathsf{B}$ in (6.5) can be obtained as

$$\mathbf{H}_{\mathsf{B},\mathrm{lin}} = \mathbf{Y}_\mathsf{B}\mathbf{W} \tag{6.27}$$

where $\mathbf{W} \in \mathbb{C}^{L\times N_\mathsf{R}}$. The LMMSE channel estimator is the optimal linear estimator that minimizes the following objective function J_{mse} [129]:

$$\begin{aligned} J_{\mathrm{mse}} &= \mathrm{E}\left(\|\mathbf{H}_\mathsf{B} - \mathbf{Y}_\mathsf{B}\mathbf{W}\|_F^2\right) \\ &= \mathrm{tr}(\mathbf{R}_\mathbf{H}) - \mathrm{tr}(\mathbf{R}_\mathbf{H}\mathbf{S}_\mathsf{B}\mathbf{W}) - \mathrm{tr}(\mathbf{W}^H\mathbf{S}_\mathsf{B}^H\mathbf{R}_\mathbf{H}) \\ &\quad + \mathrm{tr}\left(\mathbf{W}^H(\mathbf{S}_\mathsf{B}^H\mathbf{R}_\mathbf{H}\mathbf{S}_\mathsf{B} + \mathbf{R}_{\mathbf{HS}_\mathsf{A}} + \sigma_\mathsf{B}^2 N_\mathsf{B}\mathbf{I}_L)\mathbf{W}\right) \end{aligned} \tag{6.28}$$

where $\mathbf{R}_\mathbf{H} = \mathrm{E}\left(\mathbf{H}_\mathsf{B}^H\mathbf{H}_\mathsf{B}\right)$ and $\mathbf{R}_{\mathbf{HS}_\mathsf{A}} = \mathrm{E}\left(\mathbf{S}_\mathsf{A}^H\mathbf{H}_\mathsf{B}^H\mathbf{H}_\mathsf{B}\mathbf{S}_\mathsf{A}\right)$. Here we utilized the fact that $\mathbf{S}_\mathsf{A}$ and $\mathbf{N}_\mathsf{B}$ are independent with the channel $\mathbf{H}_\mathsf{B}$ and both have zero-mean. For spatially correlated channels, $\mathbf{R}_\mathbf{H}$ and $\mathbf{R}_{\mathbf{HS}_\mathsf{A}}$ contain the channel correlation information.

The optimum solution that minimizes J_{mse} is $\mathbf{W}^\star = \arg\min_\mathbf{W} J_{\mathrm{mse}}$, and it can be calculated by setting $\partial J_{\mathrm{mse}}/\partial\mathbf{W} = 0$. The solution can be explicitly expressed as

$$\mathbf{W}^\star = \left(\mathbf{S}_\mathsf{B}^H\mathbf{R}_\mathbf{H}\mathbf{S}_\mathsf{B} + \mathbf{R}_{\mathbf{HS}_\mathsf{A}} + \sigma_\mathsf{B}^2 N_\mathsf{B}\mathbf{I}_L\right)^{-1}\mathbf{S}_\mathsf{B}^H\mathbf{R}_\mathbf{H}. \tag{6.29}$$

For notational coherence with following sections, the LMMSE channel estimate is denoted as

$$\hat{\mathbf{H}}_\mathsf{B}^{(1)} = \mathbf{Y}_\mathsf{B}\left(\mathbf{S}_\mathsf{B}^H\mathbf{R}_\mathbf{H}\mathbf{S}_\mathsf{B} + \mathbf{R}_{\mathbf{HS}_\mathsf{A}} + \sigma_\mathsf{B}^2 N_\mathsf{B}\mathbf{I}_L\right)^{-1}\mathbf{S}_\mathsf{B}^H\mathbf{R}_\mathbf{H}. \tag{6.30}$$

The LMMSE channel estimator only utilizes the first and second order statistical knowledge of the channel, where in (6.30) the second-order statistical information of $\mathbf{H}_\mathsf{B}$ and $\mathbf{H}_\mathsf{B}\mathbf{S}_\mathsf{A}$ is contained in $\mathbf{R}_\mathbf{H}$ and $\mathbf{R}_{\mathbf{HS}_\mathsf{A}}$, respectively.

Similarly, the SI-aided LMMSE channel estimate in time-varying channel (6.15) is obtained as

$$\hat{\mathcal{H}}_\mathsf{B}^{(1)} = \mathrm{R}_\mathcal{H}\mathcal{S}_\mathsf{B}^H\left(\mathcal{S}_\mathsf{B}\mathrm{R}_\mathcal{H}\mathcal{S}_\mathsf{B}^H + \mathrm{R}_{\mathcal{S}_\mathsf{A}\mathcal{H}} + \sigma_\mathsf{B}^2 N_\mathsf{B}\mathbf{I}_K\right)^{-1}\mathcal{Y}_\mathsf{B} \tag{6.31}$$

where $\mathrm{R}_\mathcal{H} = \mathrm{E}\left(\mathcal{H}_\mathsf{B}\mathcal{H}_\mathsf{B}^H\right)$ and $\mathrm{R}_{\mathcal{S}_\mathsf{A}\mathcal{H}} = \mathrm{E}\left(\mathcal{S}_\mathsf{A}\mathcal{H}_\mathsf{B}\mathcal{I}\mathcal{L}_\mathsf{B}^H\mathcal{S}_\mathsf{A}^H\right)$.

6.4.2 MSE of the Channel Estimate

The MSE is a key performance measure of channel estimates. Due to space constraints, we only discuss the MSE for block-fading channels here. We assume each element of $\mathbf{H}_\mathsf{B}$ to be an i.i.d. $\mathcal{CN}(0,1)$ random variable, and define $\mathbf{\Phi}_{\mathbf{H}_\mathsf{B}} = \mathrm{E}\left((\mathrm{vec}\,\mathbf{H}_\mathsf{B})(\mathrm{vec}\,\mathbf{H}_\mathsf{B})^H\right) = \mathbf{I}_{N_\mathsf{R}} \otimes \mathbf{I}_{N_\mathsf{B}}$.

The LMMSE channel estimation error is $\tilde{\mathbf{H}}_\mathsf{B} = \mathbf{H}_\mathsf{B} - \hat{\mathbf{H}}_\mathsf{B}^{(1)}$, and the error covariance matrix is defined as $\mathbf{\Phi}_{\tilde{\mathbf{H}}_\mathsf{B}} = \mathrm{E}\left((\mathrm{vec}\,\tilde{\mathbf{H}}_\mathsf{B})(\mathrm{vec}\,\tilde{\mathbf{H}}_\mathsf{B})^H\right)$. The MSE of the estimated channel is $\sigma^2_{\tilde{\mathbf{H}}_\mathsf{B}} = \mathrm{tr}(\mathbf{\Phi}_{\tilde{\mathbf{H}}_\mathsf{B}})/(N_\mathsf{R} N_\mathsf{B})$.

The covariance matrix $\mathbf{\Phi}_{\tilde{\mathbf{H}}_\mathsf{B}}$ can be calculated according to [129] as

$$\mathbf{\Phi}_{\tilde{\mathbf{H}}_\mathsf{B}} = \mathrm{E}_{\mathbf{S}_\mathsf{B},\mathbf{H}_\mathsf{B}}\left((\mathrm{vec}\,\tilde{\mathbf{H}}_\mathsf{B})(\mathrm{vec}\,\tilde{\mathbf{H}}_\mathsf{B})^H\right) \tag{6.32}$$

$$= \mathrm{E}_{\mathbf{S}_\mathsf{B}}\left(\left(\mathbf{\Phi}_{\mathbf{H}_\mathsf{B}}^{-1} + (\mathbf{S}_\mathsf{B} \otimes \mathbf{I}_{N_\mathsf{B}})\mathbf{\Phi}_{\mathbf{V}}^{-1}(\mathbf{S}_\mathsf{B}^H \otimes \mathbf{I}_{N_\mathsf{B}})\right)^{-1}\right) \tag{6.33}$$

$$= \mathrm{E}_{\mathbf{S}_\mathsf{B}}\left(\left(\mathbf{I}_{N_\mathsf{R}} + \frac{1}{P_\mathsf{A} + \sigma_\mathsf{B}^2}\mathbf{S}_\mathsf{B}\mathbf{S}_\mathsf{B}^H\right)^{-1}\right) \otimes \mathbf{I}_{N_\mathsf{B}}. \tag{6.34}$$

In (6.32), we emphasize that the expectation is taken with respect to the distribution of $\mathbf{S}_\mathsf{B}$ and $\mathbf{H}_\mathsf{B}$. Since they are independent, we can first take expectations with respect to $\mathbf{H}_\mathsf{B}$ and obtain (6.33). To obtain (6.34), we used $\mathbf{\Phi}_\mathbf{V} = \mathrm{E}\left((\mathrm{vec}\,\mathbf{V})(\mathrm{vec}\,\mathbf{V})^H\right) = \mathrm{E}\left(\mathbf{S}_\mathsf{A}^T\mathbf{S}_\mathsf{A}^*\right) \otimes \mathbf{I}_{N_\mathsf{B}} + \mathrm{E}\left((\mathrm{vec}\,\mathbf{N}_\mathsf{B})(\mathrm{vec}\,\mathbf{N}_\mathsf{B})^H\right) = (P_\mathsf{A} + \sigma_\mathsf{B}^2)\mathbf{I}_L \otimes \mathbf{I}_{N_\mathsf{B}}$ according to (6.5). In general, the covariance matrix $\mathbf{\Phi}_{\tilde{\mathbf{H}}_\mathsf{B}}$ depends on the distribution of $\mathbf{S}_\mathsf{B}$, and therefore depends on the modulation schemes. However, due to interleaving, each entry of $\mathbf{S}_\mathsf{B}$ is i.i.d. and has power $P_\mathsf{B}/N_\mathsf{R}$. According to the law of large numbers [181], we have

$$\lim_{L\to\infty} \frac{1}{L}\mathbf{S}_\mathsf{B}\mathbf{S}_\mathsf{B}^H = \frac{P_\mathsf{B}}{N_\mathsf{R}}\mathbf{I}_{N_\mathsf{R}}. \tag{6.35}$$

That is, when the coherence interval L is large enough, the covariance matrix $\mathbf{\Phi}_{\tilde{\mathbf{H}}_\mathsf{B}}$ is

$$\mathbf{\Phi}_{\tilde{\mathbf{H}}_\mathsf{B}} \overset{L\to\infty}{=} \left(\mathbf{I}_{N_\mathsf{R}} + \frac{P_\mathsf{B} L}{N_\mathsf{R}(P_\mathsf{A} + \sigma_\mathsf{B}^2)}\mathbf{I}_{N_\mathsf{R}}\right)^{-1} \otimes \mathbf{I}_{N_\mathsf{B}} = \left(1 + \frac{P_\mathsf{B} L}{N_\mathsf{R}(P_\mathsf{A} + \sigma_\mathsf{B}^2)}\right)^{-1} \mathbf{I}_{N_\mathsf{R}} \otimes \mathbf{I}_{N_\mathsf{B}}. \tag{6.36}$$

So the MSE of the channel estimates can be expressed as

$$\sigma^2_{\tilde{\mathbf{H}}_\mathsf{B}} = \left(1 + \frac{P_\mathsf{B} L}{N_\mathsf{R}(P_\mathsf{A} + \sigma_\mathsf{B}^2)}\right)^{-1}. \tag{6.37}$$

Simulations in Section 6.6.1 show that the value of L does not have to be very large in order to satisfy (6.37). For realistic modulation schemes and practical values of L, equation (6.37) matches well with the simulated MSE of the SI-aided LMMSE channel estimates.

6.4.3 Comparison With Pure Pilot-Aided Channel Estimation

Suppose the length of the pilot sequence is L_{p} in each coherence interval, the MSE of the pure pilot-aided channel estimates using the LMMSE algorithm can be expressed as [26]

$$\sigma_{\text{p}}^2 = \left(1 + \frac{P_{\mathsf{R}} L_{\text{p}}}{N_{\mathsf{R}} \sigma_{\mathsf{B}}^2}\right)^{-1}. \tag{6.38}$$

The MSE of the SI-aided LMMSE channel estimates in (6.37) is smaller than that of the pilot-aided channel estimates when $\sigma^2_{\tilde{\mathbf{H}}_{\mathsf{B}}} < \sigma_{\text{p}}^2$. That is, when

$$L_{\text{p}} < \frac{P_{\mathsf{B}} L}{P_{\mathsf{R}}(P_{\mathsf{A}}/\sigma_{\mathsf{B}}^2 + 1)}. \tag{6.39}$$

Interestingly, the condition (6.39) does not depend on the number of antennas N_{R} and N_{B}, and only depends on the power allocation and coherence interval L. Moreover, if L is long enough, the SI-aided channel estimate will eventually outperform the pure pilot-aided channel estimates.

6.4.4 Applications to Optimized Power Allocation at Relay

When the SI-aided channel estimation is applied, the power allocation P_{A} and P_{B} at the relay determines the decoding performance at stations A and B simultaneously. In order to find the power allocation that optimizes the decoding performance, we need to first calculate the effective SNRs at the receivers after SI cancelation.

After removing the SI from the received signal matrix, the remaining signal to be decoded can be expressed as

$$\bar{\mathbf{y}}_{\mathsf{B}} = \mathbf{y}_{\mathsf{B}} - \hat{\mathbf{H}}_{\mathsf{B}}^{(1)} \mathbf{s}_{\mathsf{B}} = \hat{\mathbf{H}}_{\mathsf{B}}^{(1)} \mathbf{s}_{\mathsf{A}} + \underbrace{\tilde{\mathbf{H}}_{\mathsf{B}} \mathbf{s}_{\mathsf{B}} + \tilde{\mathbf{H}}_{\mathsf{B}} \mathbf{s}_{\mathsf{A}} + \mathbf{n}_{\mathsf{B}}}_{\mathbf{n}}. \tag{6.40}$$

Equation (6.40) describes a system with a known channel $\hat{\mathbf{H}}_{\mathsf{B}}^{(1)}$ and noise $\mathbf{n}$. Since $\hat{\mathbf{H}}_{\mathsf{B}}^{(1)}$ is the LMMSE estimation of the channel $\mathbf{H}_{\mathsf{B}}$, the channel estimation error $\tilde{\mathbf{H}}_{\mathsf{B}}$ has zero-mean entries and is uncorrelated with $\hat{\mathbf{H}}_{\mathsf{B}}^{(1)}$. According to the orthogonality principle [129], the variance of each entry in $\hat{\mathbf{H}}_{\mathsf{B}}^{(1)}$ is $\sigma^2_{\hat{\mathbf{H}}_{\mathsf{B}}^{(1)}} = \sigma^2_{\mathbf{H}_{\mathsf{B}}} - \sigma^2_{\tilde{\mathbf{H}}_{\mathsf{B}}}$, where $\sigma^2_{\mathbf{H}_{\mathsf{B}}} = 1$ is the variance of each entry in $\mathbf{H}_\mathsf{B}$. The variance of the noise term $\mathbf{n}$ in (6.40) can be calculated as

$$\sigma_\mathbf{n}^2 = \frac{1}{N_\mathsf{B}} \operatorname{tr} \mathrm{E}\left(\mathbf{n}\mathbf{n}^H\right) \tag{6.41}$$

$$\approx \frac{1}{N_\mathsf{B}} \operatorname{tr}\left(\mathrm{E}\left(\tilde{\mathbf{H}}_\mathsf{B}^H \tilde{\mathbf{H}}_\mathsf{B}\right) \mathrm{E}\left(\mathbf{s}_\mathsf{B}\mathbf{s}_\mathsf{B}^H\right)\right) + \frac{1}{N_\mathsf{B}} \operatorname{tr}\left(\mathrm{E}\left(\tilde{\mathbf{H}}_\mathsf{B}^H \tilde{\mathbf{H}}_\mathsf{B}\right) \mathrm{E}\left(\mathbf{s}_\mathsf{A}\mathbf{s}_\mathsf{A}^H\right)\right) + \sigma_\mathsf{B}^2 \tag{6.42}$$

$$= (P_\mathsf{B} + P_\mathsf{A})\sigma_{\tilde{\mathbf{H}}_\mathsf{B}}^2 + \sigma_\mathsf{B}^2 \tag{6.43}$$

where $\sigma_{\tilde{\mathbf{H}}_\mathsf{B}}^2$ in (6.43) can be calculated according to (6.37). Here we used the fact that the entries in $\mathbf{s}_\mathsf{B}$ and $\mathbf{s}_\mathsf{A}$ are uncorrelated. The approximation in (6.42) is because $\mathbf{s}_\mathsf{B}$ and $\mathbf{s}_\mathsf{A}$ contribute to the channel estimate. However, the correlation between $\tilde{\mathbf{H}}_\mathsf{B}$ and $\mathbf{s}_\mathsf{B}$ (and $\mathbf{s}_\mathsf{A}$) is very small when L is large, so we neglect the approximation errors and treat (6.43) as the accurate value of $\sigma_\mathbf{n}^2$ in the following derivations of the *approximate* effective SNR.

Treating $\mathbf{n}$ as Gaussian noise in (6.40), the effective SNR at station B can be calculated as

$$\mathrm{SNR}_{\mathsf{B,eff}} = \frac{\sigma_{\hat{\mathbf{H}}_\mathsf{B}^{(1)}}^2 P_\mathsf{A}}{\sigma_\mathbf{n}^2} = \frac{(1 - \sigma_{\tilde{\mathbf{H}}_\mathsf{B}}^2) P_\mathsf{A}}{(P_\mathsf{B} + P_\mathsf{A})\sigma_{\tilde{\mathbf{H}}_\mathsf{B}}^2 + \sigma_\mathsf{B}^2}. \tag{6.44}$$

Similarly, the effective SNR at station A can be expressed as

$$\mathrm{SNR}_{\mathsf{A,eff}} = \frac{(1 - \sigma_{\tilde{\mathbf{H}}_\mathsf{A}}^2) P_\mathsf{B}}{(P_\mathsf{B} + P_\mathsf{A})\sigma_{\tilde{\mathbf{H}}_\mathsf{A}}^2 + \sigma_\mathsf{A}^2}. \tag{6.45}$$

Since the performance at stations A and B is equally important to us, we choose *fairness* as the optimization criterion, i.e., we find P_A and P_B that maximize the minimum of the bit-error rate (BER) performance stations A and B:

$$\begin{aligned} &\text{maximize} \quad \min(\mathrm{BER}_\mathsf{A}, \mathrm{BER}_\mathsf{B}) \\ &\text{subject to} \quad P_\mathsf{B} + P_\mathsf{A} = P_\mathsf{R} \end{aligned} \tag{6.46}$$

The BER performance is a monotonically decreasing function of the effective SNR, i.e., $\mathrm{BER}_\mathsf{A} = Q_\mathsf{A}(\mathrm{SNR}_{\mathsf{A,eff}})$ and $\mathrm{BER}_\mathsf{B} = Q_\mathsf{B}(\mathrm{SNR}_{\mathsf{B,eff}})$. This problem cannot be solved analytically due to the function $\min$ in (6.46). However, if the analytical expressions of $Q_\mathsf{A}(\cdot)$ and $Q_\mathsf{B}(\cdot)$ are known, numerical methods, such as bisection method [33], can be applied to solve this problem since there is actually only one variable in the optimization problem (6.46). When the same modulation and coding schemes are applied to stations A and B, we have $Q_\mathsf{A}(\cdot) = Q_\mathsf{B}(\cdot)$, and optimizing BER performance is equivalent to optimizing the effective SNR. The optimization problem (6.46) cna be further simplified to finding the power alloca-

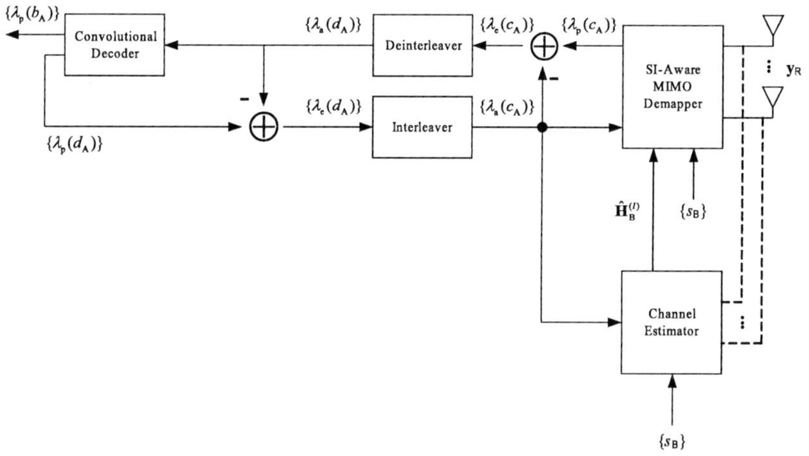

Fig. 6.5: Iterative receiver structure for channel estimation and data detection at station B.

tion P_{A} and P_{B} that maximize the minimum of the effective SNRs of stations A and B:

$$\begin{array}{ll} \text{maximize} & \min(\mathrm{SNR}_{\mathsf{B,eff}}, \mathrm{SNR}_{\mathsf{A,eff}}) \\ \text{subject to} & P_{\mathsf{B}} + P_{\mathsf{A}} = P_{\mathsf{R}} \end{array} \tag{6.47}$$

Similarly, the bisection method can be applied to solve the problem (6.47).

6.5 SI-Aided Iterative Receiver for Channel Estimation and Data Detection

In order to approach the capacity of the MIMO channel, channel coding is an integral part of MIMO transmission systems. Joint data detection and channel decoding is computationally infeasible and has to be approximated by separate iterative MIMO symbol demapping and channel decoding [107]. When channel estimation is considered, a channel estimator must be included in the receiver structure, which, together with the MIMO demapper, forms the "inner decoder" of the receiver. When pilot-aided channel estimation is applied, the channel estimator generates the channel estimates based on the pilot symbols and the feedback from the channel decoder, and pass it on to the MIMO demapper [31]. Although an optimum receiver design requires joint channel estimation and data demapping [225], the complexity of such method is very high. Compared to the separate and iterative channel estimation scheme, the joint channel estimation and data demapping scheme achieves almost the same performance unless the Doppler shift is very large [196]. Such a structure that separates channel estimation and data detection is one of the most widely used practical receiver structures [228]. Moreover, when the initial channel estimate is already available, the subsequent iterations of channel estimation and decoding is optional depending on the performance requirements at the receiver.

Similar to the pilot-aided iterative receiver structure, we can propose the SI-aided iterative receiver structure. The SI-aided iterative receiver structure for channel estimation and data detection obtains the initial channel knowledge with the assistance of the known data symbols in SI and subsequently improves the channel estimates by the decoded data. The SI-aided channel estimate obtained in Section 6.4 can be further improved utilizing the data symbol estimates fed back from the decoder. This leads to the iterative receiver structure where the SI-aided linear channel estimate serves as the initial channel estimate. The iterative receiver is composed of the channel estimator, the MIMO symbol demapper and the channel decoder. Such decomposition enables the receiver to achieve near-optimum performance [196] while allowing low-complexity implementation of each component. Corresponding to the transmitter structure in Fig. 6.2, the block diagram for the iterative receiver structure at station B is shown in Fig. 6.5. Compared to conventional iterative receivers with pilot-aided channel estimation, only minor modification is needed, except for the channel estimator, to construct the SI-aided iterative receivers. In particular, the impact of the SI has to be canceled by the MIMO demapper in the process of demapping. We consider a coded system with convolutional codes. Other channel codes can also be applied straightforwardly. In this section, we discuss each component of the iterative receiver in detail, and summarize the working process of the receiver in the end.

6.5.1 Channel Estimator

Since the initial SI-aided LMMSE channel estimation and its performance have been discussed in Section 6.4, we only discuss the decision-directed channel estimation in subsequent iterations. After the first iteration of decoding, the channel decoder feeds back the soft information that represents the reliability for each coded bit. Taking into account the soft feedback, the unknown data symbols can be reconstructed. Following similar discussions as in Section 6.4.1, the block-fading channel in (6.5) can be re-estimated in the lth iteration

$(l \geq 2)$ as:

$$\hat{\mathbf{H}}_{\mathsf{B}}^{(l)} = \mathbf{Y}_{\mathsf{B}} \left(\mathrm{E}\left(\mathbf{S}_{\mathsf{R}}^H \mathbf{R}_{\mathbf{H}} \mathbf{S}_{\mathsf{R}} \right) + \sigma_{\mathsf{B}}^2 N_{\mathsf{B}} \mathbf{I}_L \right)^{-1} \mathrm{E}\left(\mathbf{S}_{\mathsf{R}}^H \right) \mathbf{R}_{\mathbf{H}}. \tag{6.48}$$

Suppose each element s_{A} of $\mathbf{S}_{\mathsf{A}}$ is modulated within the M_{A}-ary modulation constellation set $\{S_1, \cdots, S_{M_{\mathsf{A}}}\}$, we define $\mathrm{E}(s_{\mathsf{A}}) = \sum_{m=1}^{M_{\mathsf{A}}} S_m P(s_{\mathsf{A}} = S_m)$ and $\mathrm{E}(|s_{\mathsf{A}}|^2) = \sum_{m=1}^{M_{\mathsf{A}}} |S_m|^2 P(s_{\mathsf{A}} = S_m)$. The *a priori* symbol probability $P(s_{\mathsf{A}} = S_m)$ is based on the feedback from the channel decoder. For matrices in (6.48), $\mathrm{E}(\cdot)$ can be calculated element-wise. Moreover, $\mathrm{E}(\mathbf{S}_{\mathsf{R}}) = \mathrm{E}(\mathbf{S}_{\mathsf{A}}) + \mathbf{S}_{\mathsf{B}}$, and

$$\mathrm{E}\left(\mathbf{S}_{\mathsf{R}}^H \mathbf{R}_{\mathbf{H}} \mathbf{S}_{\mathsf{R}} \right) = \mathrm{E}\left((\mathbf{S}_{\mathsf{A}} + \mathbf{S}_{\mathsf{B}})^H \mathbf{R}_{\mathbf{H}} (\mathbf{S}_{\mathsf{A}} + \mathbf{S}_{\mathsf{B}}) \right) \tag{6.49}$$

$$= \mathbf{S}_{\mathsf{B}}^H \mathbf{R}_{\mathbf{H}} \mathbf{S}_{\mathsf{B}} + \mathrm{E}\left(\mathbf{S}_{\mathsf{A}}^H \mathbf{R}_{\mathbf{H}} \mathbf{S}_{\mathsf{A}} \right) + \mathbf{S}_{\mathsf{B}}^H \mathbf{R}_{\mathbf{H}} \mathrm{E}(\mathbf{S}_{\mathsf{A}}) + \mathrm{E}(\mathbf{S}_{\mathsf{A}}^H) \mathbf{R}_{\mathbf{H}} \mathbf{S}_{\mathsf{B}}. \tag{6.50}$$

We denote $[\mathbf{X}]_{ij}$ as the entry in the ith row and jth column of the matrix $\mathbf{X}$, and $[\mathbf{X}]_i$ is the ith column of the matrix $\mathbf{X}$. Then we have

$$\left[\mathrm{E}\left(\mathbf{S}_{\mathsf{A}}^H \mathbf{R}_{\mathbf{H}} \mathbf{S}_{\mathsf{A}} \right) \right]_{ij} = \begin{cases} N_{\mathsf{B}} \mathrm{E}\left(\sum_k |[\mathbf{S}_{\mathsf{A}}]_{ki}|^2 \right), & \text{for } i = j, \\ \mathrm{E}([\mathbf{S}_{\mathsf{A}}]_i^H) \mathbf{R}_{\mathbf{H}} \mathrm{E}([\mathbf{S}_{\mathsf{A}}]_j), & \text{for } i \neq j, \end{cases} \tag{6.51}$$

where we utilized the fact that the transmitted symbols at different time slots are independent.

Similarly, for the time-varying channel, we can obtain the lth iteration channel estimate as

$$\hat{\boldsymbol{\mathcal{H}}}_{\mathsf{B}}^{(l)} = \mathbf{R}_{\mathcal{H}} \mathrm{E}(\boldsymbol{\mathcal{S}}_{\mathsf{R}}^H) \left(\mathrm{E}(\boldsymbol{\mathcal{S}}_{\mathsf{R}} \mathbf{R}_{\mathcal{H}} \boldsymbol{\mathcal{S}}_{\mathsf{R}}^H) + \sigma_{\mathsf{B}}^2 N_{\mathsf{B}} \mathbf{I}_K \right)^{-1} \boldsymbol{\mathcal{Y}}_{\mathsf{B}} \tag{6.52}$$

where $l \geq 2$, and the term $\mathrm{E}(\boldsymbol{\mathcal{S}}_{\mathsf{R}} \mathbf{R}_{\mathcal{H}} \boldsymbol{\mathcal{S}}_{\mathsf{R}}^H)$ can be calculated as follows:

$$\mathrm{E}\left(\boldsymbol{\mathcal{S}}_{\mathsf{R}}^H \mathbf{R}_{\mathcal{H}} \boldsymbol{\mathcal{S}}_{\mathsf{R}} \right) = \boldsymbol{\mathcal{S}}_{\mathsf{B}}^H \mathbf{R}_{\mathcal{H}} \boldsymbol{\mathcal{S}}_{\mathsf{B}} + \mathrm{E}\left(\boldsymbol{\mathcal{S}}_{\mathsf{A}}^H \mathbf{R}_{\mathcal{H}} \boldsymbol{\mathcal{S}}_{\mathsf{A}} \right) + \boldsymbol{\mathcal{S}}_{\mathsf{B}}^H \mathbf{R}_{\mathcal{H}} \mathrm{E}(\boldsymbol{\mathcal{S}}_{\mathsf{A}}) + \mathrm{E}(\boldsymbol{\mathcal{S}}_{\mathsf{A}}^H) \mathbf{R}_{\mathcal{H}} \boldsymbol{\mathcal{S}}_{\mathsf{B}}. \tag{6.53}$$

The entry in the ith row and the jth column of the term $\mathrm{E}\left(\boldsymbol{\mathcal{S}}_{\mathsf{A}}^H \mathbf{R}_{\mathcal{H}} \boldsymbol{\mathcal{S}}_{\mathsf{A}} \right)$ can be calculated as

$$\left[\mathrm{E}\left(\boldsymbol{\mathcal{S}}_{\mathsf{A}}^H \mathbf{R}_{\mathcal{H}} \boldsymbol{\mathcal{S}}_{\mathsf{A}} \right) \right]_{ij} = \begin{cases} \mathrm{E}\left(\sum_k |[\boldsymbol{\mathcal{S}}_{\mathsf{A}}]_{ki}|^2 \right), & \text{for } i = j, \\ \mathrm{E}([\boldsymbol{\mathcal{S}}_{\mathsf{A}}]_i^H) \mathbf{R}_{\mathcal{H}} \mathrm{E}([\boldsymbol{\mathcal{S}}_{\mathsf{A}}]_j), & \text{for } i \neq j. \end{cases} \tag{6.54}$$

6.5.2 SI-Aware MIMO Demapper

In the lth iteration demapping $(l \geq 1)$, the soft-output MIMO demapper accepts the channel estimate $\hat{\mathbf{H}}_{\mathsf{B},k}^{(l)}$ from the channel estimator and generates the log-likelihood ratio (LLR) values for each coded bits upon receiving $\mathbf{y}_{\mathsf{B},k}$ according to (6.3). For simplicity, we omit the time indices and only discuss the optimum maximum *a posteriori* probability (MAP) demapper in detail.

Since each entry of the symbol vector $\mathbf{s}_\mathsf{A}$ is modulated using 2^{m_A}-ary QAM or PSK modulation schemes, the whole symbol vector $\mathbf{s}_\mathsf{A}$ can be considered as being modulated from the $m_\mathsf{A}N_\mathsf{R}$ coded bits $c_{\mathsf{A},1}, \cdots, c_{\mathsf{A},m_\mathsf{A}N_\mathsf{R}}$. For each $c_{\mathsf{A},i}$ of the $m_\mathsf{A}N_\mathsf{R}$ coded bits associated with the received signal $\mathbf{y}_\mathsf{B}$, the soft-output MAP demapper computes the *a posteriori* LLR value $\lambda_\mathrm{p}(c_{\mathsf{A},i})$ as the output, which can be expressed as

$$\lambda_\mathrm{p}(c_{\mathsf{A},i}) = \ln \frac{p(c_{\mathsf{A},i}=1|\mathbf{y}_\mathsf{B})}{p(c_{\mathsf{A},i}=0|\mathbf{y}_\mathsf{B})} = \ln \frac{\sum_{\mathbf{s}_\mathsf{A}\in C_i^1} f(\mathbf{y}_\mathsf{B}|\mathbf{s}_\mathsf{A}) \prod_{j=1}^{m_\mathsf{A}N_\mathsf{R}} \mathrm{P}\left(c_{\mathsf{A},j}=c_j(\mathbf{s}_\mathsf{A})\right)}{\sum_{\mathbf{s}_\mathsf{A}\in C_i^0} f(\mathbf{y}_\mathsf{B}|\mathbf{s}_\mathsf{A}) \prod_{j=1}^{m_\mathsf{A}N_\mathsf{R}} \mathrm{P}\left(c_{\mathsf{A},j}=c_j(\mathbf{s}_\mathsf{A})\right)} \quad (6.55)$$

where C_i^1 and C_i^0 represent the set of transmit symbol vectors whose ith bit labeling is 1 and 0, respectively. The function $c_j(\mathbf{s}_\mathsf{A})$ denotes the jth bit associated with the labeling of $\mathbf{s}_\mathsf{A}$. Furthermore,

$$\ln f(\mathbf{y}_\mathsf{B}|\mathbf{s}_\mathsf{A}) = -\frac{\|\mathbf{y}_\mathsf{B} - \hat{\mathbf{H}}_\mathsf{B}^{(l)}(\mathbf{s}_\mathsf{A}+\mathbf{s}_\mathsf{B})\|^2}{\sigma_\mathbf{n}^2}. \quad (6.56)$$

In the calculations of (6.56), $\hat{\mathbf{H}}_\mathsf{B}^{(l)}$ is the lth iteration channel estimate from the channel estimator. In the *mismatched detection*, the MIMO demapper assumes the channel estimate to be perfect and $\sigma_\mathbf{n}^2 = \sigma_\mathsf{B}^2$. When channel estimation errors are considered, they can be treated as additional noise. Following the same derivations as (6.43), the equivalent noise variance $\sigma_\mathbf{n}^2$ can be calculated as $\sigma_\mathbf{n}^2 = (P_\mathsf{B}+P_\mathsf{A})\sigma_{\tilde{\mathbf{H}}_\mathsf{B}}^2 + \sigma_\mathsf{B}^2$, where $\sigma_{\tilde{\mathbf{H}}_\mathsf{B}}^2$ is given by (6.37) in the initial SI-aided LMMSE channel estimation; in the subsequent iterations, we have $\sigma_{\tilde{\mathbf{H}}_\mathsf{B}}^2 \approx (1+(P_\mathsf{A}+P_\mathsf{B})L/(N_\mathsf{R}\sigma_\mathsf{B}^2))^{-1}$. The probability terms in (6.55) represent the *a priori* probability and is calculated according to the channel decoder feedback $\{\lambda_\mathrm{a}(c_\mathsf{A})\}$ as

$$\mathrm{P}\left(c_{\mathsf{A},j}=c_j(\mathbf{s}_\mathsf{A})\right) = \frac{\exp(c_j(\mathbf{s}_\mathsf{A})\lambda_\mathrm{a}(c_{\mathsf{A},j}))}{1+\exp(\lambda_\mathrm{a}(c_{\mathsf{A},j}))}. \quad (6.57)$$

The detailed calculations of the *a posteriori* LLR value $\lambda_\mathrm{p}(c_{\mathsf{A},i})$ can be found in [107].

6.5.3 Convolutional Decoder

The soft-input soft-output channel decoder for convolutional codes is realized using the BCJR algorithm [14]. In every iteration, based on the output of the MIMO demapper, the channel decoder computes the *a posteriori* LLR values for each information bit and coded bit. Soft information on coded bits is fed back to the MIMO symbol demapper and the channel estimator. In order to avoid error propagation, the *a priori* information at the input of the channel decoder is subtracted from the output, and only the *extrinsic* information is fed back. In the final iteration, the decoder outputs the hard decisions on the information bits.

The overall workflow of the receiver is summarized in Algorithm 6.

Algorithm 6 Workflow of SI-aided iterative receiver

Initialize: 1. Obtain $\mathbf{H}_{\mathsf{B}}^{(1)}$ from (6.30) for block-fading channel or from (6.31) for time-varying channel.

2. Set $\{\lambda_{\mathrm{a}}(c_{\mathsf{A}})\} = \{0\}$ and $l = 0$.

repeat

Update $l = l + 1$.

In the lth iteration ($l \geq 1$):

1. The MIMO demapper calculates $\{\lambda_{\mathrm{p}}(c_{\mathsf{A}})\}$ from (6.55) using $\mathbf{H}_{\mathsf{B}}^{(l)}$.
2. Calculate $\{\lambda_{\mathrm{e}}(c_{\mathsf{A}})\} = \{\lambda_{\mathrm{p}}(c_{\mathsf{A}}) - \lambda_{\mathrm{a}}(c_{\mathsf{A}})\}$, deinterleave it and obtain $\{\lambda_{\mathrm{a}}(d_{\mathsf{A}})\}$.
3. Feed $\{\lambda_{\mathrm{a}}(d_{\mathsf{A}})\}$ to the channel decoder.
4. The channel decoder computes $\lambda_{\mathrm{p}}(d_{\mathsf{A}})$ using the BCJR algorithm [14].
5. Calculate $\{\lambda_{\mathrm{e}}(d_{\mathsf{A}})\} = \{\lambda_{\mathrm{p}}(d_{\mathsf{A}}) - \lambda_{\mathrm{a}}(d_{\mathsf{A}})\}$, interleave it and obtain $\{\lambda_{\mathrm{a}}(c_{\mathsf{A}})\}$.
6. Feed $\{\lambda_{\mathrm{a}}(c_{\mathsf{A}})\}$ to the MIMO demapper and the channel estimator.
7. The channel estimator calculates $\mathbf{H}_{\mathsf{B}}^{(l+1)}$ from (6.48) for block-fading channel or from (6.52) for time-varying channel.

until $\mathrm{BER}^{(l)} - \mathrm{BER}^{(l-1)} \leq \epsilon$, or maximum number of iteration is reached.

6.6 Simulation Results

In this section, we show the performance of the proposed SI aided channel estimation schemes in the BRC phase of two-way DF relaying systems. In particular, we compare it with the conventional pilot-aided channel estimation scheme.

6.6.1 MSE Performance in Block-Fading Channel

Fig. 6.6 shows the MSE performance of channel estimation at station B in the BRC phase, where $N_{\mathsf{A}} = N_{\mathsf{B}} = N_{\mathsf{R}} = 2$ and $P_{\mathsf{A}} = P_{\mathsf{B}} = P_{\mathsf{R}}/2$. Every entry of $\mathbf{H}_{\mathsf{B}}$ is an i.i.d. $\mathcal{CN}(0,1)$ random variable, and the system is uncoded. When pilot-aided channel estimation scheme is applied, orthogonal pilot sequences of length N_{R} obtained from the columns of Hadamard matrices are transmitted at each transmit antenna of the relay, which corresponds to the minimum length pilot sequences [26]. In addition, we consider a *genie-aided* case, where we assume the genie at the receiver of station B knows $\mathbf{S}_{\mathsf{R}}$ perfectly and calculates the channel estimate according to $\hat{\mathbf{H}}_{\mathsf{B}} = \mathbf{Y}_{\mathsf{B}}(\mathbf{S}_{\mathsf{R}}^H\mathbf{R}_{\mathbf{H}}\mathbf{S}_{\mathsf{R}} + \sigma_{\mathsf{B}}^2 N_{\mathsf{B}}\mathbf{I}_L)^{-1}\mathbf{S}_{\mathsf{R}}^H\mathbf{R}_{\mathbf{H}}$. The MSE of this genie-aided

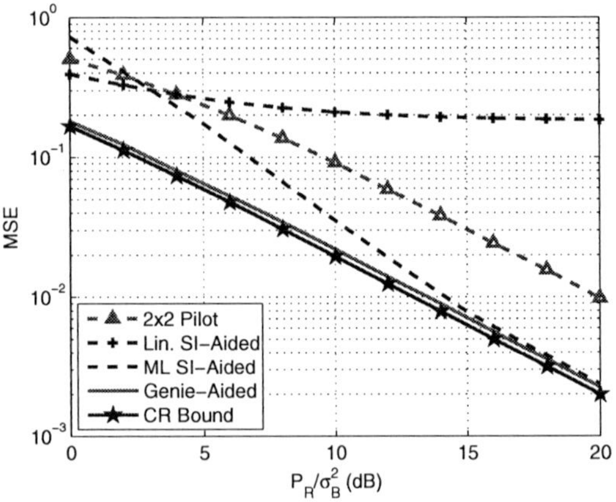

(a) Coherence interval $L = 10$

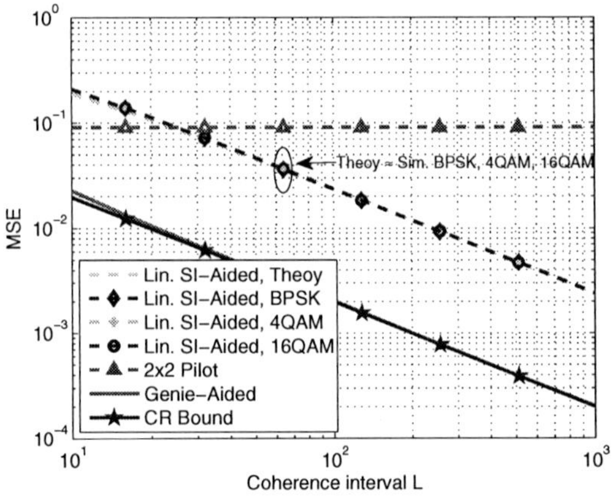

(b) $P_R/\sigma_B^2 = 10$dB

Fig. 6.6: MSE of channel estimation at station B, $N_A = N_R = N_B = 2$ and $P_A = P_B = P_R/2$, Rayleigh block-fading channel.

channel estimation serves as a lower bound for the decision-directed channel estimation in Section 6.5.1.

Fig. 6.6(a) shows how the MSE changes with P_R/σ_B^2 when $L = 10$. Each entry in s_A and s_B uses 4QAM Gray modulation. The "lin. SI-aided" channel estimate corresponds to the channel obtained from (6.30), and the "ML SI-aided" channel estimate denotes the channel estimate obtained from the joint ML channel estimation and data detection scheme in Section 6.3.1. Compared with the pilot-aided scheme, the SI-aided linear channel estimate has lower MSE when $P_R/\sigma_B^2 \leq 4$dB, but its MSE does not decrease further as P_R/σ_B^2 increases because the estimation error is mainly due to the unknown data in the received signal. The ML SI-aided channel estimate outperforms the pure pilot-aided channel estimate when $P_R/\sigma_B^2 \geq 3$dB. It approaches the genie-aided channel estimate and the Cramér-Rao lower bound when $P_R/\sigma_B^2 \geq 15$dB.

Fig. 6.6(b) shows how the MSE of the SI-aided linear channel estimation changes with L when $P_R/\sigma_B^2 = 10$dB. BPSK, 4QAM and 16QAM with Gray mapping are applied on the entries of s_A and s_B. For $L \geq 16$, the simulated SI-aided linear channel estimation MSE fits quite well with the theoretical MSE (6.37). This shows it is reasonable to calculate the SI-aided linear channel estimation MSE according to (6.37) when L is not too short. Furthermore, we observe that the SI-aided linear channel estimate outperforms the pilot-aided channel estimate when $L > 24$ in Fig. 6.6(b). This confirms our theoretical calculation in (6.39), which simplifies to $L > L_p(P_R/\sigma_B^2 + 2) = 24$ when $P_B = P_A = P_R/2$ and $L_p = 2$.

6.6.2 SI-Aided Iterative Channel Estimation in Block-Fading Channel

Fig. 6.7 shows the BER and MSE performance of the iterative receiver discussed in Section 6.5. We consider a MIMO-OFDM system with 64 subcarriers, where 50 subcarriers in each OFDM symbol transmit data and pilots. Each subcarrier corresponds to an i.i.d. Rayleigh fading channel, where the channel remains constant for $L = 32$ time slots. The transmitted bits are encoded by a rate 1/2 convolutional encoder with constraint length of 9 and a generator $(561, 753)_8$ in octal representation. The codeword length is 6400. The convolutional code is taken from the Universal Mobile Telecommunications System (UMTS) standard [62], and the coded bits are modulated to 4QAM symbols with Gray mapping. The parameters for the convolutional code and the OFDM system are summarized in Table 6.1;

The length of the pilot sequences applied at each subcarrier in each coherence interval is N_R. This wastes $2/32 = 6.25\%$ of the bandwidth. Longer pilot sequences can obtain better estimates of the channel, but at the price of more system resources. In order to make

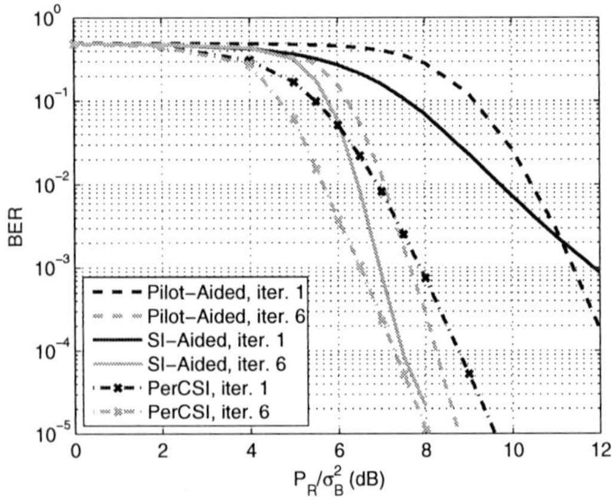

(a) BER

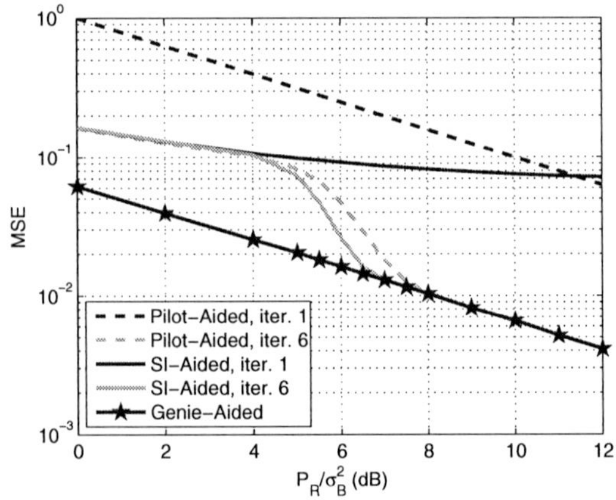

(b) MSE

Fig. 6.7: Pilot-aided and SI-aided iterative channel estimation at station B in block-fading channel under the same average power constraint and transmitting the same amount of data symbols. Coherence interval $L = 32$ time slots, $N_\mathsf{A} = N_\mathsf{R} = N_\mathsf{B} = 2$ and $P_\mathsf{A} = P_\mathsf{B} = P_\mathsf{R}/2$. The length of pilot sequences is N_R.

Table 6.1: Simulation Parameters

convolutional coding rate	1/2
convolutional encoder constraint length	9 bits
convolutional code generator polynomial	$(561, 753)_8$
interleaver type	random
codeword length	6400
number of subcarriers	64
payload (including pilots) subcarriers	50
modulation scheme	4QAM

fair comparisons between the SI-aided and the pilot-aided cases, simulations are performed under the same average transmit power constraint and transmitting the same amount of data similar to [147]. Fig. 6.7(a) shows that the BER performance of the SI-aided scheme is better than that of the pilot-aided scheme when $P_\mathsf{R}/\sigma_\mathsf{B}^2 < 11$dB in the first iteration, and outperforms that of the pilot-aided scheme in the considered BER range in the sixth iteration. When $\mathrm{BER} = 10^{-5}$, the SI-aided scheme is 0.6dB better than the pilot-aided scheme and its gap to the perfect channel state information (CSI) case is only about 0.2dB. Fig. 6.7(b) shows the MSE performance of the SI-aided and the pilot-aided iterative channel estimation. We observe that the SI-aided channel estimates have lower MSE than the pilot-aided estimates when $P_\mathsf{R}/\sigma_\mathsf{B}^2 < 11$dB in the first iteration. When $P_\mathsf{R}/\sigma_\mathsf{B}^2 > 7$dB, the SI-aided channel estimates approach the genie-aided estimates in the sixth iteration, which shows the decoding errors do not severely affect channel estimation then. The pilot-aided scheme approaches the genie-aided estimates in the sixth iteration only when $P_\mathsf{R}/\sigma_\mathsf{B}^2 > 8$dB.

6.6.3 Power Allocation in BRC Phase

Fig. 6.8 shows the optimized power allocation of $P_\mathsf{A}/P_\mathsf{R}$ in the BRC phase of two-way DF relaying systems when the SI-aided channel estimation scheme is applied. Here $P_\mathsf{R} = 10$mW and $P_\mathsf{R}/\sigma_\mathsf{A}^2 = 8$dB. The power allocation of P_A is obtained by solving (6.47). Fig. 6.8(a) shows the power allocation of P_A when $L = 32$. The figure shows that the optimized fraction $P_\mathsf{A}/P_\mathsf{R}$ decreases with $P_\mathsf{R}/\sigma_\mathsf{B}^2$ when $P_\mathsf{R}/\sigma_\mathsf{B}^2 < 9$dB, and remains constant when $P_\mathsf{R}/\sigma_\mathsf{B}^2 > 9$dB. This is because the SI-aided linear channel estimate is not sensitive to $P_\mathsf{R}/\sigma_\mathsf{B}^2$ as shown in Fig. 6.6(a). When the link to station B is weak, allocating more power to P_A is more effective to improve $\mathrm{SNR}_{\mathsf{B},\mathsf{eff}}$. However, when $P_\mathsf{R}/\sigma_\mathsf{B}^2$ and $P_\mathsf{R}/\sigma_\mathsf{A}^2$ are comparable, power allocation should be chosen by considering both the data power and the accuracy of channel estimate. In the symmetric scenario that $P_\mathsf{R}/\sigma_\mathsf{B}^2 = 8$dB, the power allocation

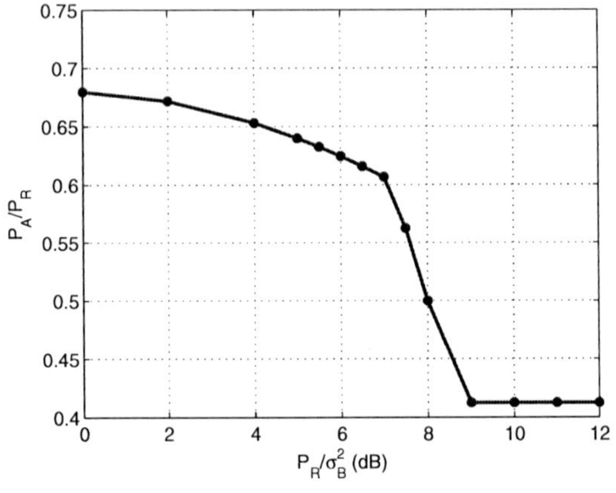

(a) Power allocation $P_{\mathsf{A}}/P_{\mathsf{R}}$, $L = 32$ time slots, $P_{\mathsf{R}}/\sigma_{\mathsf{A}}^2 = 8\text{dB}$.

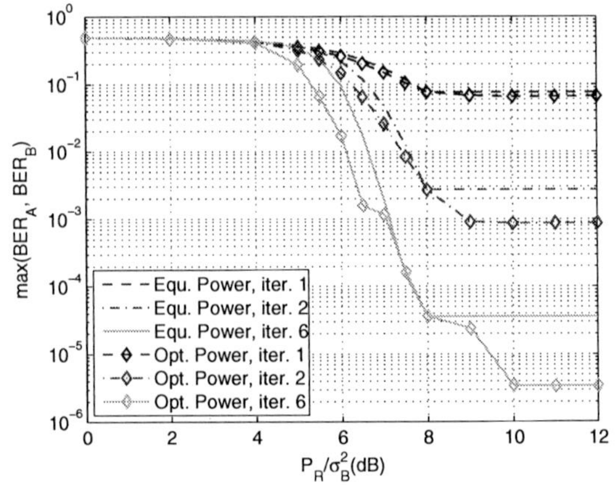

(b) BER performance, $L = 32$ time slots, $P_{\mathsf{R}}/\sigma_{\mathsf{A}}^2 = 8\text{dB}$.

Fig. 6.8: Optimized power allocation $P_{\mathsf{A}}/P_{\mathsf{R}}$ at the relay and the BER performance considering both stations A and B. $P_{\mathsf{R}}/\sigma_{\mathsf{A}}^2 = 8\text{dB}$.

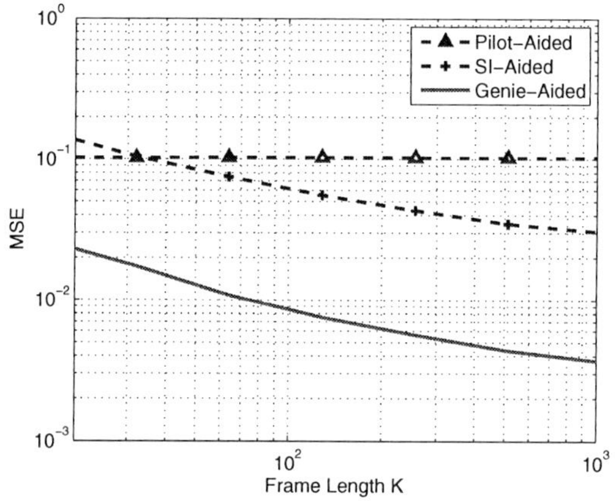

Fig. 6.9: Comparison of channel estimation MSE at station B. Time-varying channel. $N_{\mathsf{R}} = N_{\mathsf{B}} = 2$ and $P_{\mathsf{A}} = P_{\mathsf{B}} = P_{\mathsf{R}}/2$. $f_{\mathrm{D}} T_{\mathrm{s}} = 0.005$. $P_{\mathsf{R}}/\sigma_{\mathsf{B}}^2 = 8$dB. Distance between neighboring pilots is 15 time slots. Each pilot sequence has length N_{R}.

$P_{\mathsf{A}}/P_{\mathsf{R}} = 0.5$. The corresponding BER performance is shown in Fig. 6.8(b). Since we want to guarantee fairness between the two receivers, $\max(\mathrm{BER}_{\mathsf{A}}, \mathrm{BER}_{\mathsf{B}})$ is compared between the equal power allocation case and the optimized power allocation case, and the optimized power allocation has lower $\max(\mathrm{BER}_{\mathsf{A}}, \mathrm{BER}_{\mathsf{B}})$ in each iteration. This verifies our assumption that by optimizing the effective SNR the corresponding BER performance can also be optimized.

6.6.4 SI-Aided Channel Estimation in Time-Varying Channel

In the simulations for time-varying channel, we assume the channel is Rayleigh fading and follow Jakes' two dimensional isotropic scattering model [116]. h_{n_1,m_1,k_1} denotes the channel coefficient between the m_1th transmit antenna and the n_1th receive antenna at the k_1th time slot. The transmit and receive antennas are assumed to be spaced sufficiently far apart and the channel is uncorrelated across antennas. The channel gain temporal autocorrelation function is

$$\mathrm{E}(h_{n_1,m_1,k_1} h^*_{n_2,m_2,k_2}) = \delta_{n_1,n_2} \delta_{m_1,m_2} J_0(2\pi f_{\mathrm{D}} T_{\mathrm{s}}(k_1 - k_2)) \quad (6.58)$$

where $\delta_{n_1,n_2} = 1$ if $n_1 = n_2$, and $\delta_{n_1,n_2} = 0$ otherwise. J_0 is the zeroth-order Bessel function of the first kind. f_{D} is the maximum Doppler shift and T_{s} is the symbol duration. The channel

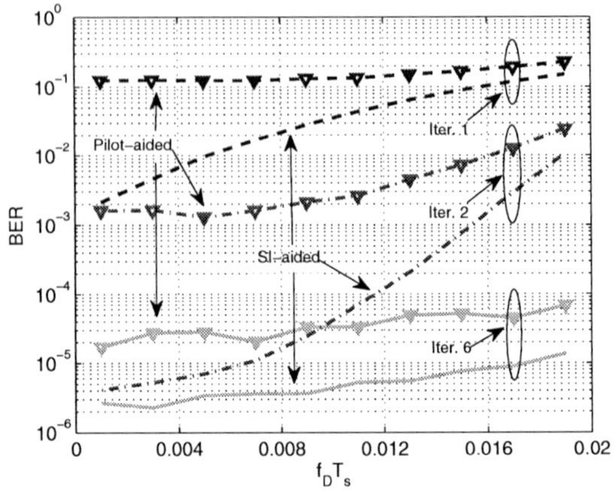

Fig. 6.10: BER comparison of pilot-aided and SI-aided iterative channel estimation at station B in time-varying channel under the same average power constraint and transmitting the same amount of data symbols. $N_{\mathsf{A}} = N_{\mathsf{R}} = N_{\mathsf{B}} = 2$ and $P_{\mathsf{A}} = P_{\mathsf{B}} = P_{\mathsf{R}}/2$. $P_{\mathsf{R}}/\sigma_{\mathsf{B}}^2 = 8\text{dB}$, and frame length $K = 128$ time slots. Pilot sequences are placed every 16 time slots, and each pilot sequence has length N_{R}.

gain temporal autocorrelation function depends on $f_{\mathrm{D}}T_{\mathrm{s}}$, which is determined by the relative velocity between the transmitter and receiver.

Fig. 6.9 compares the MSE of the SI-aided and the pilot-aided channel estimation at station B in a time-varying channel with $f_{\mathrm{D}}T_{\mathrm{s}} = 0.005$. In such a system, $P_{\mathsf{R}}/\sigma_{\mathsf{B}}^2 = 8\text{dB}$ and the distance between neighboring pilots is 15 time slots if pilot-aided channel estimation is applied. Each pilot has length $N_{\mathsf{R}} = 2$. Therefore, about $2/15 = 13.3\%$ of the bandwidth is wasted for transmitting pilots. An interesting observation is that the MSE of the SI-aided linear channel estimate decreases with the increase of frame length K. When $K = 35$, the initial SI-aided channel estimation outperforms the pilot-aided channel estimation in terms of MSE. Unlike the coherence interval in block-fading channel that is determined by the physical channel itself, the observation frame length K can be chosen by the receiver. In order to obtain better SI-aided channel estimates, larger K can be chosen. The price to pay is the higher computational complexity for channel estimation.

In Fig. 6.10, we show the BER performance vs. $f_{\mathrm{D}}T_{\mathrm{s}}$ in a system with $P_{\mathsf{R}}/\sigma_{\mathsf{B}}^2 = 8\text{dB}$ and $K = 128$ time slots. The parameters for the OFDM system and coding have already been stated in Section 6.6.2. In the pilot-aided channel estimation scheme, pilots are placed every

16 or 17 time slots of each frame in each subcarrier. This corresponds to the optimum pilot symbol spacing for the time-varying channel with $f_D T_s = 0.005$ in the turbo-coded system of [233]. The pilots altogether occupy 18 time slots in each subcarrier, which corresponds to $18/128 = 14\%$ of the bandwidth. At the sixth iteration, the SI-aided scheme shows much better BER performance compared to the pilot-aided scheme in Fig. 6.10. For example, when $f_D T_s = 0.005$, the pilot-aided case achieves BER of 2.8×10^{-5} and the BER for the SI-aided case is only 3.4×10^{-6}. For both schemes, the BER performance degrades when $f_D T_s$ is high because the pilots or SI is not sufficient to track the channel changes then. However, due to the error-correcting capacities of channel coding and the decision-directed channel estimation, there is no severe BER performance degradation in the sixth iteration.

6.7 Chapter Summary and Conclusions

We proposed two novel SI-aided channel estimation schemes for the BRC phase of the two-way DF relaying protocol. The channel estimates of the proposed schemes are obtained by exploiting the known data symbols in the SI that are inherent in the considered two-way DF relaying scenario. No pilot sequences are required to be transmitted from the relay, so the bandwidth efficiency is further improved. To the best of our knowledge, this is the first scheme that *utilizes* SI for channel estimation instead of simply canceling it out.

Besides proposing the ideas of exploiting SI to estimate the channel, we provided the whole SI-aided iterative receiver structure. The performance analysis and simulation results showed that when the coherence interval in block-fading channels or the observation frame length in time-varying channels is long enough, the SI-aided channel estimation can eventually outperform the pilot-aided channel estimation in many realistic scenarios. Our proposed scheme is particularly suitable for systems with large number of antennas and subcarriers or with high mobility stations, where the resource consumed by conventional pilot-aided channel estimation had been considered as a big hindrance for the practical implementation of the system.

6.8 Appendices

6.8.1 Proof of the Cramér-Rao Bound

Since the diagonal elements of the inverse of the Fisher information matrix (FIM) corresponds to the Cramér-Rao bounds of the parameters we wish to estimate, we first calculate the FIM. We denote $\tilde{\mathbf{s}} = \mathrm{vec}(\mathbf{S}_\mathsf{B})$ and $\mathbf{y} = \mathrm{vec}(\mathbf{Y}_\mathsf{B})$. The derivation follows that in [246]. Under the regularity conditions [129], the FIM $\mathcal{J}$ is defined as

$$\mathcal{J} = \mathrm{E}\left\{\frac{\partial \ln p(\mathbf{y},\boldsymbol{\theta})}{\partial \boldsymbol{\theta}^*}\left(\frac{\partial \ln p(\mathbf{y},\boldsymbol{\theta})}{\partial \boldsymbol{\theta}^*}\right)^H\right\} \tag{6.59}$$

where the expectation is taken over $p(\mathbf{y},\boldsymbol{\theta},\tilde{\mathbf{s}})$. The Fisher information matrix $\mathcal{J}$ can be written as

$$\mathcal{J} = \mathrm{E}\left\{\frac{\partial \ln p(\mathbf{y}\mid\boldsymbol{\theta})}{\partial \boldsymbol{\theta}^*}\left(\frac{\partial \ln p(\mathbf{y}\mid\boldsymbol{\theta})}{\partial \boldsymbol{\theta}^*}\right)^H\right\} + \mathrm{E}\left\{\frac{\partial p(\boldsymbol{\theta})}{\partial \boldsymbol{\theta}^*}\left(\frac{\partial p(\boldsymbol{\theta})}{\partial \boldsymbol{\theta}^*}\right)^H\right\}. \tag{6.60}$$

According to (6.16), the log-likelihood function can be written as

$$\begin{aligned}\ln p(\mathbf{y}\mid\boldsymbol{\theta}) &= \mathrm{Const} - \sigma_\mathsf{B}^{-2}\|\mathbf{y} - (\mathbf{I}_L \otimes \mathbf{H}_\mathsf{B})(\mathbf{s}+\tilde{\mathbf{s}})\|^2 &(6.61)\\ &= \mathrm{Const} - \sigma_\mathsf{B}^{-2}\left\|\mathbf{y} - \left((\mathbf{S}_\mathsf{A}+\mathbf{S}_\mathsf{B})^T \otimes \mathbf{I}_{N_\mathsf{B}}\right)\mathbf{h}\right\|^2. &(6.62)\end{aligned}$$

$\mathcal{J}$ can be calculated by taking partial derivatives of $\ln p(\mathbf{y}\mid\boldsymbol{\theta})$ with respect to $\mathbf{s}$ and $\mathbf{h}$ according to (6.61) and (6.62), respectively.

$$\begin{aligned}\frac{\partial \ln p(\mathbf{y}\mid\boldsymbol{\theta})}{\partial \mathbf{s}^*} &= -\sigma_\mathsf{B}^{-2}(\mathbf{I}_L \otimes \mathbf{H}_\mathsf{B})^H\left(\mathbf{y} - (\mathbf{I}_L \otimes \mathbf{H}_\mathsf{B})(\mathbf{s}+\tilde{\mathbf{s}})\right) &(6.63)\\ \frac{\partial \ln p(\mathbf{y}\mid\boldsymbol{\theta})}{\partial \mathbf{h}^*} &= -\sigma_\mathsf{B}^{-2}\left((\mathbf{S}_\mathsf{A}+\mathbf{S}_\mathsf{B})^T \otimes \mathbf{I}_{N_\mathsf{B}}\right)^H\left(\mathbf{y} - \left((\mathbf{S}_\mathsf{A}+\mathbf{S}_\mathsf{B})^T \otimes \mathbf{I}_{N_\mathsf{B}}\right)\mathbf{h}\right). &(6.64)\end{aligned}$$

Therefore,

$$\begin{aligned}\mathrm{E}\left\{\frac{\partial \ln p(\mathbf{y}\mid\boldsymbol{\theta})}{\partial \mathbf{s}^*}\left(\frac{\partial \ln p(\mathbf{y}\mid\boldsymbol{\theta})}{\partial \mathbf{s}^*}\right)^H\right\} &= \mathrm{E}\left\{\sigma_\mathsf{B}^{-2}(\mathbf{I}_L \otimes \mathbf{H}_\mathsf{B})^H(\mathbf{I}_L \otimes \mathbf{H}_\mathsf{B})\right\}\\ &= \frac{N_\mathsf{B}}{\sigma_\mathsf{B}^2}\mathbf{I}_{N_\mathsf{R}L}, &(6.65)\\ \mathrm{E}\left\{\frac{\partial \ln p(\mathbf{y}\mid\boldsymbol{\theta})}{\partial \mathbf{h}^*}\left(\frac{\partial \ln p(\mathbf{y}\mid\boldsymbol{\theta})}{\partial \mathbf{h}^*}\right)^H\right\} &= \mathrm{E}\left\{\sigma_\mathsf{B}^{-2}\left((\mathbf{S}_\mathsf{A}+\mathbf{S}_\mathsf{B})^T \otimes \mathbf{I}_{N_\mathsf{B}}\right)^H\left((\mathbf{S}_\mathsf{A}+\mathbf{S}_\mathsf{B})^T \otimes \mathbf{I}_{N_\mathsf{B}}\right)\right\}\\ &= \frac{P_\mathsf{R}L}{N_\mathsf{R}\sigma_\mathsf{B}^2}\mathbf{I}_{N_\mathsf{R}N_\mathsf{B}}, &(6.66)\\ \mathrm{E}\left\{\frac{\partial \ln p(\mathbf{y}\mid\boldsymbol{\theta})}{\partial \mathbf{s}^*}\left(\frac{\partial \ln p(\mathbf{y}\mid\boldsymbol{\theta})}{\partial \mathbf{h}^*}\right)^H\right\} &= \mathrm{E}\left\{\sigma_\mathsf{B}^{-2}(\mathbf{I}_L \otimes \mathbf{H}_\mathsf{B})^H\left((\mathbf{S}_\mathsf{A}+\mathbf{S}_\mathsf{B})^T \otimes \mathbf{I}_{N_\mathsf{B}}\right)\right\}\\ &= \mathbf{0}_{N_\mathsf{R}L\times N_\mathsf{R}N_\mathsf{B}}. &(6.67)\end{aligned}$$

In (6.67), we utilized the fact that each entry of $\mathbf{H}_\mathsf{B}$ and $(\mathbf{S}_\mathsf{A} + \mathbf{S}_\mathsf{B})$ are independent. Furthermore, according to the independence of s and $\mathbf{h}$, we have $p(\boldsymbol{\theta}) = p(\mathrm{s})p(\mathbf{h})$. Since each entry of s is modulated using QAM or PSK modulation with equal probability and each entry of $\mathbf{H}_\mathsf{B}$ is an i.i.d. $\mathcal{CN}(0,1)$ random variable, we have $p(\mathrm{s}) = \mathrm{Const}$ and $\ln p(\mathbf{h}) = \mathrm{Const} - \mathbf{h}^H\mathbf{h}$. So we obtain

$$\mathrm{E}\left\{\frac{\partial p(\boldsymbol{\theta})}{\partial \boldsymbol{\theta}^*}\left(\frac{\partial p(\boldsymbol{\theta})}{\partial \boldsymbol{\theta}^*}\right)^H\right\} = \begin{pmatrix} \mathbf{0}_{N_\mathsf{R}L\times N_\mathsf{R}L} & \mathbf{0}_{N_\mathsf{R}L\times N_\mathsf{R}N_\mathsf{B}} \\ \mathbf{0}_{N_\mathsf{R}N_\mathsf{B}\times N_\mathsf{R}L} & \mathrm{E}(\mathbf{h}\mathbf{h}^H) \end{pmatrix} = \begin{pmatrix} \mathbf{0}_{N_\mathsf{R}L\times N_\mathsf{R}L} & \mathbf{0}_{N_\mathsf{R}L\times N_\mathsf{R}N_\mathsf{B}} \\ \mathbf{0}_{N_\mathsf{R}N_\mathsf{B}\times N_\mathsf{R}L} & \mathbf{I}_{N_\mathsf{R}N_\mathsf{B}} \end{pmatrix}. \tag{6.68}$$

So the FIM can be written as

$$\mathcal{J} = \begin{pmatrix} \mathcal{J}_{1,1} & \mathcal{J}_{1,2} \\ \mathcal{J}_{1,2}^H & \mathcal{J}_{2,2} \end{pmatrix} \tag{6.69}$$

where

$$\mathcal{J}_{1,1} = \frac{N_\mathsf{B}}{\sigma_\mathsf{B}^2}\mathbf{I}_{N_\mathsf{R}L} \tag{6.70}$$

$$\mathcal{J}_{1,2} = \mathbf{0}_{N_\mathsf{R}L\times N_\mathsf{R}N_\mathsf{B}} \tag{6.71}$$

$$\mathcal{J}_{2,2} = (1 + \frac{P_\mathsf{R}L}{N_\mathsf{R}\sigma_\mathsf{B}^2})\mathbf{I}_{N_\mathsf{R}N_\mathsf{B}}. \tag{6.72}$$

The diagonal elements of the inverse of $\mathcal{J}_{1,1}$ corresponds to the Cramér-Rao bounds of data symbols s, and the diagonal elements of the inverse of $\mathcal{J}_{2,2}$ corresponds to the Cramér-Rao bounds of channel $\mathbf{h}$.

6.8.2 Calculation of *a posteriori* LLR Values

The *a posteriori* LLR value $\lambda_\mathrm{p}(c_{\mathsf{A},i})$ for ith bit $c_{\mathsf{A},i}$ of the $m_\mathsf{A}N_\mathsf{R}$ coded bits associated with the observation $\mathbf{y}_\mathsf{B}$ of the received signal can be expressed as

$$\lambda_\mathrm{p}(c_{\mathsf{A},i}) = \ln\frac{p(c_{\mathsf{A},i} = 1|\mathbf{y}_\mathsf{B})}{p(c_{\mathsf{A},i} = 0|\mathbf{y}_\mathsf{B})} \tag{6.73}$$

$$= \ln\frac{\sum_{\mathbf{s}_\mathsf{A}\in C_i^1} p(\mathbf{s}_\mathsf{A}|\mathbf{y}_\mathsf{B})}{\sum_{\mathbf{s}_\mathsf{A}\in C_i^0} p(\mathbf{s}_\mathsf{A}|\mathbf{y}_\mathsf{B})} \tag{6.74}$$

$$= \ln\frac{\sum_{\mathbf{s}_\mathsf{A}\in C_i^1} f(\mathbf{y}_\mathsf{B}|\mathbf{s}_\mathsf{A})\mathrm{P}(\mathbf{s}_\mathsf{A})}{\sum_{\mathbf{s}_\mathsf{A}\in C_i^0} f(\mathbf{y}_\mathsf{B}|\mathbf{s}_\mathsf{A})\mathrm{P}(\mathbf{s}_\mathsf{A})} \tag{6.75}$$

$$= \ln\frac{\sum_{\mathbf{s}_\mathsf{A}\in C_i^1} f(\mathbf{y}_\mathsf{B}|\mathbf{s}_\mathsf{A})\prod_{j=1}^{m_\mathsf{A}N_\mathsf{R}}\mathrm{P}\left(c_{\mathsf{A},j} = c_j(\mathbf{s}_\mathsf{A})\right)}{\sum_{\mathbf{s}_\mathsf{A}\in C_i^0} f(\mathbf{y}_\mathsf{B}|\mathbf{s}_\mathsf{A})\prod_{j=1}^{m_\mathsf{A}N_\mathsf{R}}\mathrm{P}\left(c_{\mathsf{A},j} = c_j(\mathbf{s}_\mathsf{A})\right)} \tag{6.76}$$

According to (6.57), the *a posteriori* LLR at the output of the soft demapper can be written as

$$\lambda_{\mathrm{p}}(c_{\mathsf{A},i}) = \ln\frac{p(c_{\mathsf{A},i}=1)}{p(c_{\mathsf{A},i}=0)} + \ln\frac{\sum_{\mathbf{s}_{\mathsf{A}}\in\mathcal{C}_i^1} f(\mathbf{y}_{\mathsf{B}}|\mathbf{s}_{\mathsf{A}})\prod_{j=1,j\neq i}^{m_{\mathsf{A}}N_{\mathsf{R}}} \mathrm{P}\left(c_{\mathsf{A},j}=c_j(\mathbf{s}_{\mathsf{A}})\right)}{\sum_{\mathbf{s}_{\mathsf{A}}\in\mathcal{C}_i^0} f(\mathbf{y}_{\mathsf{B}}|\mathbf{s}_{\mathsf{A}})\prod_{j=1,j\neq i}^{m_{\mathsf{A}}N_{\mathsf{R}}} \mathrm{P}\left(c_{\mathsf{A},j}=c_j(\mathbf{s}_{\mathsf{A}})\right)} \quad (6.77)$$

$$= \ln\frac{p(c_{\mathsf{A},i}=1)}{p(c_{\mathsf{A},i}=0)} + \ln\frac{\sum_{\mathbf{s}_{\mathsf{A}}\in\mathcal{C}_i^1} f(\mathbf{y}_{\mathsf{B}}|\mathbf{s}_{\mathsf{A}})\prod_{j=1,j\neq i}^{m_{\mathsf{A}}N_{\mathsf{R}}} \exp(c_j(\mathbf{s}_{\mathsf{A}})\lambda_{\mathrm{a}}(c_{\mathsf{A},j}))}{\sum_{\mathbf{s}_{\mathsf{A}}\in\mathcal{C}_i^0} f(\mathbf{y}_{\mathsf{B}}|\mathbf{s}_{\mathsf{A}})\prod_{j=1,j\neq i}^{m_{\mathsf{A}}N_{\mathsf{R}}} \exp(c_j(\mathbf{s}_{\mathsf{A}})\lambda_{\mathrm{a}}(c_{\mathsf{A},j}))} \quad (6.78)$$

$$= \lambda_{\mathrm{a}}(c_{\mathsf{A},i}) + \lambda_{\mathrm{e}}(c_{\mathsf{A},i}) \quad (6.79)$$

where $\lambda_{\mathrm{a}}(c_{\mathsf{A},i})$ and $\lambda_{\mathrm{e}}(c_{\mathsf{A},i})$ denote the *a priori* information and the *extrinsic* information for $c_{\mathsf{A},i}$, respectively. The extrinsic information $\lambda_{\mathrm{e}}(c_{\mathsf{A},i})$ in (6.79) can be calculated by using the *max-star* operator as

$$\lambda_{\mathrm{e}}(c_{\mathsf{A},i}) = \max_{\mathbf{s}_{\mathsf{A}}\in\mathcal{C}_i^1}{}^{*}\left(\ln f(\mathbf{y}_{\mathsf{B}}|\mathbf{s}_{\mathsf{A}}) + \sum_{j=1,j\neq i}^{m_{\mathsf{A}}N_{\mathsf{R}}} \exp(c_j(\mathbf{s}_{\mathsf{A}})\lambda_{\mathrm{a}}(c_{\mathsf{A},j}))\right) - \max_{\mathbf{s}_{\mathsf{A}}\in\mathcal{C}_i^0}{}^{*}\left(\ln f(\mathbf{y}_{\mathsf{B}}|\mathbf{s}_{\mathsf{A}}) + \sum_{j=1,j\neq i}^{m_{\mathsf{A}}N_{\mathsf{R}}} \exp(c_j(\mathbf{s}_{\mathsf{A}})\lambda_{\mathrm{a}}(c_{\mathsf{A},j}))\right) \quad (6.80)$$

where $\max^*(x,y) = \max(x,y) + \ln\left(1+\exp(-|x-y|)\right)$. For multiple arguments, it follows the recursive relationship $\max^*(x,y,z) = \max^*(\max^*(x,y),z)$.

Chapter 7

Achievable Rates of Bidirectional Broadcast Channels With Self-Interference Aided Channel Estimation

The self-interference (SI)-aided channel estimation has been proposed for the two-way decode-and-forward (DF) relaying protocol in Chapter 6, where we discussed the practical coding and modulation schemes and the transceiver structures. The SI, which contains the known data symbols at the receiving stations, is utilized to estimate the channel from the relay to its corresponding receiver in the broadcast (BRC) phase. In this way, the SI can play a similar role as the pilot sequences for channel estimation at the receiving stations. Such a SI-aided channel estimation scheme has been shown to be able to achieve excellent channel estimation performance in realistic scenarios. Compared to the pure pilot-aided channel estimation scheme, the SI is inherent in the two-way relaying protocols and does not occupy the system resources. The SI-aided channel estimation can achieve higher spectral efficiency compared to the pure pilot-aided channel estimation scheme. In order to quantify the spectral efficiency improvement in the BRC phase channel with SI-aided channel estimation compared to the conventional pure pilot-aided channel estimation, we calculate in this chapter the theoretical achievable rates of the bidirectional broadcast channel in the BRC phase of two-way DF relaying systems when the SI-aided channel estimation scheme is applied. The spectral efficiency improvement for systems employing this SI-aided channel estimation scheme is quantified by comparing its achievable rates to that of the conventional pilot-aided channel estimation scheme. Both the theoretical Gaussian codebook and the practical symbol modulation schemes, such as the quadrature amplitude modulation (QAM), are

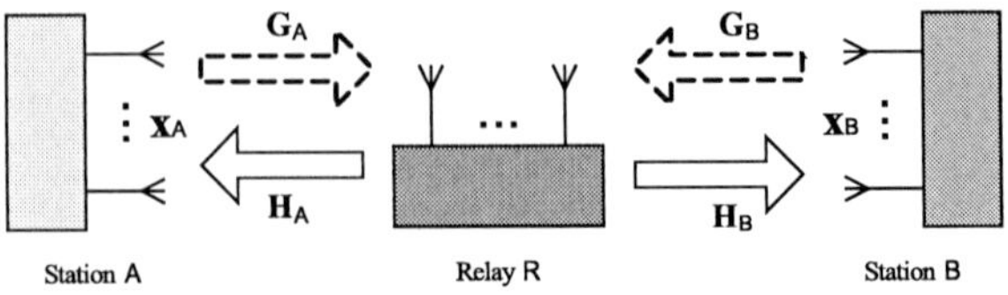

Fig. 7.1: MIMO two-way DF relaying system, where the dashed arrows represent the transmission in the MAC phase, and the solid arrows represent the transmission in the BRC phase.

considered.

7.1 Introduction

The two-way decode-and-forward (DF) relaying protocol considers the scenario that two half-duplex wireless stations, A and B, exchange data via another half-duplex wireless relay R, as shown in Fig. 7.1. Compared to traditional relaying schemes that require four phases, i.e., channel uses, to achieve the bidirectional communication, the two-way relaying protocol consists of the following two phases: the multiple access (MAC) phase and the broadcast (BRC) phase, which can be separated in time (TDD) or in frequency (FDD) (see Chapter 2). The core idea of the two-way DF relaying protocol is the *self-interference* (SI) cancellation in the broadcast (BRC) phase. That is, the relay combines the decoded data information from the two user stations. So each receiving user station knows part of the data information in the combined data from the relay. The back-propagated known data information that was from the receiving station itself in the MAC phase is called the SI, and it does not interfere with the decoding at the receivers in the BRC phase.

When the *superposition coding* (SPC) scheme is applied, the relay combines the data information on the symbol level, and the SI contains the known data symbols in this case (see Chapter 4). Instead of simply canceling the SI, we proposed a practical approach to utilize the SI for channel estimation in the BRC phase of two-way DF relaying systems in Chapter 6 when the SPC scheme is applied. There, the SI plays a similar role in the process of channel estimation, and the SI-aided channel estimation scheme is shown to be able to achieve similar bit error rate (BER) performance as traditional pilot-aided channel estimation schemes in the considered scenario. Thus, the resources occupied by pilot sequences are released, and the spectral efficiency can be further improved.

In the two-way DF relaying systems, the relay fully decode the received data from the stations A and B in the MAC phase. The transmission data rate pair of station A and B is an achievable rate pair of the two-way DF relaying system if it is achievable by both the MAC and BRC phases, i.e., we can determine the achievable rate pairs of the two-way DF relaying system by determining those for the MAC and BRC phases separately. However, the MAC phase is a conventional multiple access channel, where the optimal coding schemes are known, e.g., ref. [78]. Compared to the MAC phase, the BRC phase is a new area of research. Leaving aside the constraints of the MAC phase, we assume the messages from the stations A and B have been perfectly decoded at the relay. Thus in the BRC phase, the messages at the relay to be transmitted to station A are perfectly known to station B and vice versa for the messages intended for station B. Such a channel is called the *bidirectional broadcast channel* [170].

As we know, channel estimation is an integral part of wireless transmission schemes and it is particularly important for systems with multiple antennas. Traditional channel estimation schemes transmit orthogonal pilot sequences on different transmit antennas before sending data. The algorithms used for traditional pilot-aided channel estimation schemes are summarized in [26], where the linear least square (LS) and minimum mean square error (MMSE) approaches are the two major algorithms. The authors of [97] computed a lower bound on the capacity of a point-to-point block-fading multiple-input multiple-output (MIMO) channel that is estimated by the pilot-aided scheme. The authors of [280] also studied the block-fading MIMO channel with coherence time T. However, neither the transmitter nor the receiver has the channel state information. The asymptotic capacity of this channel at high signal-to-noise ratio (SNR) is derived. Assuming that the transmitter has M antennas and the receiver has N antennas, the corresponding capacity gain for this noncoherent channel turns out to be $M^*(1 - M^*/T)$ bits/s/Hz for every 3-dB increase in SNR, where $M^* = \min(M, N, \lfloor T/2 \rfloor)$. Compared to the capacity gain of the coherent multiple antenna channel, which is $\min(M, N)$ bits/s/Hz for every 3-dB increase in SNR, the capacity of the noncoherent channel also depends on the coherence time. The authors of [16] considered a continuously time-varying (Rayleigh fading) channel model and computed the achievable rate for a typical coded modulation transmission system operating on flat fading MIMO channels and using a perfect interleaver to combat the bursty nature of the channel. Their derivations showed the link between the LMMSE channel estimation and the average mutual information with respect to the channel dynamics. The achievable rate of the system was also optimized with respect to the amount of training information needed.

In [282], the authors proposed to transmit low-power orthogonal pilot signals concurrently with the data. A capacity lower bound for systems employing this pilot-embedding scheme has been shown in [45,46]. All those channel estimation schemes rely on an initial estimate of the MIMO channel, based on the transmission of pilot sequences.

The purpose of this chapter is to quantify the spectral efficiency improvement by computing achievable rates of the bidirectional broadcast channel when the SI-aided channel estimation scheme is applied. The achievable rates serve as theoretical performance limits of the system and enable quantitative comparisons to traditional pilot-aided schemes. Note that in order to calculate the achievable rates, the codes have to be decoded with error rate approaching zero. We consider a block-fading channel model for the BRC phase, i.e., the BRC phase contains a number of coherence intervals. Each coherence interval contains T time slots. The idea is to consider each time slot of the coherence interval as one use of T parallel channels with different signal-to-noise ratios (SNRs). The quality of the channel estimates determines the SNR for each of the parallel channels. Codes with different rates are allocated to those parallel channels according to the SNRs. We first exploit the SI to get an initial estimate of the channel and use it to decode the unknown data in the first T_s time slots of each coherence interval, where $T_s \geq N_R$. After the unknown data in those time slots are fully decoded, the data-aided approach is applied, where the decoded data are utilized to re-estimate the channel and help to provide better channel estimates for the following time slots. As the channel estimation quality improves, codes with higher rates can be allocated in subsequent time slots of the coherence interval. We calculate the achievable rates for the bidirectional broadcast channel when the relay employs Gaussian codebooks or quadrature amplitude modulation (QAM) for retransmission.

Our Contributions: The contributions of this chapter can be summarized as follows:

- We derive of the achievable rates and present the quantification of the spectral efficiency improvement of the BRC phase channel in MIMO two-way DF relaying systems from the information theoretic perspective when the SI-aided channel estimation scheme is applied. Both theoretical Gaussian codebooks and practical QAM codebooks are considered.

The remainder of this chapter is organized as follows: in Section 7.2, we introduce the system model and briefly summarize the two-way relaying technique. The SI-aided channel estimation scheme is described in Section 7.3. The achievable rates for systems employing SI-aided channel estimation schemes are compared with systems employing traditional pilot-aided schemes, and the simulation results are presented in Section 7.4. Finally, conclusions are drawn in Section 7.5.

7.2 System Model

We consider a relaying system where two wireless stations A and B exchange data via a half-duplex DF relay. We assume that there is no direct connection between station A and B (for example, due to shadowing or too large distance between them). The number of antennas at station A, the relay R and station B are denoted as N_A, N_R and N_B, respectively.

When the two-way relaying technique [193] is applied, the data of station A and B are exchanged in the MAC and BRC phases as shown in Fig. 7.1. In the MAC phase, station A and B transmit simultaneously to the relay. The data symbol vectors transmitted at station A and B in one time slot are denoted as $\mathbf{x}_\mathsf{A} \in \mathbb{C}^{N_\mathsf{A}\times 1}$ and $\mathbf{x}_\mathsf{B} \in \mathbb{C}^{N_\mathsf{B}\times 1}$, where $\mathrm{E}\{\mathbf{x}_\mathsf{A}\mathbf{x}_\mathsf{A}^H\} = Q_\mathsf{A}/N_\mathsf{A}\mathbf{I}_{N_\mathsf{A}}$ and $\mathrm{E}\{\mathbf{x}_\mathsf{B}\mathbf{x}_\mathsf{B}^H\} = Q_\mathsf{B}/N_\mathsf{B}\mathbf{I}_{N_\mathsf{B}}$. The received signal $\mathbf{y}_\mathsf{R}$ at the relay can be expressed as

$$\mathbf{y}_\mathsf{R} = \mathbf{G}_\mathsf{A}\mathbf{x}_\mathsf{A} + \mathbf{G}_\mathsf{B}\mathbf{x}_\mathsf{B} + \mathbf{v}_\mathsf{R} \tag{7.1}$$

where $\mathbf{G}_\mathsf{A} \in \mathbb{C}^{N_\mathsf{R}\times N_\mathsf{A}}$ and $\mathbf{G}_\mathsf{B} \in \mathbb{C}^{N_\mathsf{R}\times N_\mathsf{B}}$ are respectively the channel matrices from station A and B to the relay. Q_A and Q_B are the transmit power constraints at station A and B in the MAC phase. The additive noise vector at the relay is $\mathbf{v}_\mathsf{R} \sim \mathcal{CN}(0, \sigma^2\mathbf{I}_{N_\mathsf{R}})$. This is a conventional multiple access scenario. The receiver structure for decoding the data contained in $\mathbf{x}_\mathsf{A}$ and $\mathbf{x}_\mathsf{B}$ can be found in e.g., [240].

Since we focus on transmission schemes in the BRC phase, we assume the relay perfectly decodes what it receives in the MAC phase. In the BRC phase, we consider the superposition coding scheme [193]. The relay remodulates the decoded data from station A and B separately into symbol vectors $\mathbf{s}_\mathsf{A} \in \mathbb{C}^{N_\mathsf{R}\times 1}$ and $\mathbf{s}_\mathsf{B} \in \mathbb{C}^{N_\mathsf{R}\times 1}$, where $\mathbf{s}_\mathsf{A}$ and $\mathbf{s}_\mathsf{B}$ are the normalized transmit signal vectors at the relay that contain the same data as $\mathbf{x}_\mathsf{A}$ and $\mathbf{x}_\mathsf{B}$, respectively. Furthermore, we have $\mathrm{E}\{\mathbf{s}_\mathsf{A}\mathbf{s}_\mathsf{A}^H\} = \mathbf{I}_{N_\mathsf{R}}$ and $\mathrm{E}\{\mathbf{s}_\mathsf{B}\mathbf{s}_\mathsf{B}^H\} = \mathbf{I}_{N_\mathsf{R}}$. Here we assume the relay does not know the channel to the two stations in the BRC phase. This is the case in FDD systems without channel feedback from the stations. The relay then adds the two symbol vectors together and retransmits the sum vector:

$$\mathbf{s} = \sqrt{\frac{P_\mathsf{A}}{N_\mathsf{R}}}\mathbf{s}_\mathsf{A} + \sqrt{\frac{P_\mathsf{B}}{N_\mathsf{R}}}\mathbf{s}_\mathsf{B}. \tag{7.2}$$

In order to satisfy the power constraint, we require $P_\mathsf{A} + P_\mathsf{B} = P_\mathsf{R}$, where P_R is the transmit power constraint at the relay in the BRC phase. The modulation schemes and the power allocation of $\mathbf{s}_\mathsf{A}$ and $\mathbf{s}_\mathsf{B}$ used at the relay are known to both station A and B.

Note that $\mathbf{s}_\mathsf{A}$ is already known to station A, and $\mathbf{s}_\mathsf{B}$ is known to station B. The received signal part that contains the known data transmitted by the receiving station itself is the SI for the receiver. For example, the signal received at station A can be written as

$$\mathbf{y}_{\mathsf{A}} = \mathbf{H}_{\mathsf{A}}\mathbf{s} + \mathbf{v}_{\mathsf{A}} \tag{7.3}$$

$$= \underbrace{\sqrt{\frac{P_{\mathsf{A}}}{N_{\mathsf{R}}}}\mathbf{H}_{\mathsf{A}}\mathbf{s}_{\mathsf{A}}}_{\text{SI for station A}} + \sqrt{\frac{P_{\mathsf{B}}}{N_{\mathsf{R}}}}\mathbf{H}_{\mathsf{A}}\mathbf{s}_{\mathsf{B}} + \mathbf{v}_{\mathsf{A}}. \tag{7.4}$$

Similarly, the signal received at station B is

$$\mathbf{y}_{\mathsf{B}} = \mathbf{H}_{\mathsf{B}}\mathbf{s} + \mathbf{v}_{\mathsf{B}} \tag{7.5}$$

$$= \sqrt{\frac{P_{\mathsf{A}}}{N_{\mathsf{R}}}}\mathbf{H}_{\mathsf{B}}\mathbf{s}_{\mathsf{A}} + \underbrace{\sqrt{\frac{P_{\mathsf{B}}}{N_{\mathsf{R}}}}\mathbf{H}_{\mathsf{B}}\mathbf{s}_{\mathsf{B}}}_{\text{SI for station B}} + \mathbf{v}_{\mathsf{B}}. \tag{7.6}$$

Here $\mathbf{H}_{\mathsf{A}} \in \mathbb{C}^{N_{\mathsf{A}} \times N_{\mathsf{R}}}$ and $\mathbf{H}_{\mathsf{B}} \in \mathbb{C}^{N_{\mathsf{B}} \times N_{\mathsf{R}}}$ respectively denote the channel matrices from the relay to station A and B. $\mathbf{v}_{\mathsf{A}} \sim \mathcal{CN}(0, \sigma_{\mathsf{A}}^2 \mathbf{I}_{N_{\mathsf{A}}})$ and $\mathbf{v}_{\mathsf{B}} \sim \mathcal{CN}(0, \sigma_{\mathsf{B}}^2 \mathbf{I}_{N_{\mathsf{B}}})$ are the additive noise vectors at the receivers of station A and B, respectively.

The SI is “harmless” because it can be canceled at the receiver if the channel knowledge is available. The remaining part after canceling the SI only contains the unknown data transmitted from the other side. The decoding performance at the receivers is highly dependent on the accuracy of the channel knowledge. However, in reality the channel knowledge is not available at the receivers for free, and it has to be estimated. In traditional pilot-aided channel estimation schemes, the relay transmits orthogonal *pilot sequences* $\mathbf{S}_{\mathrm{t}}$ before transmitting data as shown in Fig. 7.2a. The pilot sequences occupy $T_{\mathrm{t}} \geq N_{\mathsf{R}}$ time slots, where $\mathbf{S}_{\mathrm{t}} \in \mathbb{C}^{N_{\mathsf{R}} \times T_{\mathrm{t}}}$ and $\mathbf{S}_{\mathrm{t}}\mathbf{S}_{\mathrm{t}}^H = P_{\mathsf{R}}T_{\mathrm{t}}/N_{\mathsf{R}}\mathbf{I}_{N_{\mathsf{R}}}$. The receiver correlates the received signals with the pilot sequences and obtains the channel estimates.

Since the pilot sequences do not carry data information, pilot-aided channel estimation schemes waste part of system resources. On the other hand, we observe that the SI also contains the information about the channel. In the following, we derive an achievable rate of the system in the BRC phase when the SI-aided channel estimation scheme is applied. We only calculate the achievable rate at the receiver of station A, while the same discussions also apply to the receiver of station B.

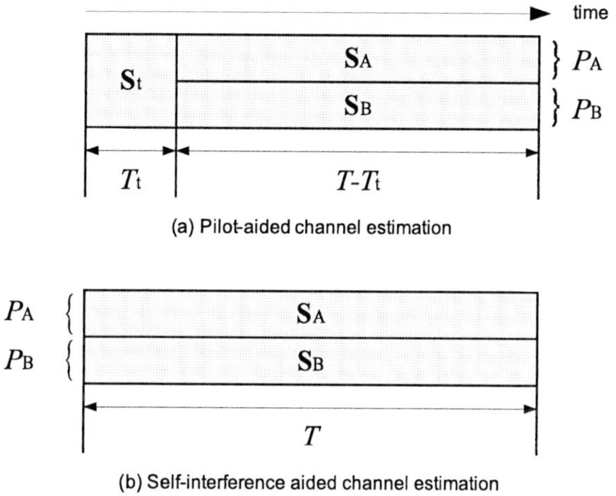

Fig. 7.2: Channel estimation schemes

7.3 Achievable Rates of SI-Aided Channel Estimation Scheme

We consider a block-fading channel model, i.e., the channels in the BRC phase remain constant for a coherence interval of T time slots and change independently between different coherence intervals. Each entry of the channel matrices $\mathbf{H}_\mathsf{A}$ and $\mathbf{H}_\mathsf{B}$ is an i.i.d. $\mathcal{CN}(0,1)$ random variable. We assume that the transmitted data symbols from the relay are uncorrelated in space and time. This is easily satisfied by most communication systems since the data bits are usually interleaved before transmission to break the correlations between neighboring data. The idea of the SI-aided channel estimation scheme is to exploit the SI to get an initial estimate of the channel, and use the initial channel estimate to decode the unknown data in the first T_s time slots, where $T_\mathrm{s} \geq N_\mathsf{R}$. After the data in the first T_s time slots are decoded, the decoded data are utilized to re-estimate the channel and help to provide better channel estimates for decoding the data in the following time slots. As the channel estimates improve, higher rate codes can be allocated in subsequent time slots of the coherence interval.

7.3.1 Initial Channel Estimate Using SI

In the SI-aided channel estimation scheme as shown in Fig. 7.2b, no pilot sequence is used. To emphasize the time reference, we rewrite the received signal at station A as

$$\mathbf{y}_{\mathsf{A},k} = \sqrt{\frac{P_{\mathsf{A}}}{N_{\mathsf{R}}}}\mathbf{H}_{\mathsf{A}}\mathbf{s}_{\mathsf{A},k} + \sqrt{\frac{P_{\mathsf{B}}}{N_{\mathsf{R}}}}\mathbf{H}_{\mathsf{A}}\mathbf{s}_{\mathsf{B},k} + \mathbf{v}_{\mathsf{A},k} \tag{7.7}$$

where the index k denotes the kth time slot, $k \leq T_{\mathrm{s}}$. In order to estimate the channel $\mathbf{H}_{\mathsf{A}}$ for decoding the data $\mathbf{s}_{\mathsf{B},k}$ at time slot k, we utilize the received signals in the remaining time slots $i \neq k, i \in \{1, \cdots, T\}$. As we will see later, estimating the channel only based on the received signals of the remaining time slots avoids the correlation between the estimated channel and the data $\mathbf{s}_{\mathsf{B},k}$, which facilitates the derivation of achievable rates of the system.

The signal received in the remaining time slots, i.e., the signals of the whole coherence interval except the kth time slot, can be expressed as

$$\bar{\mathbf{Y}}_{\mathsf{A}} = \mathbf{H}_{\mathsf{A}}\left(\sqrt{\frac{P_{\mathsf{A}}}{N_{\mathsf{R}}}}\bar{\mathbf{S}}_{\mathsf{A}} + \sqrt{\frac{P_{\mathsf{B}}}{N_{\mathsf{R}}}}\bar{\mathbf{S}}_{\mathsf{B}}\right) + \bar{\mathbf{V}}_{\mathsf{A}}, \tag{7.8}$$

$$= \sqrt{\frac{P_{\mathsf{A}}}{N_{\mathsf{R}}}}\mathbf{H}_{\mathsf{A}}\bar{\mathbf{S}}_{\mathsf{A}} + \underbrace{\sqrt{\frac{P_{\mathsf{B}}}{N_{\mathsf{R}}}}\mathbf{H}_{\mathsf{A}}\bar{\mathbf{S}}_{\mathsf{B}} + \bar{\mathbf{V}}_{\mathsf{A}}}_{\mathbf{W}}. \tag{7.9}$$

Here $\bar{\mathbf{Y}}_{\mathsf{A}} = [\mathbf{y}_{\mathsf{A},1}, \cdots, \mathbf{y}_{\mathsf{A},k-1}, \mathbf{y}_{\mathsf{A},k+1}, \cdots, \mathbf{y}_{\mathsf{A},T}]$ is the $N_{\mathsf{A}} \times (T-1)$ received signal matrix. Similarly, $\bar{\mathbf{S}}_{\mathsf{A}}$, $\bar{\mathbf{S}}_{\mathsf{B}}$ and $\bar{\mathbf{V}}_{\mathsf{A}}$ are $N_{\mathsf{R}} \times (T-1)$ matrices denoting the signals transmitted from the relay and the noise matrices at station A in the whole coherence interval except the kth time slot, respectively.

In (7.9), $\bar{\mathbf{S}}_{\mathsf{A}}$ and $\bar{\mathbf{Y}}_{\mathsf{A}}$ are both known to the receiver of station A. Also based on the statistics of $\mathbf{W}$, we can get a first estimate of $\mathbf{H}_{\mathsf{A}}$ and use it for decoding $\mathbf{s}_{\mathsf{B},k}$ in (7.7). By treating the unknown noisy symbol matrix $\mathbf{W}$ as noise and using the Bayesian Gauss-Markov Theorem in [129] (see Appendix 7.6.1), we can obtain a linear minimum mean square error (LMMSE) estimate of the channel as

$$\bar{\mathbf{H}}_{\mathsf{A}} = \bar{\mathbf{Y}}_{\mathsf{A}}\left[\frac{P_{\mathsf{A}}}{N_{\mathsf{R}}}\bar{\mathbf{S}}_{\mathsf{A}}^{H}\mathbf{R}_{\mathbf{h}}\bar{\mathbf{S}}_{\mathsf{A}} + \mathbf{R}_{\mathbf{w}}\right]^{-1}\left(\sqrt{\frac{P_{\mathsf{A}}}{N_{\mathsf{R}}}}\bar{\mathbf{S}}_{\mathsf{A}}^{H}\right)\mathbf{R}_{\mathbf{h}}$$

$$= \sqrt{\frac{N_{\mathsf{R}}}{P_{\mathsf{A}}}}\bar{\mathbf{Y}}_{\mathsf{A}}\left[\bar{\mathbf{S}}_{\mathsf{A}}^{H}\bar{\mathbf{S}}_{\mathsf{A}} + \frac{N_{\mathsf{R}}(P_{\mathsf{B}} + \sigma_{\mathsf{A}}^{2})}{P_{\mathsf{A}}}\mathbf{I}_{T-1}\right]^{-1}\bar{\mathbf{S}}_{\mathsf{A}}^{H}. \tag{7.10}$$

Here we have $\mathbf{R}_{\mathbf{h}} = \mathrm{E}\left\{\mathbf{h}\mathbf{h}^{H}\right\} = \mathbf{I}_{N_{\mathsf{R}}}$ and $\mathbf{R}_{\mathbf{w}} = \mathrm{E}\left\{\mathbf{w}\mathbf{w}^{H}\right\} = (P_{\mathsf{B}} + \sigma_{\mathsf{A}}^{2})\mathbf{I}_{T-1}$, where $\mathbf{h}^{H}$

and $\mathbf{w}^H$ represent the rows of $\mathbf{H}_\mathsf{A}$ and $\mathbf{W}$, respectively. In (7.10), we used the property that $\mathbf{H}_\mathsf{A}$ and $\bar{\mathbf{S}}_\mathsf{B}$ are independent. In addition, we denote the estimation error as $\tilde{\mathbf{H}}_\mathsf{A} = \mathbf{H}_\mathsf{A} - \bar{\mathbf{H}}_\mathsf{A}$. Let $\tilde{\mathbf{h}}^H$ denote the rows of $\tilde{\mathbf{H}}_\mathsf{A}$. The covariance matrix of $\tilde{\mathbf{h}}$ is [129]

$$\mathbf{R}_{\tilde{\mathbf{h}}} = \mathrm{E}_{\mathbf{H}_\mathsf{A},\bar{\mathbf{S}}_\mathsf{A}}\left\{\tilde{\mathbf{h}}\tilde{\mathbf{h}}^H\right\} \tag{7.11}$$

$$= \mathrm{E}_{\bar{\mathbf{S}}_\mathsf{A}}\left\{\left[\mathbf{R}_{\mathbf{h}}^{-1} + \frac{P_\mathsf{A}}{N_\mathsf{R}}\bar{\mathbf{S}}_\mathsf{A}\mathbf{R}_{\mathbf{w}}^{-1}\bar{\mathbf{S}}_\mathsf{A}^H\right]^{-1}\right\} \tag{7.12}$$

$$= \mathrm{E}_{\bar{\mathbf{S}}_\mathsf{A}}\left\{\left[\mathbf{I}_{N_\mathsf{R}} + \frac{P_\mathsf{A}}{N_\mathsf{R}(P_\mathsf{B}+\sigma_\mathsf{A}^2)}\bar{\mathbf{S}}_\mathsf{A}\bar{\mathbf{S}}_\mathsf{A}^H\right]^{-1}\right\}. \tag{7.13}$$

The mean square error (MSE) of the estimated channel is $\sigma^2_{\tilde{\mathbf{H}}_\mathsf{A}} = \mathrm{tr}\,\mathbf{R}_{\tilde{\mathbf{h}}}/N_\mathsf{R}$. When the entries of $\bar{\mathbf{S}}_\mathsf{A}$ and $\bar{\mathbf{S}}_\mathsf{B}$ are i.i.d. Gaussian, i.e., when we use a Gaussian codebook for transmitting data symbols, the MSE of the channel estimate can be expressed as [224]

$$\sigma^2_{\tilde{\mathbf{H}}_\mathsf{A}} = \frac{1}{N_\mathsf{R}}\,\mathrm{tr}\,\mathbf{R}_{\tilde{\mathbf{h}}}$$

$$= \frac{1}{N_\mathsf{R}}\int_0^\infty\left[1 + \frac{P_\mathsf{A}}{N_\mathsf{R}(P_\mathsf{B}+\sigma_\mathsf{A}^2)}x\right]^{-1} \times \sum_{k=0}^{N_\mathsf{R}-1}\frac{k!\left[L_k^{T-1-N_\mathsf{R}}(x)\right]^2}{(k+T-1-N_\mathsf{R})!}x^{T-1-N_\mathsf{R}}e^{-x}dx \tag{7.14}$$

where $L_j^i(x)$ are the associated Laguerre polynomials.

After we get the channel estimate $\bar{\mathbf{H}}_\mathsf{A}$, we remove the SI from the received signal matrix as if the channel estimate $\bar{\mathbf{H}}_\mathsf{A}$ is the real channel matrix. The remaining signal at time slot k can be expressed as

$$\bar{\mathbf{y}}_{\mathsf{A},k} = \mathbf{y}_{\mathsf{A},k} - \sqrt{\frac{P_\mathsf{A}}{N_\mathsf{R}}}\bar{\mathbf{H}}_\mathsf{A}\mathbf{s}_{\mathsf{A},k} \tag{7.15}$$

$$= \sqrt{\frac{P_\mathsf{B}}{N_\mathsf{R}}}\bar{\mathbf{H}}_\mathsf{A}\mathbf{s}_{\mathsf{B},k} + \underbrace{\sqrt{\frac{P_\mathsf{A}}{N_\mathsf{R}}}\tilde{\mathbf{H}}_\mathsf{A}\mathbf{s}_{\mathsf{A},k} + \sqrt{\frac{P_\mathsf{B}}{N_\mathsf{R}}}\tilde{\mathbf{H}}_\mathsf{A}\mathbf{s}_{\mathsf{B},k} + \mathbf{v}_{\mathsf{A},k}}_{\mathbf{n}}. \tag{7.16}$$

Since $\bar{\mathbf{H}}_\mathsf{A}$ is the LMMSE estimation of the channel $\mathbf{H}$, the channel estimation error $\tilde{\mathbf{H}}_\mathsf{A}$ has zero-mean entries and is uncorrelated with $\bar{\mathbf{H}}_\mathsf{A}$ and $\mathbf{s}_{\mathsf{B},k}$. Thus, the noise term $\mathbf{n}$ in (7.16) is uncorrelated with the data $\mathbf{s}_{\mathsf{B},k}$ and has zero-mean entries. The noise variance at time slot k

is

$$\begin{aligned}
\sigma_{\mathbf{n}}^2 &= \frac{1}{N_{\mathsf{A}}} \operatorname{tr} \mathrm{E}\left\{\mathbf{n}\mathbf{n}^H\right\} \\
&= \frac{1}{N_{\mathsf{A}}} \mathrm{E} \operatorname{tr}\left[\frac{P_{\mathsf{A}}}{N_{\mathsf{R}}} \tilde{\mathbf{H}}_{\mathsf{A}}^H \tilde{\mathbf{H}}_{\mathsf{A}} \mathbf{s}_{\mathsf{A},k} \mathbf{s}_{\mathsf{A},k}^H\right] + \frac{1}{N_{\mathsf{A}}} \mathrm{E} \operatorname{tr}\left[\frac{P_{\mathsf{B}}}{N_{\mathsf{R}}} \tilde{\mathbf{H}}_{\mathsf{A}}^H \tilde{\mathbf{H}}_{\mathsf{A}} \mathbf{s}_{\mathsf{B},k} \mathbf{s}_{\mathsf{B},k}^H\right] + \frac{1}{N_{\mathsf{A}}} \mathrm{E} \operatorname{tr}\left[\mathbf{v}_{\mathsf{A},k} \mathbf{v}_{\mathsf{A},k}^H\right] \\
&= \frac{P_{\mathsf{A}}}{N_{\mathsf{A}} N_{\mathsf{R}}} \operatorname{tr}\left[\mathrm{E}\left(\tilde{\mathbf{H}}_{\mathsf{A}}^H \tilde{\mathbf{H}}_{\mathsf{A}}\right) \mathrm{E}\left(\mathbf{s}_{\mathsf{A},k} \mathbf{s}_{\mathsf{A},k}^H\right)\right] + \frac{P_{\mathsf{B}}}{N_{\mathsf{A}} N_{\mathsf{R}}} \operatorname{tr}\left[\mathrm{E}\left(\tilde{\mathbf{H}}_{\mathsf{A}}^H \tilde{\mathbf{H}}_{\mathsf{A}}\right) \mathrm{E}\left(\mathbf{s}_{\mathsf{B},k} \mathbf{s}_{\mathsf{B},k}^H\right)\right] + \sigma_{\mathsf{A}}^2 \\
&= (P_{\mathsf{A}} + P_{\mathsf{B}}) \sigma_{\tilde{\mathbf{H}}_{\mathsf{A}}}^2 + \sigma_{\mathsf{A}}^2.
\end{aligned} \tag{7.17}$$

Here we used the fact that the entries in $\mathbf{s}_{\mathsf{A},k}$ and $\mathbf{s}_{\mathsf{B},k}$ are uncorrelated and also $\tilde{\mathbf{H}}_{\mathsf{A}}$ is uncorrelated with $\mathbf{s}_{\mathsf{A},k}$ and $\mathbf{s}_{\mathsf{B},k}$. This is because $\bar{\mathbf{H}}_{\mathsf{A}}$ is obtained using the received signals in time slots $i \neq k, i \in \{1, \cdots, T\}$. Eq. (7.16) describes a system with a known channel $\bar{\mathbf{H}}_{\mathsf{A}}$ and noise $\mathbf{n}$ with variance $\sigma_{\mathbf{n}}^2$. According to [97, Theorem 1] (see Appendix 7.6.2), the worst case noise $\mathbf{n}$ has a zero-mean Gaussian distribution. An achievable rate of such a system can be calculated by substituting $\mathbf{n}$ with a Gaussian noise with the same variance $\sigma_{\mathbf{n}}^2$.

7.3.2 Improving Channel Estimation by Data-Aided Approach

The initial channel estimation using SI is subject to the residual error due to the unknown data part $\bar{\mathbf{S}}_{\mathsf{B}}$. Thus the MSE of the channel estimate $\bar{\mathbf{H}}_{\mathsf{A}}$ can still be high. On the other hand, after decoding the symbols in the first several time slots, we can re-estimate the channel by exploiting the decoded data. This is the commonly used *data-aided* approach. Note that the data-aided approach can only be started after the unknown data in the first $T_{\mathrm{s}} \geq N_{\mathsf{R}}$ time slots have been decoded. An initial channel estimate is indispensable to the data-aided approach.

Suppose the data in the first $k-1$ time slots have been decoded. Let

$$\mathbf{S} = \sqrt{\frac{P_{\mathsf{A}}}{N_{\mathsf{R}}}} \mathbf{S}_{\mathsf{A}[k-1]} + \sqrt{\frac{P_{\mathsf{B}}}{N_{\mathsf{R}}}} \mathbf{S}_{\mathsf{B}[k-1]} \tag{7.18}$$

where $\mathbf{S}_{\mathsf{A}[k-1]} = [\mathbf{s}_{\mathsf{A},1}, \cdots, \mathbf{s}_{\mathsf{A},k-1}]$ and $\mathbf{S}_{\mathsf{B}[k-1]} = [\mathbf{s}_{\mathsf{B},1}, \cdots, \mathbf{s}_{\mathsf{B},k-1}]$ denote the matrices composed of the transmitted signals in the first $k-1$ time slots. The received signal matrix in the first $k-1$ time slots is denoted as $\mathbf{Y} = [\mathbf{y}_{\mathsf{A},1}, \cdots, \mathbf{y}_{\mathsf{A},k-1}]$. The MMSE channel estimate based on $\mathbf{S}$ and $\mathbf{Y}$ can be written as

$$\hat{\mathbf{H}}_{\mathsf{A}} = \mathbf{Y}\left(\sigma_{\mathsf{A}}^2 \mathbf{I}_{k-1} + \mathbf{S}^H \mathbf{S}\right)^{-1} \mathbf{S}^H. \tag{7.19}$$

We also define the channel estimation error as $\check{\mathbf{H}}_{\mathsf{A}} = \mathbf{H}_{\mathsf{A}} - \hat{\mathbf{H}}_{\mathsf{A}}$. Let $\check{\mathbf{h}}^H$ denote the rows of

$\check{\mathbf{H}}_{\mathsf{A}}$. The covariance matrix $\mathbf{R}_{\check{\mathbf{h}}}$ of $\check{\mathbf{h}}$ is

$$\mathbf{R}_{\check{\mathbf{h}}} = \mathrm{E}\left\{\check{\mathbf{h}}\check{\mathbf{h}}^H\right\} = \mathrm{E}\left\{\left(\mathbf{I}_{N_{\mathsf{R}}} + \frac{1}{\sigma_{\mathsf{A}}^2}\mathbf{S}\mathbf{S}^H\right)^{-1}\right\} \tag{7.20}$$

where we used the property that each entry in $\mathbf{S}_{\mathsf{A}[k-1]}$ and $\mathbf{S}_{\mathsf{B}[k-1]}$ is uncorrelated in space and time. So the MSE of the channel estimation error is

$$\sigma_{\check{\mathbf{H}}_{\mathsf{A}}}^2 = \frac{1}{N_{\mathsf{R}}}\operatorname{tr}\mathrm{E}\left\{\check{\mathbf{h}}\check{\mathbf{h}}^H\right\}. \tag{7.21}$$

After canceling the SI, the received signal vector at time slot k is

$$\hat{\mathbf{y}}_{\mathsf{A},k} = \sqrt{\frac{P_{\mathsf{B}}}{N_{\mathsf{R}}}}\hat{\mathbf{H}}_{\mathsf{A}}\mathbf{s}_{\mathsf{B},k} + \underbrace{\sqrt{\frac{P_{\mathsf{A}}}{N_{\mathsf{R}}}}\check{\mathbf{H}}_{\mathsf{A}}\mathbf{s}_{\mathsf{A},k} + \sqrt{\frac{P_{\mathsf{B}}}{N_{\mathsf{R}}}}\check{\mathbf{H}}_{\mathsf{A}}\mathbf{s}_{\mathsf{B},k} + \mathbf{v}_{\mathsf{A},k}}_{\check{\mathbf{n}}}. \tag{7.22}$$

This describes a system with known channel $\hat{\mathbf{H}}_{\mathsf{A}}$ and noise term $\check{\mathbf{n}}$. We have the covariance matrix of the noise vector $\check{\mathbf{n}}$ as

$$\mathrm{E}\left(\check{\mathbf{n}}\check{\mathbf{n}}^H\right) = \left[(P_{\mathsf{A}} + P_{\mathsf{B}})\sigma_{\check{\mathbf{H}}_{\mathsf{A}}}^2 + \sigma_{\mathsf{A}}^2\right]\mathbf{I}_{N_{\mathsf{A}}}. \tag{7.23}$$

So the noise variance is

$$\sigma_{\check{\mathbf{n}}}^2 = (P_{\mathsf{A}} + P_{\mathsf{B}})\sigma_{\check{\mathbf{H}}_{\mathsf{A}}}^2 + \sigma_{\mathsf{A}}^2. \tag{7.24}$$

Again, for a system described in (7.22), the worst case for noise term $\check{\mathbf{n}}$ is when $\check{\mathbf{n}}$ is Gaussian. An achievable rate can be calculated by assuming $\check{\mathbf{n}} \sim \mathcal{CN}\left(0, \sigma_{\check{\mathbf{n}}}^2\mathbf{I}_{N_{\mathsf{A}}}\right)$.

7.3.3 Achievable Rates of Bidirectional Broadcast Channel

In this section, we derive achievable rates of the bidirectional broadcast channel when the SI-aided channel estimation scheme is applied. The idea is that we can allocate codes with different rates on different time slots of the coherence intervals, as shown in Fig. 7.3. Each time slot of a coherence interval can be considered as one use of T parallel channels with different SNRs. At the beginning of each coherence interval, only SI is available for channel estimation and the system model is shown in (7.16). Here the effective SNR is low. Low rate codes are allocated at those time slots so that they can be fully decoded after many channel uses, i.e., coherence intervals. After the data at the beginning time slots of each coherence interval are decoded, the data-aided approach can be used to improve the channel

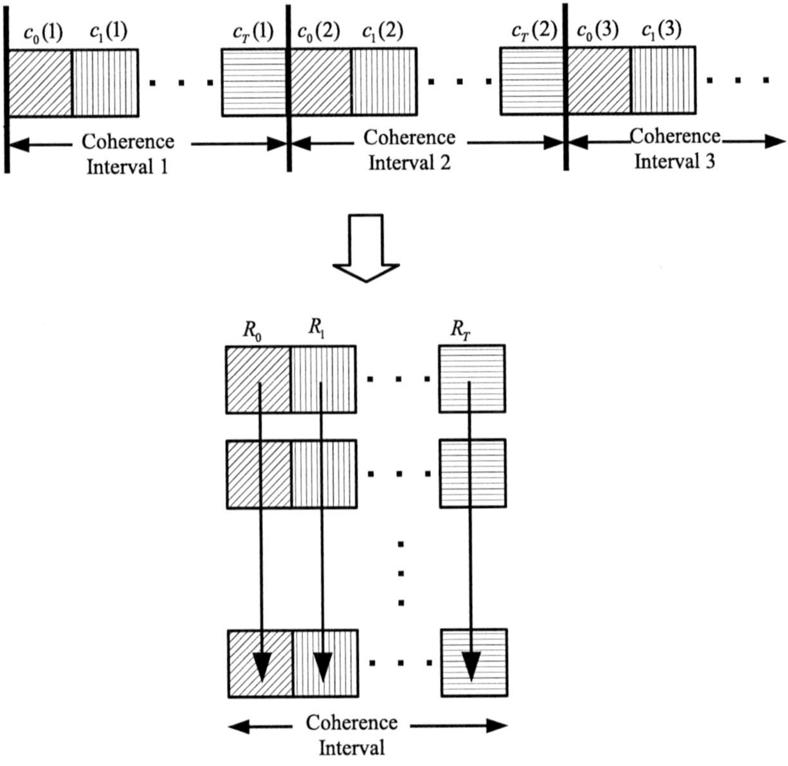

Fig. 7.3: Rate allocation in the time slots. Different hatching represents different code rate. Note coding is spread across many different coherence intervals.

estimation and the effective SNR, where the system model is shown in (7.22). Thus higher rate codes can be allocated at those time slots. Since we assume the channel changes independently between different coherence intervals, each coherence interval can be considered as a realization of the T parallel channels. According to channel coding theorem in a fast fading channel, those codes can be decoded *without error* after many independent realizations of those parallel channels. This method is also used in [46] to derive an achievable rate of the pilot-embedding schemes when data-aided approach is applied. Note this decoding scheme is different from [279], where we were interested in a practical scheme for improving the BER performance and the coded bits were spread across the whole coherence interval. There we utilized the error-correcting capability of convolutional codes in each iteration to correct the errors contained in the unknown data, and the decoded data might contain errors in each iteration.

Assuming the initial SI estimated channel $\bar{\mathbf{H}}_{\mathsf{A}}$ is used to decode the data in the first T_{s} time slots, the following expression gives an average achievable rate expression for the data decoded at station A in the BRC phase

$$C_{\mathsf{A}} = \frac{0.5}{T}\left\{\sum_{k=1}^{T_{\mathrm{s}}} I(\mathbf{y}_{\mathsf{A},k};\mathbf{s}_{\mathsf{B},k}|\bar{\mathbf{H}}_{\mathsf{A}}) + \sum_{k=T_{\mathrm{s}}+1}^{T} I(\mathbf{y}_{\mathsf{A},k};\mathbf{s}_{\mathsf{B},k}|\hat{\mathbf{H}}_{\mathsf{A}})\right\}$$

where the factor 0.5 is due to the fact that the rate can only be achieved after both the MAC and BRC phases, and the length of the two phases are equal. In the expression of C_{A}, the first term is the mutual information conveyed by the data symbols decoded by using the channel estimated purely by SI, and the second term represents the mutual information conveyed by the data symbols decoded by using the data-aided approach. When a Gaussian codebook is used, we can write

$$C_{\mathsf{A}}^{\mathrm{Gau}} = \frac{0.5T_{\mathrm{s}}}{T}\mathrm{E}\log\det\left(\mathbf{I}_{N_{\mathsf{A}}} + \rho_0\frac{\bar{\mathbf{H}}_{\mathsf{A}}\bar{\mathbf{H}}_{\mathsf{A}}^{H}}{N_{\mathsf{R}}}\right) + \frac{0.5}{T}\sum_{k=T_{\mathrm{s}}+1}^{T}\mathrm{E}\log\det\left(\mathbf{I}_{N_{\mathsf{A}}} + \rho_k\frac{\hat{\mathbf{H}}_{\mathsf{A}}\hat{\mathbf{H}}_{\mathsf{A}}^{H}}{N_{\mathsf{R}}}\right) \quad (7.25)$$

where $\rho_0 = P_{\mathsf{B}}/\sigma_{\bar{n}}^2$ and $\rho_k = P_{\mathsf{B}}/\sigma_{\check{n}}^2$ according to (7.17) and (7.24). However, the optimum choice of codebook of $\mathbf{s}_{\mathsf{A},k}$ and $\mathbf{s}_{\mathsf{B},k}$ is still an open question. This is because Gaussian codebooks maximize the mutual information only if the channel is perfectly known at the receiver. When the channel has to be estimated, Gaussian codebooks do not necessarily lead to the lowest $\sigma_{\tilde{\mathbf{H}}_{\mathsf{A}}}^2$ in the channel estimation in (7.16).

For QAM modulations, no explicit expression for mutual information is available. However, it can be obtained by Monte Carlo simulations by using the following expression

$$C_{\mathsf{A}}^{\mathrm{QAM}} = \frac{0.5T_{\mathrm{s}}}{T}\mathrm{E}\left\{\log\frac{p(\mathbf{y}_{\mathsf{A},k}|\mathbf{s}_{\mathsf{B},k},\bar{\mathbf{H}}_{\mathsf{A}})}{\sum_{\mathbf{s}_{\mathsf{B},k}} p(\mathbf{y}_{\mathsf{A},k}|\mathbf{s}_{\mathsf{B},k},\bar{\mathbf{H}}_{\mathsf{A}})\cdot p(\mathbf{s}_{\mathsf{B},k})}\right\}$$
$$+\frac{0.5}{T}\sum_{k=T_{\mathrm{s}}+1}^{T}\mathrm{E}\left\{\log\frac{p(\mathbf{y}_{\mathsf{A},k}|\mathbf{s}_{\mathsf{B},k},\hat{\mathbf{H}}_{\mathsf{A}})}{\sum_{\mathbf{s}_{\mathsf{B},k}} p(\mathbf{y}_{\mathsf{A},k}|\mathbf{s}_{\mathsf{B},k},\hat{\mathbf{H}}_{\mathsf{A}})\cdot p(\mathbf{s}_{\mathsf{B},k})}\right\}.$$

Here it is assumed that $\mathbf{s}_{\mathsf{B},k}$ is chosen from a QAM constellation with equal probabilities. According to (7.16) and (7.22), we have

$$p(\mathbf{y}_{\mathsf{A},k}|\mathbf{s}_{\mathsf{B},k},\bar{\mathbf{H}}_{\mathsf{A}}) = \frac{1}{(\pi\sigma_{\mathbf{n}}^2)^{N_{\mathsf{A}}}}\exp\left(-\frac{1}{\sigma_{\mathbf{n}}^2}\|\mathbf{y}_{\mathsf{A},k}-\bar{\mathbf{H}}_{\mathsf{A}}\mathbf{s}_{\mathsf{B},k}\|^2\right),$$
$$p(\mathbf{y}_{\mathsf{A},k}|\mathbf{s}_{\mathsf{B},k},\hat{\mathbf{H}}_{\mathsf{A}}) = \frac{1}{(\pi\sigma_{\check{\mathbf{n}}}^2)^{N_{\mathsf{A}}}}\exp\left(-\frac{1}{\sigma_{\check{\mathbf{n}}}^2}\|\mathbf{y}_{\mathsf{A},k}-\hat{\mathbf{H}}_{\mathsf{A}}\mathbf{s}_{\mathsf{B},k}\|^2\right).$$

Here we choose the noise vectors $\mathbf{n}\sim\mathcal{CN}(0,\sigma_{\mathbf{n}}^2\mathbf{I}_{N_{\mathsf{A}}})$ and $\check{\mathbf{n}}\sim\mathcal{CN}(0,\sigma_{\check{\mathbf{n}}}^2\mathbf{I}_{N_{\mathsf{A}}})$ in order to calculate the achievable rates following the discussions in Section 7.3.

7.4 Simulation Results

In this section, we compare the average achievable rates of the SI-aided channel estimation scheme with that of the traditional pilot-aided channel estimation scheme by using Monte Carlo simulations. The data-aided approach is utilized to improve the channel estimation in both cases. We consider a two-way relaying system where $N_{\mathsf{A}} = N_{\mathsf{R}} = N_{\mathsf{B}} = 2$. We only consider the BRC phase and equal power is allocated for transmission to station A and B, i.e., $P_{\mathsf{A}} = P_{\mathsf{B}} = P_{\mathsf{R}}/2$. The first N_{R} rows of the Hadamard matrix are taken as the pilot sequences in the pilot-aided scheme. In Fig. 7.4, SNR $= P_{\mathsf{R}}/\sigma_{\mathsf{A}}^2 = P_{\mathsf{R}}/\sigma_{\mathsf{B}}^2 = 10$dB, and we plot the average sum rate of station A and B. We can observe that the achievable rate increases with the coherence time. On the one hand, this is because the initial channel estimate gets better in the SI-aided channel estimation when the coherence interval gets longer; on the other hand, better channel estimate can be obtained at the end of each coherence interval in the data-aided approach as the coherence time increases. In Fig. 7.5, we show how the average achievable rate changes with the SNR for $T = 30$, where SNR $= P_{\mathsf{R}}/\sigma_{\mathsf{A}}^2 = P_{\mathsf{R}}/\sigma_{\mathsf{B}}^2$. We can observe that the gains of the SI-aided channel estimation scheme remain nearly as constants in the considered SNR range. This is due to the fact that the received power of both SI and the noise term $\mathbf{W}$ in (7.9) increase as SNR increases. In 4QAM modulations, we can also observe that the SI-aided channel estimation can achieve nearly the same rates at high SNR

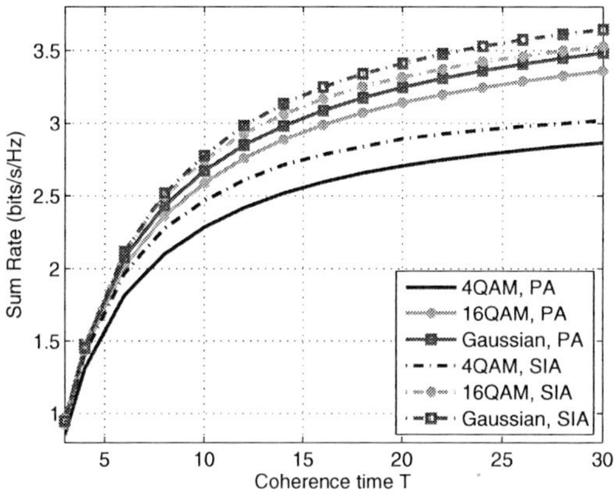

Fig. 7.4: Average achievable sum rates in the BRC phase of two-way relaying systems ($N_A = N_R = N_B = 2$, SNR $= P_R/\sigma_A^2 = P_R/\sigma_B^2 = 10$dB). "PA" and "SIA" denote pilot-aided and SI-aided channel estimation, respectively.

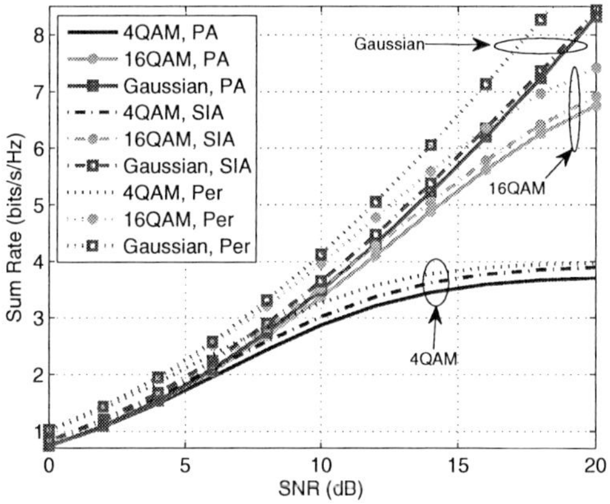

Fig. 7.5: Average achievable sum rates in the BRC phase of two-way relaying systems ($N_A = N_R = N_B = 2$, the coherence time $T = 30$). "Per" denotes the achievable rates when channel knowledge is perfectly available at the receivers.

as when perfect channel knowledge is available at the receivers.

7.5 Chapter Summary and Conclusions

We derived achievable rate expressions for bidirectional broadcast channels of two-way relaying systems when the SI-aided channel estimation scheme is applied. This scheme utilizes the SI to obtain an initial estimate of the channel and use data-aided approaches to improve the channel estimates. Simulation results showed that our scheme can achieve higher rates compared to traditional pilot-aided channel estimation schemes. The gain is relatively high in the low SNR regime. In high SNR regime, the performance of the SI-aided channel estimation scheme for 4QAM modulations is nearly as good as when perfect channel knowledge is available at the receivers.

7.6 Appendices

7.6.1 Bayesian Gauss-Markov Theorem

Theorem 7.6.1 ([129] P391). *If the data are described by the Bayesian linear model from*

$$\mathbf{x} = \mathbf{H}\boldsymbol{\theta} + \mathbf{w} \tag{7.26}$$

where $\mathbf{x}$ *is an* $N \times 1$ *data vector,* $\mathbf{H}$ *is a known* $N \times p$ *observation matrix,* $\boldsymbol{\theta}$ *is a* $p \times 1$ *random vector of parameters whose realization is to be estimated and has mean* $\mathrm{E}\{\boldsymbol{\theta}\}$ *and covariance matrix* $\mathbf{C}_{\theta\theta}$*, and* $\mathbf{w}$ *is an* $N \times 1$ *random vector with zero mean and covariance matrix* $\mathbf{C}_{\mathbf{w}}$ *and is uncorrelated with* $\boldsymbol{\theta}$ *(the joint PDF* $p(\mathbf{w}, \boldsymbol{\theta})$ *is otherwise arbitrary), then the LMMSE estimator of* $\boldsymbol{\theta}$ *is*

$$\hat{\boldsymbol{\theta}} = \mathrm{E}\{\boldsymbol{\theta}\} + \mathbf{C}_{\theta\theta}\mathbf{H}^T(\mathbf{H}\mathbf{C}_{\theta\theta}\mathbf{H}^T + \mathbf{C}_{\mathbf{w}})^{-1}(\mathbf{x} - \mathbf{H}\mathrm{E}\{\boldsymbol{\theta}\}) \tag{7.27}$$

$$= \mathrm{E}\{\boldsymbol{\theta}\} + (\mathbf{C}_{\theta\theta}^{-1} + \mathbf{H}^T\mathbf{C}_{\mathbf{w}}^{-1}\mathbf{H})^{-1}\mathbf{H}^T\mathbf{C}_{\mathbf{w}}^{-1}(\mathbf{x} - \mathbf{H}\mathrm{E}\{\boldsymbol{\theta}\}). \tag{7.28}$$

The performance of the estimator is measured by the error $\boldsymbol{\epsilon} = \boldsymbol{\theta} - \hat{\boldsymbol{\theta}}$ *whose mean is zero and whose covariance matrix is*

$$\mathbf{C}_{\epsilon} = \mathrm{E}_{\mathbf{x},\boldsymbol{\theta}}(\boldsymbol{\epsilon}\boldsymbol{\epsilon}^T) \tag{7.29}$$

$$= \mathbf{C}_{\theta\theta} - \mathbf{C}_{\theta\theta}\mathbf{H}^T(\mathbf{H}\mathbf{C}_{\theta\theta}\mathbf{H}^T + \mathbf{C}_{\mathbf{w}})^{-1}\mathbf{H}\mathbf{C}_{\theta\theta} \tag{7.30}$$

$$= (\mathbf{C}_{\theta\theta}^{-1} + \mathbf{H}^T\mathbf{C}_{\mathbf{w}}^{-1}\mathbf{H})^{-1}. \tag{7.31}$$

The error covariance matrix is also the minimum MSE matrix $\mathbf{M}_{\hat{\boldsymbol{\theta}}}$ *whose diagonal elements yield the minimum Bayesian MSE*

$$
\begin{aligned}
[\mathbf{M}_{\hat{\boldsymbol{\theta}}}]_{ii} &= [\mathbf{C}_{\boldsymbol{\epsilon}}]_{ii} && (7.32)\\
&= Bmse(\hat{\boldsymbol{\theta}}_i). && (7.33)
\end{aligned}
$$

7.6.2 Worst-Case Uncorrelated Additive Noise

Theorem 7.6.2 ([97]). *Consider the matrix-valued additive noise known channel*

$$\mathbf{x} = \sqrt{\frac{\rho}{M}}\mathbf{s}\mathbf{H} + \mathbf{v}, \tag{7.34}$$

where $\mathbf{H} \in \mathbb{C}^{M\times N}$ *is the known channel, and where the signal* $\mathbf{s} \in \mathbb{C}^{1\times M}$ *and the additive noise* $\mathbf{v} \in \mathbb{C}^{1\times N}$ *satisfy the power constraints*

$$\mathrm{E}\left\{\frac{1}{M}\mathbf{s}\mathbf{s}^H\right\} = 1 \quad \textit{and} \quad \mathrm{E}\left\{\frac{1}{N}\mathbf{v}\mathbf{v}^H\right\} = 1 \tag{7.35}$$

and are uncorrelated:

$$\mathrm{E}\left\{\mathbf{s}^H\mathbf{v}\right\} = \mathbf{0}_{M\times N}. \tag{7.36}$$

Let $R_{\mathbf{v}} = \mathrm{E}\left\{\mathbf{v}^H\mathbf{v}\right\}$ *and* $R_{\mathbf{s}} = \mathrm{E}\left\{\mathbf{s}^H\mathbf{s}\right\}$*. Then the worst-case noise has a zero-mean Gaussian distribution,* $\mathbf{v} \sim \mathcal{CN}(0, R_{\mathbf{v},\mathrm{opt}})$*, where* $R_{\mathbf{v},\mathrm{opt}}$ *is the minimizing noise covariance in*

$$C_{worst} = \min_{R_{\mathbf{v}},\mathrm{tr}(R_{\mathbf{v}})=N}\ \max_{R_{\mathbf{s}},\mathrm{tr}(R_{\mathbf{s}})=M} \mathrm{E}\log_2\det\left(\mathbf{I}_N + \frac{\rho}{M}R_{\mathbf{v}}^{-1}\mathbf{H}^H R_{\mathbf{s}}\mathbf{H}\right). \tag{7.37}$$

We also have the minimax property

$$I_{\mathbf{v}\sim\mathcal{CN}(0,R_{\mathbf{v},\mathrm{opt}}),\mathbf{s}}(\mathbf{x};\mathbf{s}) \le I_{\mathbf{v}\sim\mathcal{CN}(0,R_{\mathbf{v},\mathrm{opt}}),\mathbf{s}\sim\mathcal{CN}(0,R_{\mathbf{s},\mathrm{opt}})}(\mathbf{x};\mathbf{s}) = C_{worst} \le I_{\mathbf{v},\mathbf{s}\sim\mathcal{CN}(0,R_{\mathbf{s},\mathrm{opt}})}(\mathbf{x};\mathbf{s}), \tag{7.38}$$

where $R_{\mathbf{s},\mathrm{opt}}$ *is the maximizing signal covariance matrix in* (7.37)*. When the distribution on* $\mathbf{H}$ *is left rotationally invariant, i.e., when* $p(\Theta\mathbf{H}) = p(\mathbf{H})$ *for all* Θ *such that* $\Theta\Theta^H = \Theta^H\Theta = \mathbf{I}_M$*, then*

$$R_{\mathbf{s},\mathrm{opt}} = \mathbf{I}_M. \tag{7.39}$$

When the distribution on $\mathbf{H}$ *is right rotationally invariant, i.e. when* $p(\mathbf{H}\Theta) = p(\mathbf{H})$ *for all* Θ *such that* $\Theta\Theta^H = \Theta^H\Theta = \mathbf{I}_N$*, then*

$$R_{\mathbf{v},\mathrm{opt}} = \mathbf{I}_N. \tag{7.40}$$

Chapter 8

Asymmetric Data Rate Transmission in Two-Way Relaying Systems With Network Coding

In two-way decode-and-forward (DF) relaying systems, the major problem faced by the network coding scheme is how to transmit with asymmetric data rates to the user stations according to their individual link qualities in the broadcast (BRC) phase. This chapter proposes a novel transmission strategy to solve this problem. The idea is to *utilize* the bit-level self-interference (SI) when the network coding scheme is applied. In the proposed scheme, the weaker link receiver exploits the *a priori* bit information in each received data symbol, so that it only needs to decode on a subset of the signal constellation. Subject to the same bit error rate constraint, the weaker link receiver can decode at lower signal-to-noise ratio (SNR) compared to the stronger link. The signal labeling used for mapping bits to symbols at the relay is shown to be crucial for the performance at the receivers, and we provide the criteria and methods for finding the optimized signal labeling schemes. Simulations show that the proposed transmission scheme can be applied to practical scenarios with asymmetric channel qualities, and the optimized labeling schemes are able to significantly outperform conventional ones at both receivers.

8.1 Introduction

The *two-way relaying* protocol [193] considers the scenario that two half-duplex wireless stations exchange data via another half-duplex wireless relay. Such a relaying protocol has been shown to be able to compensate for a large portion of the spectral efficiency loss that is due to the half-duplex constraint of practical relays. We consider two-way decode-and-forward (DF) relaying systems in this chapter, where the data from the two stations are exchanged in two phases: the multiple-access (MAC) phase and the broadcast (BRC) phase. In the MAC phase, the two stations transmit their data to the relay and the relay decodes the received signal; the decoded data are combined at the relay and are retransmitted to the two stations in the BRC phase (see Chapter 2). The basic idea of the two-way relaying protocol is that the back-propagated known data in the received signals at the user stations, which is called self-interference (SI), can be canceled and does not degrade the decoding performance at the receivers. There are two major practical data-combining schemes, i.e., the *superposition coding* (SPC) scheme [193] and the *network coding* scheme [260]. The SPC scheme combines the decoded data on the symbol level at the relay, and the network coding scheme combines the decoded data on the bit level (see Chapter 4). Traditional two-way relaying schemes simply *cancel* the SI instead of *utilize* it. In Chapter 6, we proposed a channel estimation scheme to utilize the SI when the SPC scheme is applied in two-way DF relaying systems, where we showed that the SI can play a similar role as pilot sequences for channel estimation. In the SPC scheme, the SI is on the symbol level. How to utilize the bit-level SI when the network coding scheme is applied at the relay is the topic of this chapter.

When the network coding scheme is applied in the BRC phase of two-way DF relaying systems, the relay combines the data on the bit level using the XOR operation before modulation. Compared to the SPC scheme, the network coding scheme does not split the power for transmitting the two sets of data, which avoids the signal-to-noise ratio (SNR) degradation after the SI cancellation [88]. However, the network coding scheme requires that both receiving stations decode the combined data bits from the same transmit symbols in the BRC phase. It was shown in [170] that network coding is optimal for transmitting the same amount of data to both stations in symmetric channel conditions, i.e., when the channel qualities from the relay to the two stations are equal. However, asymmetric channel conditions are very common in reality, where the channel quality from the relay to one user station is much better than the channel to the other. In such a situation, it is not preferable for the relay to transmit at a data rate according to the weak link channel since it sacrifices the strong link user. When the channel qualities to the two stations are asymmetric, how to transmit data, so that the data rates from the relay are not limited by the weaker link of the two stations, is an important problem for practical systems. From the information theoretic aspects, it was shown in [170] and [264] that by using random coding approaches, it is possible for the relay to transmit information rates equal to the individual link capacities simultaneously to the two receiving stations. Moreover, the authors of [265] and [15] respectively proposed schemes of combining channel coding for binary transmission and lattice coding with network coding in two-way relaying systems and derived the achievable rate regions for each case. However, real-world applications call for practical and low-complexity transmission schemes, especially for multi-antenna systems.

On the other hand, the data traffic in real-world two-way relaying systems may not always be symmetric. For example, consider a scenario that a dedicated relay station facilitates the exchange of data between a base station and a mobile station in a cellular system. Usually the mobile station has less data to transmit to the base station in the uplink than the base station has for the downlink. In this case, combining equal amount of data at the relay is not preferable in practice either.

In this chapter, we propose a novel transmission scheme for the BRC phase of two-way DF relaying systems when network coding is applied. In the proposed scheme, the data rates transmitted by the relay to the two receiving stations can be adjusted according to their individual link qualities subject to certain criterion, such as the bit error rate (BER) constraints. We call it *asymmetric data rate transmission*. The proposed scheme has low complexity and can be applied to systems with single or multiple antennas. The core idea is to utilize the bit-level SI when the network coding scheme is applied. In the proposed scheme, the relay combines the data in such a way that some bits in each transmit symbol are *a priori* known to the weaker link receiver. That receiver can hence exploit the known bits and only needs to decode on a subset of the transmit signal constellation. Therefore, the *a priori* bit information is translated into the coding gain, and enables the weaker link receiver to achieve the same decoding performance as the stronger link. We show that the *signal labeling* at the relay, i.e., the assignment of bit patterns to each symbol in the signal constellation, plays an important role in the system performance. The criteria and methods for finding the optimized labeling schemes are also proposed. Furthermore, we show an example of systems with 8PSK constellation on each transmit antenna. The optimization criteria are the mutual information and the error floor when the received SNR is high. We provide the optimized labeling schemes with their performance results. To the best of our knowledge, this is the first scheme that exploits the *a priori* bit information on the symbol level for network coding schemes in two-way DF relaying systems.

Our Contributions: The contributions of this chapter can be summarized as follows.

- We propose a practical coding and modulation scheme for the BRC phase transmission in two-way DF relaying systems when the network coding scheme is applied at the relay. The proposed scheme is able to transmit with asymmetric data rates from the

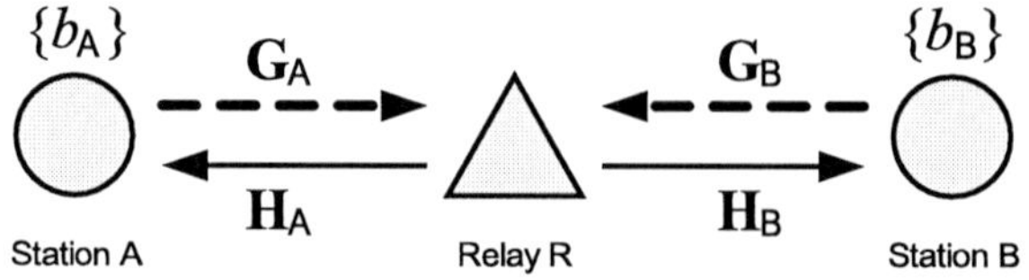

Fig. 8.1: Two-way DF relaying system. The dashed arrows represent the transmission in the MAC phase, and the solid arrows represent the transmission in the BRC phase.

relay to the receivers according to their individual link qualities. It overcomes the traditional problem that the transmission rates of the network coding scheme is limited by the weakest link.

- We provide the criteria and methods for finding the optimized signal labeling schemes that map bits to symbols at the relay, which is crucial for the performance of the proposed scheme.
- To the best of our knowledge, this is the first scheme that exploits the bit-level SI for network coding schemes in two-way DF relaying systems.

The rest of this chapter is organized as follows: the two-way DF relaying protocol with network coding is recapitulated in Section 8.2. The details of the proposed transmission scheme are discussed in Section 8.3, where we provide the transceiver structures and show how they work. The criterion and method for designing optimized labeling schemes are discussed in Section 8.4. Simulation results that compare the decoding performance of the optimized labeling with that of conventional ones are presented in Section 8.5. Conclusions are drawn in Section 8.6.

8.2 System Model

We consider a relaying system where two wireless stations A and B exchange data via a half-duplex relay as shown in Fig. 8.1. We assume that there is no direct connection between stations A and B (e.g., due to shadowing). The number of antennas at station A, the relay R and station B are N_A, N_R and N_B, respectively. $\mathbf{G}_\mathsf{A} \in \mathbb{C}^{N_\mathsf{R} \times N_\mathsf{A}}$ and $\mathbf{G}_\mathsf{B} \in \mathbb{C}^{N_\mathsf{R} \times N_\mathsf{B}}$ respectively denote the channel matrices from stations A and B to the relay in the MAC phase. $\mathbf{H}_\mathsf{A} \in \mathbb{C}^{N_\mathsf{A} \times N_\mathsf{R}}$ and $\mathbf{H}_\mathsf{B} \in \mathbb{C}^{N_\mathsf{B} \times N_\mathsf{R}}$ denote the channel matrices from the relay to stations A and B in the BRC phase, respectively. Station A wants to send the bit sequence $\{b_\mathsf{A}\}$ to station B, and station B wants to send the bit sequence $\{b_\mathsf{B}\}$ to station A.

When the two-way DF relaying protocol is applied, the bit sequences $\{b_{\mathsf{A}}\}$ and $\{b_{\mathsf{B}}\}$ are respectively modulated and transmitted to the relay by stations A and B in the MAC phase. The receiver structure at the relay can be found in, e.g., [240].[1] In the following, we focus on the BRC phase and assume the MAC phase has been completed, i.e., the bit sequences $\{b_{\mathsf{A}}\}$ and $\{b_{\mathsf{B}}\}$ have already been transmitted to the relay.

In the BRC phase, we apply network coding [260] at the relay to retransmit the data. The basic idea is that the relay combines the decoded bit sequences on the bit level using the XOR operation, and remodulates the combined bit sequence into transmit symbols, i.e.,

$$\{b_{\mathsf{A}} \oplus b_{\mathsf{B}}\} = \{b_{\mathsf{R}}\} \longmapsto \{\mathbf{s}_{\mathsf{R}}\}. \tag{8.1}$$

At one time slot, the received signals at station A and B are

$$\mathbf{y}_{\mathsf{A}} = \mathbf{H}_{\mathsf{A}}\mathbf{s}_{\mathsf{R}} + \mathbf{n}_{\mathsf{A}} \tag{8.2}$$

$$\mathbf{y}_{\mathsf{B}} = \mathbf{H}_{\mathsf{B}}\mathbf{s}_{\mathsf{R}} + \mathbf{n}_{\mathsf{B}} \tag{8.3}$$

where $\mathbf{n}_{\mathsf{A}} \sim \mathcal{CN}(0, \sigma_{\mathsf{A}}^2 \mathbf{I}_{N_{\mathsf{A}}})$ and $\mathbf{n}_{\mathsf{B}} \sim \mathcal{CN}(0, \sigma_{\mathsf{B}}^2 \mathbf{I}_{N_{\mathsf{B}}})$ are the additive noise vectors at stations A and B, respectively. The two stations demodulate the received signals and reveal the unknown data bits by XOR-ing the decoded data $\{\hat{b}_{\mathsf{R}}\}$ with their own transmitted data on the bit level. That is,

$$\{\hat{b}_{\mathsf{B}}\} = \{\hat{b}_{\mathsf{R}} \oplus b_{\mathsf{A}}\}, \quad \text{at station A;}$$

$$\{\hat{b}_{\mathsf{A}}\} = \{\hat{b}_{\mathsf{R}} \oplus b_{\mathsf{B}}\}, \quad \text{at station B.}$$

Since both receiving stations have to decode the data contained in the symbol $\mathbf{s}_{\mathsf{R}}$, it was conventionally thought that the relay must transmit at a data rate that can be supported by both links. This sacrifices the stronger link, and is not desirable in practice. In the following, we propose a practical scheme that can transmit with asymmetric data rates simultaneously from the relay to the two stations according to their individual link qualities in the BRC phase.

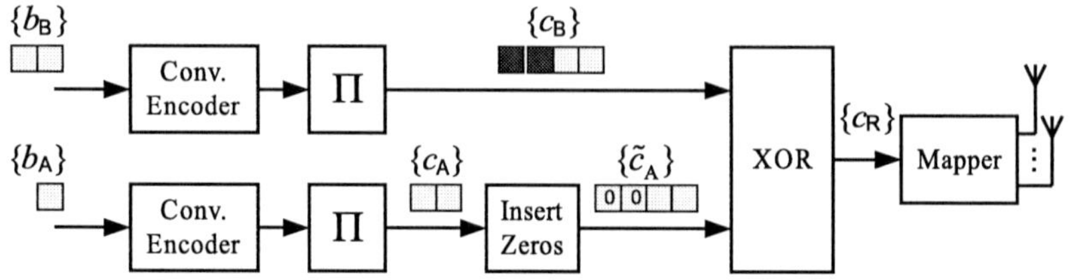

(a) Transmitter at Relay R

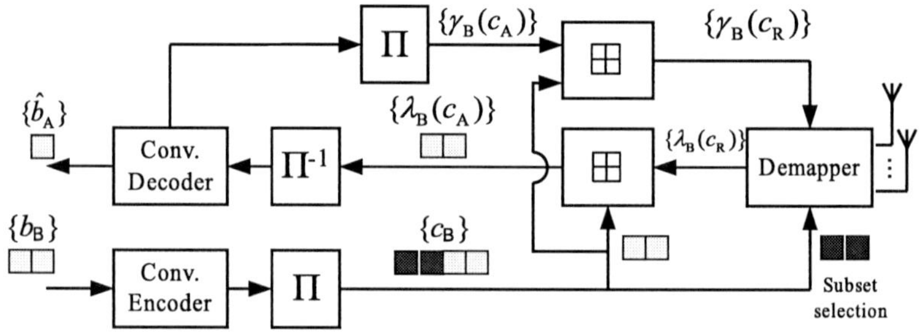

(b) Receiver at Station B (weaker link)

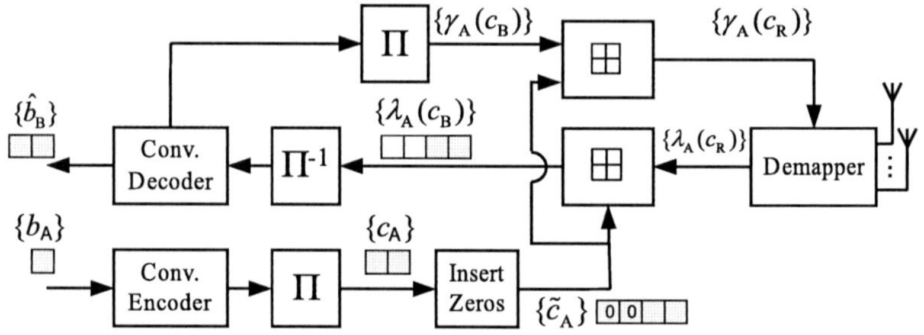

(c) Receiver at Station A (stronger link)

Fig. 8.2: Transmitter and receiver diagrams for asymmetric data rate transmission in the BRC phase. The box-plus "⊞" module at the receiver side is defined in (8.13).

8.3 Transceiver Structures for Asymmetric Data Rate Transmission

The transmitter and receiver diagrams of the proposed scheme are shown in Fig. 8.2. We assume the link from the relay to station A has better channel quality, e.g., higher signal-to-noise ratio (SNR), than the link to station B in the BRC phase. The aim of the proposed scheme is to utilize the stronger link to transmit more data bits per channel use to station A, while at the same time transmitting to station B at a data rate that can be supported by its link.[2] We assume the channel matrices $\mathbf{H}_{\mathsf{A}}$ and $\mathbf{H}_{\mathsf{B}}$ are respectively known to stations A and B. The relay only knows the receive SNRs at the two stations. The system diagram in Fig. 8.2 applies bit-interleaved coded modulation with iterative decoding (BICM-ID). However, the proposed idea can actually be applied to both coded and uncoded systems.

8.3.1 Transmission Strategy at the Relay

Fig. 8.2(a) shows the proposed transmitter structure at the relay R. The information bit sequences $\{b_{\mathsf{A}}\}$ and $\{b_{\mathsf{B}}\}$ decoded in the MAC phase are encoded individually by a convolutional encoder with coding rate r. Here we assume the two sequences are encoded by the same encoder for ease of implementation at the relay. Without loss of generality, we consider a spatial-multiplexing structure with N_{R} independent data streams. Similar discussions apply to less data streams, e.g., transmitting $\min(N_{\mathsf{R}}, N_{\mathsf{A}}, N_{\mathsf{B}})$ streams to enable efficient decoding. In each transmission, the relay sends $r \cdot mN_{\mathsf{R}}$ information bits to station A and $r \cdot nN_{\mathsf{R}}$ information bits to station B simultaneously, where $n < m$. The relay determines m and n according to the knowledge of the average receive SNRs at A and B. The transmit symbol on each relay antenna belongs to the M-ary QAM or PSK symbol alphabets, where $M = 2^m$.

The transmitter works as follows: the output bits of the convolutional encoders are bitwise interleaved to form the code sequences $\{c_{\mathsf{A}}\}$ and $\{c_{\mathsf{B}}\}$, where $c_{\mathsf{A}}, c_{\mathsf{B}} \in \{0, 1\}$. Then the bit-interleaved codewords are respectively partitioned into groups of nN_{R} and mN_{R} bits. Each pair of the two corresponding bit groups is denoted as $\mathbf{c}_{\mathsf{A}} = [c_{\mathsf{A}}^{nN_{\mathsf{R}}}, \cdots, c_{\mathsf{A}}^{1}]^T$ and $\mathbf{c}_{\mathsf{B}} =$

[1]A simpler, albeit suboptimal, approach is to transmit $\{b_{\mathsf{A}}\}$ and $\{b_{\mathsf{B}}\}$ separately to the relay as in [260].

[2]This requires more data bits $\{d_{\mathsf{B}}\}$ to be available at the relay for the BRC phase transmission. Hence, it may require station B to use more temporal or spectral resources, e.g., more subcarriers in OFDMA systems, to transmit those data bits to the relay in the MAC phase. This can be motivated by e.g., the relay has a buffer and accumulated the data decoded from the MS. Another way to motivate this is that in a OFDMA system, the MS use more subcarriers to transmit to the relay while the BS use less subcarriers to transmit to the relay in the MAC phase.

$[c_{\mathsf{B}}^{mN_{\mathsf{R}}}, \cdots, c_{\mathsf{B}}^{1}]^T$, respectively. For each $\mathbf{c}_{\mathsf{A}}$, we insert $(m-n)N_{\mathsf{R}}$ *dummy* zeros in it and obtain

$$\tilde{\mathbf{c}}_{\mathsf{A}} = [\underbrace{0, \cdots, 0}_{(m-n)N_{\mathsf{R}}}, c_{\mathsf{A}}^{nN_{\mathsf{R}}}, \cdots, c_{\mathsf{A}}^{1}]^T.$$

Those dummy zeros contain no information. Their positions are fixed and known to both stations A and B. After inserting zeros, the corresponding bits of $\tilde{\mathbf{c}}_{\mathsf{A}}$ and $\mathbf{c}_{\mathsf{B}}$ are combined into $\mathbf{c}_{\mathsf{R}} = [c_{\mathsf{R}}^{mN_{\mathsf{R}}}, \cdots, c_{\mathsf{R}}^{1}]^T$ using the XOR operation, i.e.,

$$\mathbf{c}_{\mathsf{R}} = \mathbf{c}_{\mathsf{B}} \oplus \tilde{\mathbf{c}}_{\mathsf{A}} \tag{8.4}$$

$$= [\underbrace{c_{\mathsf{B}}^{mN_{\mathsf{R}}}, \cdots, c_{\mathsf{B}}^{nN_{\mathsf{R}}+1}}_{(m-n)N_{\mathsf{R}}}, c_{\mathsf{B}}^{nN_{\mathsf{R}}} \oplus c_{\mathsf{A}}^{nN_{\mathsf{R}}}, \cdots, c_{\mathsf{B}}^{1} \oplus c_{\mathsf{A}}^{1}]^T. \tag{8.5}$$

Due to the dummy zeros in $\tilde{\mathbf{c}}_{\mathsf{A}}$, $[c_{\mathsf{B}}^{mN_{\mathsf{R}}}, \cdots, c_{\mathsf{B}}^{nN_{\mathsf{R}}+1}]$ are kept unchanged after the XOR operation when $\mathbf{c}_{\mathsf{R}}$ is generated, and those bits are known to station B.

Each of the combined bit group $\mathbf{c}_{\mathsf{R}}$ is mapped to an N_{R} dimensional complex symbol vector $\mathbf{s}_{\mathsf{R}} = [s_{\mathsf{R},1}, \cdots, s_{\mathsf{R},N_{\mathsf{R}}}]^T = \mu(\mathbf{c}_{\mathsf{R}})$ on the relay antennas by the mapper, where $\mu(\cdot)$ denotes the mapping function. Each element $s_{\mathsf{R},i}$, $i \in \{1, \cdots, N_{\mathsf{R}}\}$, belongs to the M-ary QAM or PSK symbol alphabets $\mathcal{A} = \{a_1, \cdots, a_M\}$, where $M = 2^m$. The N_{R} dimensional signal constellation is denoted as $\mathcal{X}$, i.e.,

$$\mathcal{X} = \{\mathbf{s} \mid \mathbf{s} = \mu(\mathbf{c}), \forall \mathbf{c} \in \{0,1\}^{mN_{\mathsf{R}}}\} = \mathcal{A}^{N_{\mathsf{R}}}$$

and $|\mathcal{X}| = 2^{mN_{\mathsf{R}}}$. Furthermore, $\mathrm{E}(\mathbf{s}_{\mathsf{R}}\mathbf{s}_{\mathsf{R}}^H) = P_{\mathsf{R}}/N_{\mathsf{R}}\mathbf{I}_{N_{\mathsf{R}}}$, where P_{R} is the average transmit power constraint at the relay in the BRC phase. Here we allocate equal power on each data symbol at the relay antennas. Both the encoding scheme and the mapping scheme $\mu(\cdot)$ at the relay are known to stations A and B. Note we only require there to be one-to-one mapping between each bit block $\mathbf{c} \in \{0,1\}^{mN_{\mathsf{R}}}$ and the symbol vector $\mathbf{s} \in \mathcal{X}$, which is different from the usual schemes that map bits separately to symbols on each antenna.

8.3.2 Decoding Strategies at the Receivers

Upon receiving $\mathbf{y}_{\mathsf{A}}$ and $\mathbf{y}_{\mathsf{B}}$ as in (8.2) and (8.3), stations A and B demodulate the received signals, and reveal the unknown data based on the bits contained in $\mathbf{s}_{\mathsf{R}}$ and their own data bits. The receiver structures are shown in Fig. 8.2(b) and Fig. 8.2(c). There are two important issues in the design of the receivers: firstly, in order to make it possible for the weaker link receiver B to decode at lower SNR, we must *exploit* its *a priori* known bits contained in $\mathbf{c}_{\mathsf{R}}$ in the demapping process; secondly, we use the box-plus "$\boxplus$" module defined in Section 8.3.2.2 to convert the log-likelihood ratios (LLRs) of $\{c_\mathsf{R}\}$ to that of $\{c_\mathsf{A}\}$ or $\{c_\mathsf{B}\}$ between the demapper and the channel decoder, so that the reliability information for the decoded bits is kept intact in the iterative process.

8.3.2.1 Exploiting a priori bit information at station B

The transmit symbol vector $\mathbf{s}_\mathsf{R}$ contains different *useful* bits for stations A and B: in order to obtain $\mathbf{c}_\mathsf{B}$, every bit $c_\mathsf{R}^i, \forall i \in \{1, \cdots, mN_\mathsf{R}\}$, is useful for station A according to (8.5) and needs to be decoded, whereas the useful bits for station B are c_R^i, $i \in \{1, \cdots, nN_\mathsf{R}\}$, because the $(m-n)N_\mathsf{R}$ bits $[c_\mathsf{B}^{mN_\mathsf{R}}, \cdots, c_\mathsf{B}^{nN_\mathsf{R}+1}]$ in (8.5) are *a priori* known at its receiver. Instead of decoding every bit in $\mathbf{c}_\mathsf{R}$ and discarding the known bits, we propose to exploit this *a priori* bit information, so that the receiver B only needs to demap on the subset of the transmit signal constellation whose labels contain $[c_\mathsf{B}^{mN_\mathsf{R}}, \cdots, c_\mathsf{B}^{nN_\mathsf{R}+1}]$ at the corresponding positions.

An motivating example with 8PSK ($m = 3$) transmission is given in Fig. 8.3, where $N_\mathsf{R} = 1$ and two labeling schemes for the transmit symbol s_R are shown: the Gray labeling in Fig. 8.3(a) and the set partitioning (SP) labeling [231] in Fig. 8.3(b). We assume $n = 2$ and $c_\mathsf{B}^3 = 0$. Since station B knows $c_\mathsf{B}^3 = 0$, it only needs to consider the symbols whose 3rd bit is 0 (indicated by circles in Fig. 8.3) for the demapping process. Given the known bits $[c_\mathsf{B}^{mN_\mathsf{R}}, \cdots, c_\mathsf{B}^{nN_\mathsf{R}+1}]$, we define the subset of symbol constellation, whose labels contain those known bits at the corresponding positions, as $\mathcal{S}(c_\mathsf{B}^{mN_\mathsf{R}}, \cdots, c_\mathsf{B}^{nN_\mathsf{R}+1}) \subset \mathcal{X}$, i.e.,

$$\mathcal{S}(c_\mathsf{B}^{mN_\mathsf{R}}, \cdots, c_\mathsf{B}^{nN_\mathsf{R}+1}) = \left\{\mathbf{s} \mid c_{mN_\mathsf{R}}(\mathbf{s}) = c_\mathsf{B}^{mN_\mathsf{R}}, \cdots, c_{nN_\mathsf{R}+1}(\mathbf{s}) = c_\mathsf{B}^{nN_\mathsf{R}+1}, \mathbf{s} \in \mathcal{X}\right\} \quad (8.6)$$

where $c_j(\mathbf{s})$ denotes the jth bit associated with the label of symbol $\mathbf{s}$. Given $c_\mathsf{B}^3 = 0$ in Fig. 8.3, the subset to be demapped at station B can be denoted as $\mathcal{S}(0)$. Fig. 8.3 also shows that different labeling schemes lead to different subsets for the given *a priori* bits and influence the decoding performance at the receivers. Given $c_\mathsf{B}^3 = 0$ for SP labeling, the components in $\mathcal{S}(0)$ are same as those of QPSK, and the minimum Euclidean distance (MED) between symbols in $\mathcal{S}(0)$ is increased compared to that of the original 8PSK. This leads to better decoding performance at station B when it demaps only on $\mathcal{S}(0)$ instead of on $\mathcal{X}$. However, the MED of $\mathcal{S}(0)$ for Gray labeling is not increased, and simulations show that demapping only on $\mathcal{S}(0)$ in Gray labeling does not improve the decoding performance. How to find the optimized labeling schemes will be discussed in Section 8.4.

The demappers at stations A and B work as follows. In each iteration, the soft-output demapper of station A calculates the *a posteriori* LLR values $\Lambda_\mathsf{A}(c_\mathsf{R}^i)$ for each of the coded

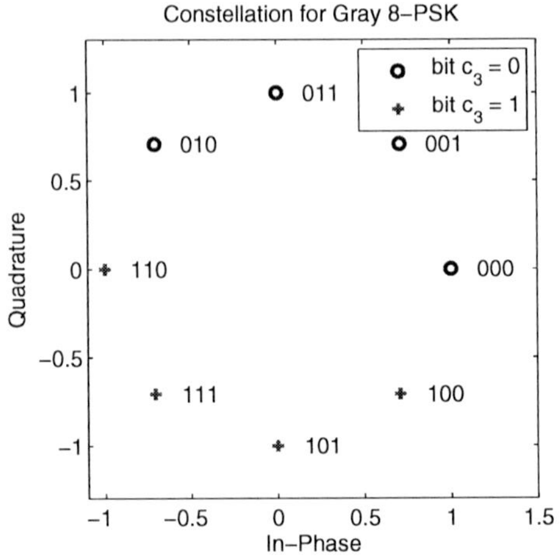

(a) 8PSK with Gray labeling

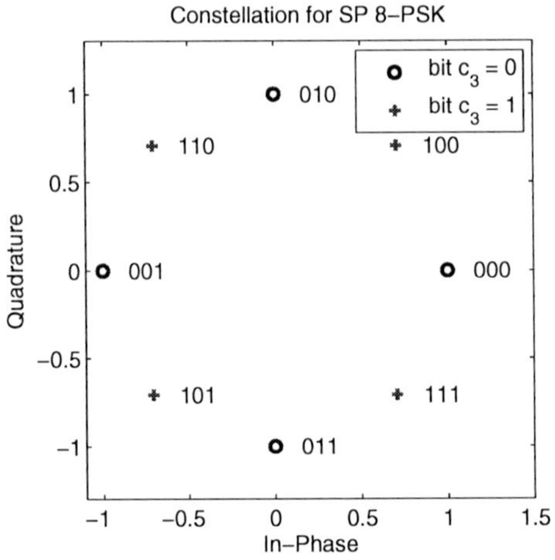

(b) 8PSK with SP labeling

Fig. 8.3: 8PSK labeling schemes, single transmit antenna case.

bit c_{R}^i, $i \in \{1, \cdots, mN_{\mathsf{R}}\}$ associated with $\mathbf{y}_{\mathsf{A}}$ [107]:

$$\Lambda_{\mathsf{A}}(c_{\mathsf{R}}^i) = \ln \frac{p(c_{\mathsf{R}}^i = 1|\mathbf{y}_{\mathsf{A}})}{p(c_{\mathsf{R}}^i = 0|\mathbf{y}_{\mathsf{A}})} \tag{8.7}$$

$$\approx \min_{\mathbf{s}_{\mathsf{R}} \in \mathcal{X}_i^0} \frac{\|\mathbf{y}_{\mathsf{A}} - \mathbf{H}_{\mathsf{A}}\mathbf{s}_{\mathsf{R}}\|^2}{\sigma_{\mathsf{A}}^2} - \min_{\mathbf{s}_{\mathsf{R}} \in \mathcal{X}_i^1} \frac{\|\mathbf{y}_{\mathsf{A}} - \mathbf{H}_{\mathsf{A}}\mathbf{s}_{\mathsf{R}}\|^2}{\sigma_{\mathsf{A}}^2} \tag{8.8}$$

where $\mathcal{X}_i^1$ and $\mathcal{X}_i^0$ represent the sets of transmit symbol vectors whose ith bit labeling is 1 and 0, respectively. Similarly, the *a posteriori* LLR values $\Lambda_{\mathsf{B}}(c_{\mathsf{R}}^i)$ for the coded bits c_{R}^i, $i \in \{1, \cdots, nN_{\mathsf{R}}\}$ are calculated at the demapper B as

$$\Lambda_{\mathsf{B}}(c_{\mathsf{R}}^i) = \ln \frac{p(c_{\mathsf{R}}^i = 1|\mathbf{y}_{\mathsf{B}})}{p(c_{\mathsf{R}}^i = 0|\mathbf{y}_{\mathsf{B}})} \tag{8.9}$$

$$\approx \min_{\mathbf{s}_{\mathsf{R}} \in \mathcal{S}_i^0} \frac{\|\mathbf{y}_{\mathsf{B}} - \mathbf{H}_{\mathsf{B}}\mathbf{s}_{\mathsf{R}}\|^2}{\sigma_{\mathsf{B}}^2} - \min_{\mathbf{s}_{\mathsf{R}} \in \mathcal{S}_i^1} \frac{\|\mathbf{y}_{\mathsf{B}} - \mathbf{H}_{\mathsf{B}}\mathbf{s}_{\mathsf{R}}\|^2}{\sigma_{\mathsf{B}}^2} \tag{8.10}$$

where $\mathcal{S}_i^1$ and $\mathcal{S}_i^0$ represent the sets of transmit symbol vectors whose ith bit labeling is 1 and 0 in the constellation subset $\mathcal{S}(c_{\mathsf{B}}^{mN_{\mathsf{R}}}, \cdots, c_{\mathsf{B}}^{nN_{\mathsf{R}}+1})$, respectively.

In order to avoid error propagation, the *a priori* LLRs $\gamma_{\mathsf{A}}(c_{\mathsf{R}}^i)$ and $\gamma_{\mathsf{B}}(c_{\mathsf{R}}^i)$ from the feedback of the channel decoders are subtracted from $\Lambda_{\mathsf{A}}(c_{\mathsf{R}}^i)$ and $\Lambda_{\mathsf{B}}(c_{\mathsf{R}}^i)$ to generate the *extrinsic* LLRs $\lambda_{\mathsf{A}}(c_{\mathsf{R}}^i)$ and $\lambda_{\mathsf{B}}(c_{\mathsf{R}}^i)$ as:

$$\lambda_{\mathsf{A}}(c_{\mathsf{R}}^i) = \Lambda_{\mathsf{A}}(c_{\mathsf{R}}^i) - \gamma_{\mathsf{A}}(c_{\mathsf{R}}^i), \quad i \in \{1, \cdots, mN_{\mathsf{R}}\}, \tag{8.11}$$

$$\lambda_{\mathsf{B}}(c_{\mathsf{R}}^i) = \Lambda_{\mathsf{B}}(c_{\mathsf{R}}^i) - \gamma_{\mathsf{B}}(c_{\mathsf{R}}^i), \quad i \in \{1, \cdots, nN_{\mathsf{R}}\}. \tag{8.12}$$

8.3.2.2 LLR values of $\{c_{\mathsf{A}}\}$ and $\{c_{\mathsf{B}}\}$

The output of the demappers are the LLR values for $\{c_{\mathsf{R}}\}$. They must be converted to the LLR values for $\{c_{\mathsf{A}}\}$ and $\{c_{\mathsf{B}}\}$ for channel decoding. This is accomplished by the "$\boxplus$" module.

The sign of each LLR value shows the estimate that the corresponding bit is 1 or 0, and its absolute value represents the reliability of such estimation. For $i \in \{1, \cdots, nN_{\mathsf{R}}\}$, we have $c_{\mathsf{B}}^i \oplus c_{\mathsf{A}}^i = c_{\mathsf{R}}^i$. The bit c_{B}^i (resp. c_{A}^i) differs with c_{R}^i only when $c_{\mathsf{A}}^i = 1$ (resp. $c_{\mathsf{B}}^i = 1$). Given the LLR value λ of c_{R}^i and the known bit c (i.e., c_{A}^i or c_{B}^i), the LLR value of the unknown bit (i.e., c_{B}^i or c_{A}^i) can be calculated as

$$\lambda \boxplus c = \begin{cases} \lambda, & \text{if } c = 0, \\ -\lambda, & \text{if } c = 1. \end{cases} \tag{8.13}$$

That is, the "⊞" module flips the sign of λ according to the corresponding input bit c. Since each bit c is perfectly known, it does not change the reliability of the decoded bits.

After the "⊞" module, the LLR values for $\{c_\mathsf{A}\}$ and $\{c_\mathsf{B}\}$ are given to the input of the convolutional decoder, where the BCJR algorithm [14] is applied. Similar to the demapper, *extrinsic* LLR values for the coded bits are generated at the output of convolutional decoders. The feedback LLR values for $\{c_\mathsf{R}\}$ are calculated according to the outputs of the channel decoder and the bit sequence $\{\tilde{c}_\mathsf{A}\}$ and $\{c_\mathsf{B}\}$ again using the "⊞" module. In the final iteration, the decoder outputs the hard decisions on the information bits. The overall workflow of the iterative receivers is summarized in Algorithm 7.

Algorithm 7 Workflow of receivers at stations A and B

Initialize: 1. Obtain $\{\tilde{c}_\mathsf{A}\}$ and $\{c_\mathsf{B}\}$ at A and B.
2. Set $\{\gamma_\mathsf{A}(c_\mathsf{R})\} = \{0\}$ and $\{\gamma_\mathsf{B}(c_\mathsf{R})\} = \{0\}$.
3. Set $l = 0$.

repeat

Update $l = l + 1$; In the lth iteration ($l \geq 1$):
1. Construct $\mathcal{S}(c_\mathsf{B}^{mN_\mathsf{R}}, \cdots, c_\mathsf{B}^{nN_\mathsf{R}+1})$ according to (8.6) at station B;
2. Calculate $\lambda_\mathsf{A}(c_\mathsf{R}^i)$ and $\lambda_\mathsf{B}(c_\mathsf{R}^i)$ according to (8.7)–(8.12) at stations A and B;
3. Calculate $\lambda_\mathsf{A}(c_\mathsf{B}^i) = \lambda_\mathsf{A}(c_\mathsf{R}^i) \boxplus \tilde{c}_\mathsf{A}^i$, $\forall i \in \{1, \cdots, mN_\mathsf{R}\}$ and $\lambda_\mathsf{B}(c_\mathsf{A}^i) = \lambda_\mathsf{B}(c_\mathsf{R}^i) \boxplus c_\mathsf{B}^i$, $\forall i \in \{1, \cdots, nN_\mathsf{R}\}$ according to (8.13);
4. Deinterleave $\{\lambda_\mathsf{A}(c_\mathsf{B})\}$ and $\{\lambda_\mathsf{B}(c_\mathsf{A})\}$; feed them to channel decoders;
5. Interleave the channel decoder outputs;
6. Calculate $\gamma_\mathsf{A}(c_\mathsf{R}^i) = \gamma_\mathsf{A}(c_\mathsf{B}^i) \boxplus \tilde{c}_\mathsf{A}^i$, $\forall i \in \{1, \cdots, mN_\mathsf{R}\}$ and $\gamma_\mathsf{B}(c_\mathsf{R}^i) = \gamma_\mathsf{B}(c_\mathsf{B}^i) \boxplus c_\mathsf{B}^i$, $\forall i \in \{1, \cdots, nN_\mathsf{R}\}$ according to (8.13);
7. Feed $\gamma_\mathsf{A}(c_\mathsf{R})$ and $\gamma_\mathsf{B}(c_\mathsf{R})$ to the demappers;

until $\mathrm{BER}^{(l)} - \mathrm{BER}^{(l-1)} \leq \epsilon$, or maximum number of iteration is reached.

Generate the output information bits $\{\hat{b}_\mathsf{A}\}$ and $\{\hat{b}_\mathsf{B}\}$.

8.4 Optimized Signal Labeling

This section discusses the criteria and methods for finding the optimized signal labeling of the transmit symbol constellation at the relay. We consider two criteria: one is based on the error bound of the decoding performance at receivers in high SNR regime, and the other is based on the mutual information in coded modulations. Both single and multiple antenna transmissions are considered. For simplicity, we assume $N_\mathsf{A} = N_\mathsf{B} \triangleq N$.

8.4.1 Optimization Criterion Based on Error Bounds

For BICM-ID systems, the asymptotic decoding performance at the receivers in high SNR regime can be characterized by the *pairwise error probability* (PEP). Let c and $\hat{\mathrm{c}}$ denote two transmit codewords with Hamming distance d. $\mathrm{P}(\mathrm{c} \to \hat{\mathrm{c}})$ denotes the PEP that the decoder chooses $\hat{\mathrm{c}}$ instead of the transmitted bit sequence c. Assuming perfect interleaving and averaged over all symbols and bit positions, $\mathrm{P}(\mathrm{c} \to \hat{\mathrm{c}})$ at receiver A can be bounded using the union bound by [36]

$$\mathrm{P}(\mathrm{c} \to \hat{\mathrm{c}}) \leq E^d \tag{8.14}$$

$$= \left\{ \frac{1}{mN_{\mathsf{R}} \cdot 2^{mN_{\mathsf{R}}}} \sum_{i=1}^{mN_{\mathsf{R}}} \sum_{b=0}^{1} \sum_{\mathbf{s}_{\mathsf{R}} \in \mathcal{X}_i^{(b)}} \sum_{\hat{\mathbf{s}}_{\mathsf{R}} \in \mathcal{X}_i^{(\bar{b})}} \mathrm{P}(\mathbf{s}_{\mathsf{R}} \to \hat{\mathbf{s}}_{\mathsf{R}}) \right\}^d \tag{8.15}$$

where $\bar{b} = 1 - b$ for $b \in \{0, 1\}$, and $\mathrm{P}(\mathbf{s}_{\mathsf{R}} \to \hat{\mathbf{s}}_{\mathsf{R}})$ denotes the PEP between symbol vector $\mathbf{s}_{\mathsf{R}}$ and $\hat{\mathbf{s}}_{\mathsf{R}}$ in the symbol constellation. $\mathcal{X}_i^{(b)}$ and $\mathcal{X}_i^{(\bar{b})}$ in (8.15) denote the two sets of symbol vectors that only differ on their ith bit labeling. In high SNR regime and after a sufficient number of iterations, we assume the channel decoder feeds back perfect information about the other unknown bits of each symbol vector. That is why we only consider the PEP between symbol vectors $\mathbf{s}_{\mathsf{R}}$ and $\hat{\mathbf{s}}_{\mathsf{R}}$ that only differ by one bit in their labeling.

According to the signal model at stations A in (8.2), the conditional the PEP that $\hat{\mathbf{s}}_{\mathsf{R}}$ is selected when $\mathbf{s}_{\mathsf{R}}$ is transmitted for a given channel matrix realization $\mathbf{H}_{\mathsf{A}}$ can be expressed as

$$\mathrm{P}(\mathbf{s}_{\mathsf{R}} \to \hat{\mathbf{s}}_{\mathsf{R}} \mid \mathbf{H}_{\mathsf{A}}) = Q\left(\sqrt{\frac{1}{2\sigma_{\mathsf{A}}^2}\|\mathbf{H}_{\mathsf{A}}(\mathbf{s}_{\mathsf{R}} - \hat{\mathbf{s}}_{\mathsf{R}})\|^2}\right) \tag{8.16}$$

The expression in (8.16) can be bounded by the Chernoff bound as

$$\mathrm{P}(\mathbf{s}_{\mathsf{R}} \to \hat{\mathbf{s}}_{\mathsf{R}} \mid \mathbf{H}_{\mathsf{A}}) \leq \exp\left(-\frac{1}{4\sigma_{\mathsf{A}}^2}\|\mathbf{H}_{\mathsf{A}}(\mathbf{s}_{\mathsf{R}} - \hat{\mathbf{s}}_{\mathsf{R}})\|^2\right) \tag{8.17}$$

We can now average this conditional PEP over the statistics of the channel matrix $\mathbf{H}_{\mathsf{A}}$. Assuming that the channel path coefficients in $\mathbf{H}_{\mathsf{A}}$ are i.i.d. $\mathcal{CN}(0, 1)$ random variables, i.e., complex-valued zero-mean Gaussian spatially white channel, the average of the PEP in (8.17) over the statistics of the channel path coefficients yields the upper bound on the average PEP as [222]

$$\mathrm{P}(\mathbf{s}_\mathsf{R} \to \hat{\mathbf{s}}_\mathsf{R}) \le \mathrm{E}_{\mathbf{H}_\mathsf{A}} \left\{ \exp\left(-\frac{1}{4\sigma_\mathsf{A}^2} \|\mathbf{H}_\mathsf{A}(\mathbf{s}_\mathsf{R} - \hat{\mathbf{s}}_\mathsf{R})\|^2 \right) \right\} \tag{8.18}$$

$$= \frac{1}{\left\{ \det\left[\mathbf{I}_{N_\mathsf{R}} + \frac{1}{4\sigma_\mathsf{A}^2} (\mathbf{s}_\mathsf{R} - \hat{\mathbf{s}}_\mathsf{R})(\mathbf{s}_\mathsf{R} - \hat{\mathbf{s}}_\mathsf{R})^H \right] \right\}^{N_\mathsf{A}}} \tag{8.19}$$

$$= \left(1 + \frac{1}{4\sigma_\mathsf{A}^2} \|\mathbf{s}_\mathsf{R} - \hat{\mathbf{s}}_\mathsf{R}\|^2 \right)^{-N_\mathsf{A}} \tag{8.20}$$

At high SNR regime, i.e., when $\|\mathbf{s}_\mathsf{R} - \hat{\mathbf{s}}_\mathsf{R}\|^2/(4\sigma_\mathsf{A}^2) \gg 1$, the PEP in (8.18) may be upper bounded by

$$\mathrm{P}(\mathbf{s}_\mathsf{R} \to \hat{\mathbf{s}}_\mathsf{R}) \le (4\sigma_\mathsf{A}^2)^{N_\mathsf{A}} \cdot \frac{1}{\|\mathbf{s}_\mathsf{R} - \hat{\mathbf{s}}_\mathsf{R}\|^{2N_\mathsf{A}}} \tag{8.21}$$

By substituting (8.21) in (8.15), we can show that E may be upper bounded in i.i.d. Rayleigh fading channels by the Chernoff bound as,

$$E \le c_\mathrm{D} \cdot D \tag{8.22}$$

where c_D is a constant that is not related to the labeling, and

$$D = \frac{1}{mN_\mathsf{R} \cdot 2^{mN_\mathsf{R}}} \sum_{i=1}^{mN_\mathsf{R}} \sum_{b=0}^{1} \sum_{\mathbf{s}_\mathsf{R} \in \mathcal{X}_i^{(b)}} \sum_{\hat{\mathbf{s}}_\mathsf{R} \in \mathcal{X}_i^{(\bar{b})}} \frac{1}{\|\mathbf{s}_\mathsf{R} - \hat{\mathbf{s}}_\mathsf{R}\|^{2N_\mathsf{A}}} \tag{8.23}$$

where D provides a measure for the decoding error bound in high SNR regime. By minimizing D in (8.23), we correspondingly minimize the error bound of $\mathrm{P}(\mathbf{c} \to \hat{\mathbf{c}})$ in high SNR at receiver A in (8.15).

Following the same discussions as for the error bounds at receiver A, a similar measure for the decoding performance at receiver B in high SNR regime can be derived. For given n and the *a priori* known bits $\mathbf{c}^{(\mathrm{a})} = [c_\mathsf{B}^{mN_\mathsf{R}}, \cdots, c_\mathsf{B}^{nN_\mathsf{R}+1}]$, the demapping at station B is performed on $\mathcal{S}(\mathbf{c}^{(\mathrm{a})})$, which is a subset of the constellation $\mathcal{X}$. We define

$$D_\mathcal{S}(\mathbf{c}^{(\mathrm{a})}) = \frac{1}{nN_\mathsf{R} \cdot 2^{nN_\mathsf{R}}} \sum_{i=1}^{nN_\mathsf{R}} \sum_{b=0}^{1} \sum_{\mathbf{s}_\mathsf{R} \in \mathcal{S}_i^{(b)}} \sum_{\hat{\mathbf{s}}_\mathsf{R} \in \mathcal{S}_i^{(\bar{b})}} \frac{1}{\|\mathbf{s}_\mathsf{R} - \hat{\mathbf{s}}_\mathsf{R}\|^{2N_\mathsf{B}}} \tag{8.24}$$

where $\mathcal{S}_i^{(b)}$ and $\mathcal{S}_i^{(\bar{b})}$ in (8.24) denote the two sets of symbol vectors that only differ on their ith bit labeling within the subset $\mathcal{S}(\mathbf{c}^{(\mathrm{a})})$. Since the bits $[c_\mathsf{B}^{mN_\mathsf{R}}, \cdots, c_\mathsf{B}^{nN_\mathsf{R}+1}]$ are random in the system, all of their possible realizations must be considered. For given n, the worst-case

PEP bound, which determines the decoding performance in high SNR regime at station B, can be minimized by minimizing the following measure

$$D_{\mathcal{S}}^{n} = \max_{\mathbf{c}^{(\mathrm{a})}} D_{\mathcal{S}}(\mathbf{c}^{(\mathrm{a})}), \tag{8.25}$$

for all possible realizations of $[c_{\mathsf{B}}^{mN_{\mathsf{R}}}, \cdots, c_{\mathsf{B}}^{nN_{\mathsf{R}}+1}]$, where $c_{\mathsf{B}}^{i} \in \{0,1\}$, $i = nN_{\mathsf{R}} + 1, \cdots, mN_{\mathsf{R}}$.

8.4.2 Optimization Criterion Based on Mutual Information

Another criterion for optimizing the signal labeling at the relay is based on the *mutual information* between the transmitted signal and the received signal, which is also called the *constellation constrained capacity*. For a given constellation and the number of *a priori* known bits n in each transmit symbol, our goal is to maximize the mutual information between the transmit symbol vector and the received signals at stations A and B. Unlike the criterion based on the error bounds, the mutual information criterion is an information theoretical criterion. In order to achieve the mutual information, capacity-achieving codes, such as turbo codes, are usually required to be applied in the system. Nevertheless, the mutual information provides an optimization criterion for the system performance limit.

According to the signal model at stations A in (8.2), the mutual information between the transmitted symbol vector $\mathbf{s}_{\mathsf{R}}$ at the relay and the received signal vector $\mathbf{y}_{\mathsf{A}}$ with the given channel matrix $\mathbf{H}_{\mathsf{A}}$ can be expressed as

$$I(\mathbf{s}_{\mathsf{R}}; \mathbf{y}_{\mathsf{A}}|\mathbf{H}_{\mathsf{A}}) = H(\mathbf{s}_{\mathsf{R}}|\mathbf{H}_{\mathsf{A}}) - H(\mathbf{s}_{\mathsf{R}}|\mathbf{y}_{\mathsf{A}}, \mathbf{H}_{\mathsf{A}}) \tag{8.26}$$

$$= H(\mathbf{s}_{\mathsf{R}}) - H(\mathbf{s}_{\mathsf{R}}|\mathbf{y}_{\mathsf{A}}, \mathbf{H}_{\mathsf{A}}) \tag{8.27}$$

where $H(\mathbf{s}_{\mathsf{R}})$ is the entropy of the transmitted symbol vector. Under the assumption that each symbol vector is transmitted with equal probability, we have $H(\mathbf{s}_{\mathsf{R}}) = \log|\mathcal{X}| = mN_{\mathsf{R}}$, where $|\mathcal{X}|$ denotes the cardinality of the constellation $\mathcal{X}$. By using Bayes' rule and observing that the a priori probabilities for all symbol vectors $\mathbf{s}_{\mathsf{R}}$ are equal, we have

$$H(\mathbf{s}_{\mathsf{R}}|\mathbf{y}_{\mathsf{A}}, \mathbf{H}_{\mathsf{A}}) = \mathrm{E}_{\mathbf{s}_{\mathsf{R}},\mathbf{y}_{\mathsf{A}}}\left[\log \frac{1}{p(\mathbf{s}_{\mathsf{R}}|\mathbf{y}_{\mathsf{A}}, \mathbf{H}_{\mathsf{A}})}\right] \tag{8.28}$$

$$= \mathrm{E}_{\mathbf{s}_{\mathsf{R}},\mathbf{y}_{\mathsf{A}}}\left[\log \frac{\sum_{\check{\mathbf{s}}_{\mathsf{R}}\in\mathcal{X}} p(\mathbf{y}_{\mathsf{A}}|\check{\mathbf{s}}_{\mathsf{R}}, \mathbf{H}_{\mathsf{A}})}{p(\mathbf{y}_{\mathsf{A}}|\mathbf{s}_{\mathsf{R}}, \mathbf{H}_{\mathsf{A}})}\right]. \tag{8.29}$$

So for the given channel $\mathbf{H}_\mathsf{A}$, we have

$$I(\mathbf{s}_\mathsf{R};\mathbf{y}_\mathsf{A}|\mathbf{H}_\mathsf{A}) = H(\mathbf{s}_\mathsf{R}) - H(\mathbf{s}_\mathsf{R}|\mathbf{y}_\mathsf{A},\mathbf{H}_\mathsf{A}) \tag{8.30}$$

$$= mN_\mathsf{R} - \mathrm{E}_{\mathbf{s}_\mathsf{R},\mathbf{y}_\mathsf{A}}\left[\log\frac{\sum_{\check{\mathbf{s}}_\mathsf{R}\in\mathcal{X}} p(\mathbf{y}_\mathsf{A}|\check{\mathbf{s}}_\mathsf{R},\mathbf{H}_\mathsf{A})}{p(\mathbf{y}_\mathsf{A}|\mathbf{s}_\mathsf{R},\mathbf{H}_\mathsf{A})}\right] \tag{8.31}$$

Here the noise vector at station A is Gaussian, i.e., $\mathbf{n}_\mathsf{A} \sim \mathcal{CN}(0,\sigma_\mathsf{A}^2\mathbf{I}_{N_\mathsf{A}})$. So we have

$$p(\mathbf{y}_\mathsf{A}|\mathbf{s}_\mathsf{R},\mathbf{H}_\mathsf{A}) = \frac{1}{(\pi\sigma_\mathsf{A}^2)^{N_\mathsf{A}}}\exp\left(-\frac{1}{\sigma_\mathsf{A}^2}\|\mathbf{y}_\mathsf{A}-\mathbf{H}_\mathsf{A}\mathbf{s}_\mathsf{R}\|^2\right). \tag{8.32}$$

For a fast fading channel, the ergodic mutual information $I(\mathbf{s}_\mathsf{R};\mathbf{y}_\mathsf{A})$ can be expressed as

$$I(\mathbf{s}_\mathsf{R};\mathbf{y}_\mathsf{A}) = \mathrm{E}_{\mathbf{H}_\mathsf{A}}\{I(\mathbf{s}_\mathsf{R};\mathbf{y}_\mathsf{A}|\mathbf{H}_\mathsf{A})\} \tag{8.33}$$

$$= mN_\mathsf{R} - \mathrm{E}_{\mathbf{s}_\mathsf{R},\mathbf{y}_\mathsf{A},\mathbf{H}_\mathsf{A}}\left[\log\frac{\sum_{\check{\mathbf{s}}_\mathsf{R}\in\mathcal{X}} p(\mathbf{y}_\mathsf{A}|\check{\mathbf{s}}_\mathsf{R},\mathbf{H}_\mathsf{A})}{p(\mathbf{y}_\mathsf{A}|\mathbf{s}_\mathsf{R},\mathbf{H}_\mathsf{A})}\right] \tag{8.34}$$

Here the expectation is taken with respect to the equally generated symbol vector $\mathbf{s}_\mathsf{R}$, the Rayleigh fading channel $\mathbf{H}_\mathsf{A}$ and the correspondingly received signal $\mathbf{y}_\mathsf{A}$ according to (8.2). For a given constellation, there is no closed-form expression for the mutual information. However, the mutual information can be evaluated by Monte Carlo simulations according to (8.34). In order to find the signal labeling that maximize the mutual information $I(\mathbf{s}_\mathsf{R};\mathbf{y}_\mathsf{A})$, we can equivalently define the cost function as

$$D = \mathrm{E}_{\mathbf{s}_\mathsf{R},\mathbf{y}_\mathsf{A},\mathbf{H}_\mathsf{A}}\left[\log\frac{\sum_{\check{\mathbf{s}}_\mathsf{R}\in\mathcal{X}} p(\mathbf{y}_\mathsf{A}|\check{\mathbf{s}}_\mathsf{R},\mathbf{H}_\mathsf{A})}{p(\mathbf{y}_\mathsf{A}|\mathbf{s}_\mathsf{R},\mathbf{H}_\mathsf{A})}\right], \tag{8.35}$$

which is to be minimized. Note the mutual information and the corresponding cost function D depends on the noise variance σ_A^2, i.e., the SNR at station A. Since we generate the signal $\mathbf{s}_\mathsf{R}$ equally likely, we can write

$$\mathrm{E}_{\mathbf{s}_\mathsf{R},\mathbf{y}_\mathsf{A},\mathbf{H}_\mathsf{A}}[\cdot] = \frac{1}{|\mathcal{X}|}\sum_{\mathbf{s}_\mathsf{R}\in\mathcal{X}}\mathrm{E}_{\mathbf{y}_\mathsf{A},\mathbf{H}_\mathsf{A}}[\cdot]. \tag{8.36}$$

where $\mathbf{H}_\mathsf{A}$ is generated according to an i.i.d. Rayleigh fading channel.

Following the same discussions as for the mutual information between the transmitted symbol and the received signal at receiver A, the mutual information between the transmitted symbol and the received signal at receiver B can also be derived. For given n and the *a priori* known bits $\mathbf{c}^{(\mathsf{a})} = [c_\mathsf{B}^{mN_\mathsf{R}},\cdots,c_\mathsf{B}^{nN_\mathsf{R}+1}]$, the mutual information between the transmitted symbol $\mathbf{s}_\mathsf{R}$ and the received signal $\mathbf{y}_\mathsf{B}$ at receiver B can be expressed as

$$I(\mathbf{s}_\mathsf{R};\mathbf{y}_\mathsf{B}|\mathbf{c}^{(a)}) = H(\mathbf{s}_\mathsf{R}|\mathbf{c}^{(a)}) - H(\mathbf{s}_\mathsf{R}|\mathbf{y}_\mathsf{B},\mathbf{c}^{(a)}) \tag{8.37}$$

$$= nN_\mathsf{R} - \mathrm{E}_{\mathbf{s}_\mathsf{R},\mathbf{y}_\mathsf{B},\mathbf{H}_\mathsf{B}}\left[\log\frac{\sum_{\check{\mathbf{s}}_\mathsf{R}\in\mathcal{S}(\mathbf{c}^{(a)})} p(\mathbf{y}_\mathsf{B}|\check{\mathbf{s}}_\mathsf{R},\mathbf{H}_\mathsf{B})}{p(\mathbf{y}_\mathsf{B}|\mathbf{s}_\mathsf{R},\mathbf{H}_\mathsf{B})}\right] \tag{8.38}$$

Here the noise vector at station B is Gaussian, i.e., $\mathbf{n}_\mathsf{B} \sim \mathcal{CN}(0,\sigma_\mathsf{B}^2\mathbf{I}_{N_\mathsf{A}})$. So we have

$$p(\mathbf{y}_\mathsf{B}|\mathbf{s}_\mathsf{R},\mathbf{H}_\mathsf{B}) = \frac{1}{(\pi\sigma_\mathsf{B}^2)^{N_\mathsf{B}}}\exp\left(-\frac{1}{\sigma_\mathsf{B}^2}\|\mathbf{y}_\mathsf{B}-\mathbf{H}_\mathsf{B}\mathbf{s}_\mathsf{R}\|^2\right). \tag{8.39}$$

The realizations of $\mathbf{c}^{(a)}$ happens with equal probability, the mutual information $I(\mathbf{s}_\mathsf{R};\mathbf{y}_\mathsf{B}|\mathbf{c}^{(a)})$ averaged over all the possible realizations of $\mathbf{c}^{(a)}$ can be expressed as

$$I(\mathbf{s}_\mathsf{R};\mathbf{y}_\mathsf{B}) = \mathrm{E}_{\mathbf{c}^{(a)}}\left\{I(\mathbf{s}_\mathsf{R};\mathbf{y}_\mathsf{B}|\mathbf{c}^{(a)})\right\} \tag{8.40}$$

$$= \frac{1}{2^{(m-n)N_\mathsf{R}}}\sum_{\mathbf{c}^{(a)}}\left[I(\mathbf{s}_\mathsf{R};\mathbf{y}_\mathsf{B}|\mathbf{c}^{(a)})\right]. \tag{8.41}$$

In order to maximize the mutual information $I(\mathbf{s}_\mathsf{R};\mathbf{y}_\mathsf{B})$, we can equivalently define the cost function

$$D_\mathcal{S}^n = \mathrm{E}_{\mathbf{c}^{(a)},\mathbf{s}_\mathsf{R},\mathbf{y}_\mathsf{B},\mathbf{H}_\mathsf{B}}\left[\log\frac{\sum_{\check{\mathbf{s}}_\mathsf{R}\in\mathcal{S}(\mathbf{c}^{(a)})} p(\mathbf{y}_\mathsf{B}|\check{\mathbf{s}}_\mathsf{R},\mathbf{H}_\mathsf{B})}{p(\mathbf{y}_\mathsf{B}|\mathbf{s}_\mathsf{R},\mathbf{H}_\mathsf{B})}\right] \tag{8.42}$$

where the transmit symbol vector $\mathbf{s}_\mathsf{R}$ is chosen in the constellation subset $\mathcal{S}(\mathbf{c}^{(a)})$ with equal probability, and

$$\mathrm{E}_{\mathbf{c}^{(a)},\mathbf{s}_\mathsf{R},\mathbf{y}_\mathsf{A},\mathbf{H}_\mathsf{A}}[\cdot] = \frac{1}{|\mathcal{S}(\mathbf{c}^{(a)})|}\sum_{\mathbf{s}_\mathsf{R}\in\mathcal{S}(\mathbf{c}^{(a)})}\mathrm{E}_{\mathbf{y}_\mathsf{A},\mathbf{H}_\mathsf{A},\mathbf{c}^{(a)}}[\cdot]. \tag{8.43}$$

8.4.3 Optimized Labeling

8.4.3.1 Optimized Labeling for Criterion Based on Error Bounds

In a two-way relaying system, the decoding performance at stations A and B have to be both considered. This is a *multi-objective optimization* problem. For given m and n, we propose to find labeling schemes that minimize the cost function $D + w \cdot D_\mathcal{S}^n$ for the criterion based on the error bounds, where D and $D_\mathcal{S}^n$ are defined in (8.23) and (8.25), respectively. $w > 0$ is the weighting factor. When the constellation size is large, search exhaustively for the optimum labeling that minimizes the cost function becomes impossible. So we applied the binary switching algorithm [200] to search for the optimized labeling (see Appendix 8.7.1).

Table 8.1: Optimized 8PSK labeling scheme, $N_\mathsf{R} = 2$

$0 \mapsto (x_3, x_4)$	$1 \mapsto (x_6, x_0)$	$2 \mapsto (x_7, x_7)$	$3 \mapsto (x_1, x_3)$
$4 \mapsto (x_7, x_0)$	$5 \mapsto (x_2, x_4)$	$6 \mapsto (x_4, x_3)$	$7 \mapsto (x_6, x_7)$
$8 \mapsto (x_1, x_1)$	$9 \mapsto (x_4, x_4)$	$10 \mapsto (x_5, x_3)$	$11 \mapsto (x_1, x_7)$
$12 \mapsto (x_5, x_5)$	$13 \mapsto (x_0, x_0)$	$14 \mapsto (x_1, x_6)$	$15 \mapsto (x_4, x_2)$
$16 \mapsto (x_7, x_1)$	$17 \mapsto (x_2, x_5)$	$18 \mapsto (x_3, x_3)$	$19 \mapsto (x_5, x_0)$
$20 \mapsto (x_3, x_5)$	$21 \mapsto (x_6, x_1)$	$22 \mapsto (x_7, x_6)$	$23 \mapsto (x_2, x_3)$
$24 \mapsto (x_5, x_6)$	$25 \mapsto (x_0, x_1)$	$26 \mapsto (x_1, x_0)$	$27 \mapsto (x_5, x_4)$
$28 \mapsto (x_1, x_2)$	$29 \mapsto (x_4, x_5)$	$30 \mapsto (x_5, x_2)$	$31 \mapsto (x_0, x_7)$
$32 \mapsto (x_5, x_1)$	$33 \mapsto (x_0, x_4)$	$34 \mapsto (x_1, x_4)$	$35 \mapsto (x_4, x_7)$
$36 \mapsto (x_0, x_5)$	$37 \mapsto (x_4, x_0)$	$38 \mapsto (x_3, x_7)$	$39 \mapsto (x_0, x_2)$
$40 \mapsto (x_7, x_5)$	$41 \mapsto (x_2, x_0)$	$42 \mapsto (x_3, x_0)$	$43 \mapsto (x_7, x_3)$
$44 \mapsto (x_2, x_1)$	$45 \mapsto (x_6, x_4)$	$46 \mapsto (x_6, x_3)$	$47 \mapsto (x_3, x_6)$
$48 \mapsto (x_1, x_5)$	$49 \mapsto (x_4, x_1)$	$50 \mapsto (x_5, x_7)$	$51 \mapsto (x_0, x_3)$
$52 \mapsto (x_3, x_1)$	$53 \mapsto (x_0, x_6)$	$54 \mapsto (x_7, x_2)$	$55 \mapsto (x_4, x_6)$
$56 \mapsto (x_3, x_2)$	$57 \mapsto (x_6, x_5)$	$58 \mapsto (x_7, x_4)$	$59 \mapsto (x_2, x_7)$
$60 \mapsto (x_6, x_6)$	$61 \mapsto (x_2, x_2)$	$62 \mapsto (x_2, x_6)$	$63 \mapsto (x_6, x_2)$

Here w is set to be 1.

Considering the case that the relay employs 8PSK constellation on each antenna, we selected the labeling schemes that work well for both $n = 2$ and 1. When $N_\mathsf{R} = 1$, we found the optimized labeling scheme as shown in Fig. 8.4(a). When $N_\mathsf{R} = 2$, the 8PSK symbol on each antenna is indicated as in Fig. 8.4(b), and the optimized labeling assignment is shown in Table 8.1, which shows how $\mathbf{c}_\mathsf{R}$ in decimal format is mapped to the symbol vector $[x_i, x_j]^T$ on the two antennas. For example, $\mathbf{c}_\mathsf{R} = 000000$ is mapped to the symbol vector $\mathbf{s}_\mathsf{R} = [x_3, x_4]^T$.

8.4.3.2 Optimized Labeling for Criterion Based on Mutual Information

For a given transmit symbol vector constellation $\mathcal{X}$ at the relay, the mutual information $I(\mathbf{s}_\mathsf{R}; \mathbf{y}_\mathsf{A})$ between the transmit symbol vector $\mathbf{s}_\mathsf{R}$ and the received signal $\mathbf{y}_\mathsf{A}$ does not depend on the signal labeling. For the given number of *a priori* known bits n in each transmit symbol vector, we only need to find the signal labeling that leads to the minimum of the cost function D_S^n in (8.42). When the constellation is large, it is difficult to search exhaustively for the optimum labeling that minimizes the cost function D_S^n in (8.42). So we applied the binary switching algorithm [200] to search for the optimized labeling (see Appendix 8.7.1). Since it only finds local minimum of cost functions, several round of random initialization is required to search for the global optimum. It is found that set-partitioning (SP) labeling (as shown in Fig. 8.3(b)) is optimum for the 8PSK constellation, i.e., it has higher mutual

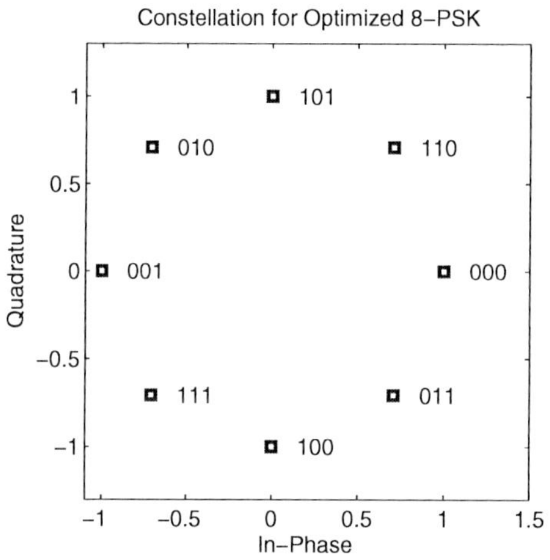

(a) Optimized labeling, $N_R = 1$

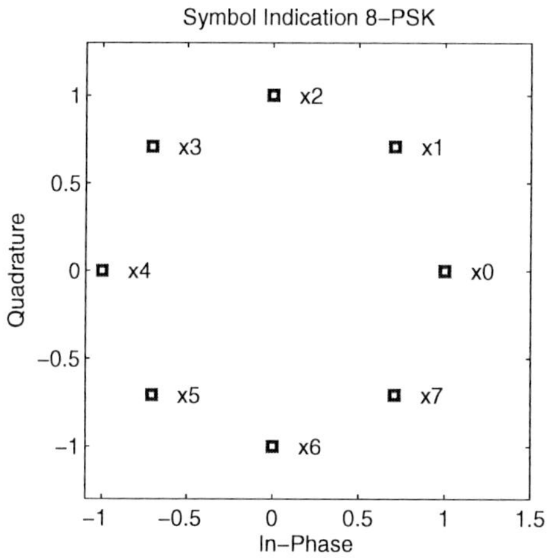

(b) Symbol indication on each antenna

Fig. 8.4: 8PSK labeling scheme and symbol indication

Table 8.2: Required SNR for decoding at BER $= 10^{-4}$

Labeling, $N_\mathsf{R} = 1$	A, 1.5bpt	B, 1bpt	B, 0.5bpt
Gray	15.3dB	15.9dB	17dB
SP	13.5dB	10.3dB	8.7dB
Opt.	10dB	9dB	8.7dB
Labeling, $N_\mathsf{R} = 2$	A, 3bpt	B, 2bpt	B, 1bpt
Gray	12.6dB	13dB	14dB
SP	10.7dB	7.3dB	5.5dB
Opt.	6.7dB	5dB	3.1dB

information than other labeling schemes when $n = 1, 2$ for the 8PSK constellation. In fact, SP labeling itself is a local minimum of the cost function.

8.5 Simulation Results

In this section, we show the performance of the proposed asymmetric data rate transmission scheme. In particular, we compare the performance of the optimized labeling schemes with conventional ones (Gray and SP labeling on each antenna). At the relay transmitter, we use a convolutional encoder with coding rate $r = 1/2$ and generator $(4, 7)_8$ in octal representation. The interleaver length is 12000 bits. The data from the relay is transmitted using the OFDM technique with 1024 subcarriers. Each subcarrier corresponds to a Rayleigh fading channel and the channel of each subcarrier remains constant for two OFDM symbols.

The simulated BER performance is shown in Fig. 8.5. The transmission on each relay antenna employs 8PSK symbols ($m = 3$). Since the transmissions to stations A and B do not interfere with each other, we show their performance in the same figures. The "SNR" on the x-axis represents $P_\mathsf{R}/\sigma_\mathsf{A}^2$ for transmission to Station A and $P_\mathsf{R}/\sigma_\mathsf{B}^2$ for transmission to Station B. The comparison of the required SNR at BER $= 10^{-4}$ is also shown in Table 8.2. Note the optimized labeling is found by minimizing its error bound at the high SNR regime.

Fig. 8.5(a) considers the case $N_\mathsf{A} = N_\mathsf{R} = N_\mathsf{B} = 1$, where the information data rate to Station A is $r \cdot mN_\mathsf{R} = 1.5$bits/transmission (bpt), and the data rate to Station B is 1bpt ($n = 2$) in Fig. 8.5(a). With the SP and the optimized labeling, Fig. 8.5(a) shows that Station B can decode at lower SNR compared to Station A by exploiting the *a priori* bit information subject to certain BER constraints. When BER $= 10^{-4}$, the required SNR at Station B is 3.2dB lower than that of Station A for SP labeling. The optimized labeling outperforms the SP labeling by achieving lower BER at the high SNR regime (see Fig. 8.5(a)). However,

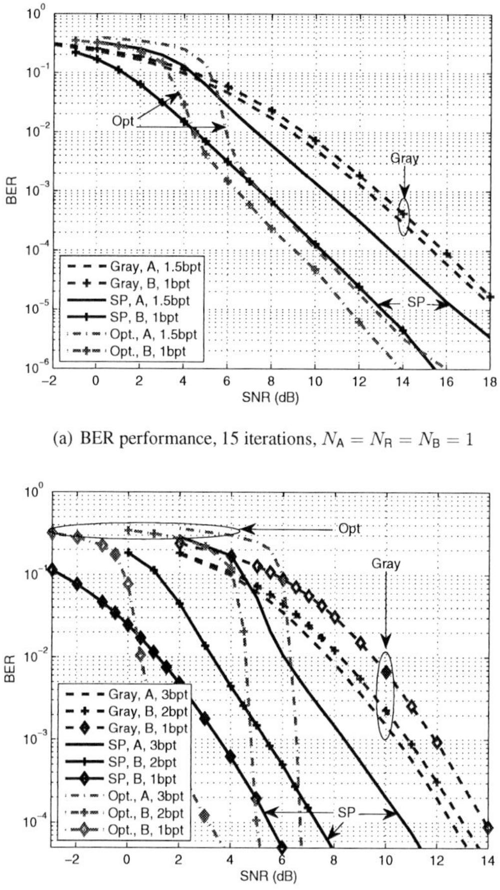

(a) BER performance, 15 iterations, $N_A = N_R = N_B = 1$

(b) BER performance, 15 iterations, $N_A = N_R = N_B = 2$

Fig. 8.5: BER performance at stations A and B, $N_R = 1$ or 2. The "bpt" stands for bit/transmission.

it may lead to worse BER in low SNR. With Gray labeling, the decoding performance at Station B does not improve compared to Station A . This is because the average Euclidean distance between $\mathcal{X}_i^{(0)}$ and $\mathcal{X}_i^{(1)}$ is larger when $i = 2$ and 3 than when $i = 1$ for Gray labeling.

Fig. 8.5(b) shows the BER performance at stations A and B when $N_{\mathsf{A}} = N_{\mathsf{R}} = N_{\mathsf{B}} = 2$. The information data rate to Station A is $r \cdot mN_{\mathsf{R}} = 3$bpt, and the data rate to Station B is 2bpt ($n = 2$) or 1bpt ($n = 1$). Similar to Fig. 8.5(a), Gray labeling does not provide decoding benefits at Station B compared to Station A. Compared to Fig. 8.5(a), the advantage of the optimized labeling to the SP labeling is more obvious. Subject to BER $= 10^{-4}$, the optimized labeling achieves coding gains of 4dB to the SP labeling for transmission to Station A. The BER bounds of 2bpt and 1bpt transmission for the optimized labeling converge at sufficiently high SNR in Fig. 8.5(b), which are lower than that of SP and Gray labeling.

The comparison of the mutual information for different labeling schemes using 8PSK constellation is shown in Fig. 8.6. Here we assume the channels $\mathbf{H}_{\mathsf{A}}$ and $\mathbf{H}_{\mathsf{B}}$ are i.i.d. Rayleigh fading. For the transmission to station A, the mutual information only depends on the transmit symbol constellation and not on the signal labeling. For the transmission to station B, the SP labeling outperforms other signal labeling schemes in the simulations. Fig. 8.6 compares the mutual information the SP labeling scheme and the Gray labeling for $n = 1$ and 2. For the SISO case ($N_{\mathsf{R}} = N_{\mathsf{B}} = 1$) as in Fig. 8.6(b), the SP labeling outperforms the Gray labeling by about 4dB when $n = 2$ and the mutual information equals 1 bit. When $n = 1$ and the mutual information equals 0.5 bits, the SP labeling outperforms the Gray labeling by about 8dB for the SISO case ($N_{\mathsf{R}} = N_{\mathsf{B}} = 1$). Similar observations can be obtained for the MIMO case ($N_{\mathsf{R}} = 2$) in Fig. 8.6(b).

8.6 Chapter Summary and Conclusions

In this chapter, we proposed a novel asymmetric data rate transmission scheme for the BRC phase of two-way DF relaying systems when network coding is applied. The idea is to exploit the *a priori* bit information in the transmit symbols at the weaker link so that it can decode at lower SNR compared to the stronger link. The signal labeling is shown the be crucial for the proposed scheme. We presented two criteria: one is based on the decoding error bound at high SNR, and the other is based on the mutual information to the weaker link receiver in the BRC phase. The methods to find the optimized signal labeling were also presented for systems with single or multiple antennas. We also showed that the optimized labeling can significantly outperform the conventional ones in such scenario.

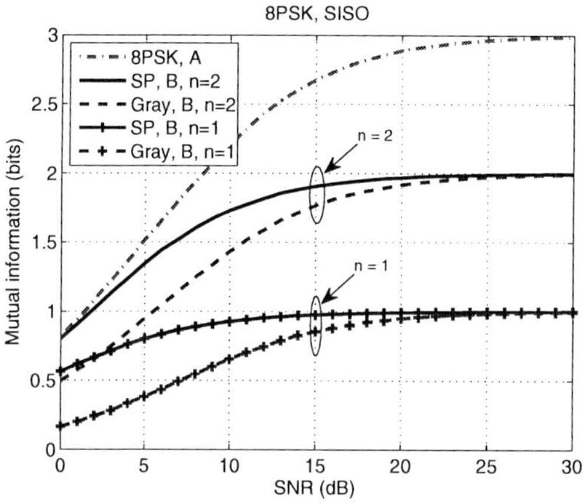

(a) Mutual information, 8PSK, $N_A = N_R = N_B = 1$

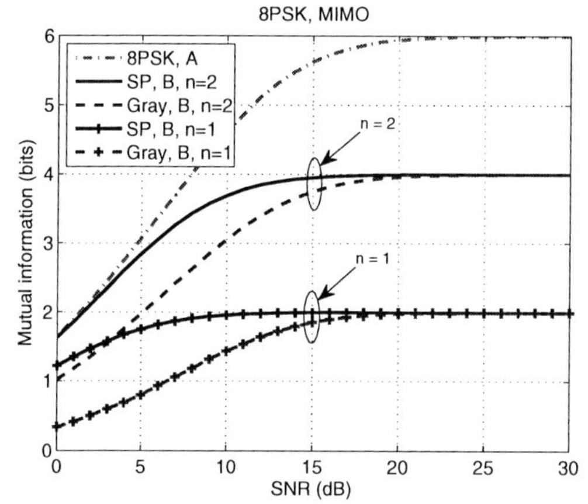

(b) Mutual information, 8PSK, $N_A = N_R = N_B = 2$

Fig. 8.6: Mutual information comparison at stations A and B, $N_R = 1$ or 2. For station B, the number of unknown bits in each symbol at each transmit antenna $n = 1$ or 2.

8.7 Appendices

8.7.1 Binary Switching Algorithm

When the number of constellation points is large, the optimization of index assignments for a given constellation based on some criterion may be intractable if an exhaustive search has to be applied. The binary switching algorithm (BSA) was proposed for index optimization in vector quantization [273]. In the communication systems, the BSA can be applied for the index assignment problem so that an average cost is optimized [201]. However, since the BSA only finds a local optimum that depends on the initial mapping given to the input of the algorithm, several executions of the algorithm with random initial mappings as input are required to generate the presumed global optimum.

The BSA works as follows: First give an initial signal labeling for a given constellation to the BSA as an input. Calculate the cost of each symbol and the total cost for the signal labeling according to the cost functions defined in Section 8.4 based on the error bounds or the mutual information. Generate an ordered list of the symbols, sorted by decreasing costs. Pick the symbol with the highest cost in the list, which has the strongest contribution to a "bad" performance. Try to switch the index of this symbol with the index of another symbol, such that the decrease of the total cost due to switch is as large as possible. If no switch partner can be found for the symbol with the highest cost, the symbol with the second highest cost will be tried to be switched next. This process continues for symbols in the list with decreasing costs until a symbol is found that allows a switch that lowers the total cost. After a switch is performed, a new ordered list of symbols is generated, and the algorithm continues as described above until no further reduction of the total cost is possible. Since the BSA only finds a local optimum, several algorithm executions with random initial mappings may yield the presumed global optimum.

The BSA algorithm for finding the optimized labeling based on the error bounds is summarized in Algorithm 8. The BSA algorithm for finding the optimized labeling based on the mutual information is summarized in Algorithm 9. We only discuss the BSA for finding the optimized labeling according to the cost function at station B.

Algorithm 8 Binary switching algorithm for finding the optimized labeling based on the error bounds

Input m, n, N_{R}, N_{B}, the constellation points and initial labeling.

repeat

1. Calculate $d_k = \sum_{\hat{\mathbf{s}}_{\mathsf{R}}} ||\mathbf{s}_{\mathsf{R}} - \hat{\mathbf{s}}_{\mathsf{R}}||^{-2N_{\mathsf{B}}}$, where $\mathbf{s}_{\mathsf{R}} \in \mathcal{X}$ is the kth $(1 \leq k \leq 2^{mN_{\mathsf{R}}})$ symbol vector in the constellation according to the current labeling assignment. $\hat{\mathbf{s}}_{\mathsf{R}}$ is the symbol vector whose labeling differs with that of $\mathbf{s}_{\mathsf{R}}$ by one bit (in the bit position $1, \ldots, nN_{\mathsf{R}}$).

2. Calculate $d = \sum_k d_k$.

3. Sort d_k in descending order. Let $\tilde{k} = \pi(k)$ represent the ordered list of indices (i.e., $\tilde{1} = \pi(1)$ is the index of the biggest d_k and $\widetilde{2^m} = \pi(2^m)$ is the index of the smallest d_k)

for $k = 1 : 2^{mN_{\mathsf{R}}}$ **do**

Try: Swap the symbol vectors corresponding to the labeling of $\tilde{k}$ and k', $\forall k' \neq \tilde{k}$, and recalculate $d_j = \sum_{\hat{\mathbf{s}}_{\mathsf{R}}} ||\mathbf{s}_{\mathsf{R}} - \hat{\mathbf{s}}_{\mathsf{R}}||^{-2N_{\mathsf{B}}}$ using the new labeling for each $\mathbf{s}_{\mathsf{R}} \in \mathcal{X}$. Calculate $d(k') = \sum_j d_j$.

if $\min_{k'}(d(k')) < d$ **then**

Swap the symbol vectors corresponding to the labeling of $\tilde{k}$ and $k_{\min}$, where $k_{\min} = \arg\min_{k'} d(k')$. Use it as the new labeling scheme.

jump out of the **for** loop.

end if

end for

until $k = 2^{mN_{\mathsf{R}}}$ and $\min_{k'}(d(k')) > d$ (i.e., we come to the smallest d_k and there is no labeling to be switched for decreasing the total cost.)

Algorithm 9 Binary switching algorithm for finding the optimized labeling based on the mutual information

Input m, n, N_R, N_B, the constellation points and initial labeling.

repeat

1. Calculate $d_k = \mathrm{E}_{\mathbf{y}_\mathsf{B},\mathbf{H}_\mathsf{B}}\left[\log \frac{\sum_{\check{\mathbf{s}}_\mathsf{R}} p(\mathbf{y}_\mathsf{B}|\check{\mathbf{s}}_\mathsf{R},\mathbf{H}_\mathsf{B})}{p(\mathbf{y}_\mathsf{B}|\mathbf{s}_\mathsf{R},\mathbf{H}_\mathsf{B})}\right]$, where $\mathbf{s}_\mathsf{R} \in \mathcal{X}$ is the kth ($1 \leq k \leq 2^{mN_\mathsf{R}}$) symbol vector in the constellation according to the current labeling assignment. $\check{\mathbf{s}}_\mathsf{R}$ is the symbol vector whose labeling differs with that of $\mathbf{s}_\mathsf{R}$ only in the bit position $1, \ldots, nN_\mathsf{R}$.

2. Calculate $d = \sum_k d_k$.

3. Sort d_k in descending order. Let $\tilde{k} = \pi(k)$ represent the ordered list of indices (i.e., $\tilde{1} = \pi(1)$ is the index of the biggest d_k and $\widetilde{2^m} = \pi(2^m)$ is the index of the smallest d_k)

for $k = 1 : 2^{mN_\mathsf{R}}$ **do**

  Try: Swap the symbol vectors corresponding to the labeling of $\tilde{k}$ and k', $\forall k' \neq \tilde{k}$, and recalculate $d_j = \mathrm{E}_{\mathbf{y}_\mathsf{B},\mathbf{H}_\mathsf{B}}\left[\log \frac{\sum_{\check{\mathbf{s}}_\mathsf{R}} p(\mathbf{y}_\mathsf{B}|\check{\mathbf{s}}_\mathsf{R},\mathbf{H}_\mathsf{B})}{p(\mathbf{y}_\mathsf{B}|\mathbf{s}_\mathsf{R},\mathbf{H}_\mathsf{B})}\right]$ using the new labeling for each $\mathbf{s}_\mathsf{R} \in \mathcal{X}$.

  Calculate $d(k') = \sum_j d_j$.

  if $\min_{k'}(d(k')) < d$ **then**

    Swap the symbol vectors corresponding to the labeling of $\tilde{k}$ and $k_{\min}$, where $k_{\min} = \arg\min_{k'} d(k')$. Use it as the new labeling scheme.

    jump out of the **for** loop.

  end if

end for

until $k = 2^{mN_\mathsf{R}}$ and $\min_{k'}(d(k')) > d$ (i.e., we come to the smallest d_k and there is no labeling to be switched for decreasing the total cost.)

Chapter 9

Conclusions and Outlooks

This dissertation has considered the communication technologies in relaying systems with multiple antennas, especially in the MIMO two-way relaying systems. Both information-theoretic aspects and practical communication strategies have been proposed and analyzed. For the information-theoretic analysis, an analytical framework for the coverage of MIMO relaying systems based on an outage capacity criterion has been proposed. For MIMO two-way relaying systems, different data combining schemes at the relay have been compared based on their achievable rates. In addition, optimal time-division (TD) strategies for MIMO two-way decode-and-forward (DF) relaying systems have been proposed, which considerably increases the achievable rate regions of the system compared to the equal TD strategy. For the practical transmission schemes, we proposed the self-interference (SI) aided channel estimation and data detection schemes for the broadcast phase of two-way DF relaying systems. Such schemes exploit the SI in two-way DF relaying systems when the superposition coding (SPC) scheme is applied. When the network coding scheme is applied in two-way DF relaying systems, we proposed an asymmetric data rate transmission scheme that utilizes the known data bits at the receivers. Such a scheme exploits the *a priori* known bits at the weak link receiver in the BRC phase.

9.1 Conclusions

After introducing the motivation of the dissertation in Chapter 1, the state-of-the-art summary of recent developments in modern wireless communications has been presented in Chapter 2, where we focused on three topics, namely, multi-antenna communications, relay communications and the two-way relaying communication. Those topics laid the foundation for the following discussions on MIMO relaying systems and MIMO two-way relaying systems in the rest of the dissertation.

Extending the coverage of cellular systems by placing dedicated relay stations has been an important motivation for incorporating relaying communication into cellular networks. An analytical framework for investigating the coverage extension in cellular systems using MIMO relays has been presented in Chapter 3, where we proposed the concepts of *coverage angle* and *circular coverage range* to describe the relation between the number of relay and the coverage extension in a cellular relaying network. The quality of service (QoS) requirement that determines the circular coverage range is based on an achievable outage capacity at the mobile stations. According to the QoS requirement, we provided upper and lower bounds, as well as an approximation, for the maximum circular coverage range when uniform power is allocated at the transmitters. Those proposals were verified by simulations. In general, there are two ways of extending the coverage of cellular network: the first is to use more relay stations and the second is to place more antennas at those stations. Our simulations also showed that if the first hop channel is already very good, as is the case for most relaying systems, placing more antennas at the base station does not provide substantial additional coverage extension. In addition, if the channel knowledge is available at the transmitters for MIMO relaying systems, additional coverage range can be achieved by using the optimum transmit signal covariance matrices.

In modern wireless communication systems, e.g., cellular systems and wireless local area networks (WLAN), bidirectional information flows are common transmission scenarios where the two-way relaying protocols can be applied. The transmission strategies and the achievable rates for MIMO two-way amplify-and-forward (AF) and decode-and-forward (DF) relaying protocols have been discussed in Chapter 4. Equal time (or frequency) resources are allocated to the multiple access (MAC) and broadcast (BRC) phases. For the MIMO two-way DF relaying protocol, we discussed the different data combining schemes at the relay in the BRC phase, i.e., the superposition coding (SPC) scheme and the network coding scheme. Furthermore, we presented a method for characterizing the capacity region in the BRC phase channel, i.e., the bidirectional broadcast channel, and calculating the maximum achievable sum rate of MIMO two-way DF relaying systems. We showed that the two-way relaying protocol achieves substantial improvement in the spectral efficiency compared to conventional unidirectional relaying protocols whether or not the channel knowledge is available at the relay. By comparing the achievable sum rates when the channel knowledge is and is not available at the relay, we found that their difference increases with increasing ratio between number of relay antennas and number of node antennas. We further showed that the network coding scheme achieves nearly the optimal sum rate when the transmit channel knowledge is available at the relay. Whether the transmit channel knowledge is or is not available at the relay, the two-way DF relaying protocol can significantly outperforms the two-way AF relaying protocol.

For the two-way DF relaying protocol, the MAC and BRC phase share the time (or frequency) resources. The allocation of the time (or frequency) resources among the MAC and BRC phases is called the *time-division* (TD) strategies. Chapter 4 considered equal TD strategies. In Chapter 5, the optimum TD strategies for MIMO two-way DF relaying systems were discussed. The methods that maximizes the achievable rate regions under the peak power constraint and the average power constraint were proposed. Simulation results showed that the achievable rate region can be significantly increased by choosing the optimum TD strategies. At high signal-to-noise ratio (SNR), the optimum TD strategies improve the ergodic sum rate when the MAC phase is the bottleneck of the system, e.g., when the sum of the number of antennas at the user stations is larger than that of the relay. Otherwise, the gain in ergodic sum rate by using the optimum TD strategies is small. However, the average achievable rate of one user given the QoS requirement of the other can still be increased a lot by using optimum TD strategies under both peak power and average power constraints.

In wireless communications, we are not only interested in the performance limits described from the information-theoretic aspects, but also practical transmission schemes that can be applied to real-world systems. The received data at the user stations contain the known data called *self-interference* (SI). The SI is unique in the two-way relaying protocols. The SI may be in the form of data symbols for the SPC scheme, or it may be in the form of data bits when the network coding scheme is applied. When the SPC scheme is applied in the two-way DF relaying protocol, we proposed the novel SI-aided channel estimation schemes for the BRC phase in Chapter 6. The channel estimates of the proposed schemes are obtained by exploiting the known data symbols in the SI that are inherent in the considered two-way DF relaying scenario. The proposed SI-aided channel estimation can be applied without pilots, which achieves higher bandwidth efficiency. On the other hand, the SI can also be used together with pilots to offer superior channel estimation performances than schemes that purely based on pilots. To the best of our knowledge, this is the first scheme that *utilizes* SI for channel estimation. Besides proposing the ideas of exploiting SI to estimate the channel, we provided the whole SI-aided iterative receiver structure. The performance analysis and simulation results showed that when the coherence interval in block-fading channels or the observation frame length in time-varying channels is long enough, the SI-aided channel estimation can eventually outperform the pure pilot-aided channel estimation in realistic scenarios. Our proposed scheme is particularly suitable for systems with large number of antennas and subcarriers or with high mobility stations, where the resource consumed by conventional pilot-aided channel estimation had been considered as a big hindrance for the practical implementation of the system. We showed how SI-aided linear channel estimator could be integrated in commonly used receivers, e.g., the iterative receiver structure for channel estimation and data detection. Considering SI, we showed that only small modifications to existing receiver structures are required. The proposed scheme is particularly interesting for multi-carrier systems because SI can track the channel in all subcarriers, and there is no need of doing interpolations in time and frequency as for pilot-aided schemes.

In Chapter 7, we derived achievable rate expressions for bidirectional broadcast channels of two-way relaying systems when the SI-aided channel estimation scheme is applied. This scheme utilizes the SI to obtain an initial estimate of the channel and use data-aided approaches to improve the channel estimates. Simulation results showed that our scheme can achieve higher rates compared to traditional pilot-aided channel estimation schemes. The gain is relatively high in the low SNR regime. In high SNR regime, the performance of the SI-aided channel estimation scheme for 4QAM modulations is nearly as good as when perfect channel knowledge is available at the receivers.

We proposed a novel asymmetric data rate transmission scheme for the BRC phase of two-way DF relaying systems when network coding is applied in Chapter 8. The idea is to exploit the *a priori* bit information in the transmit symbols at the weaker link so that it can decode at lower SNR compared to the stronger link. We also showed that the optimized labeling can significantly outperform the conventional ones in such scenario. We discussed two optimization criteria in this chapter: one is based on the error bound when the receive SNR is asymptotically high, and the other is based on the mutual information when practical modulation schemes are applied. For the error bound criterion, we proposed optimized symbol labeling for 8PSK modulations considering both SISO and MIMO systems found by using the binary switching algorithm. For the mutual information criterion, we showed that the set-partitioning (SP) labeling is optimal compared to other labeling schemes.

9.2 Future Work

Many aspects of the analysis and design of communication techniques in spectrally efficient MIMO relaying systems have been elaborated in this dissertation. However, much work remains to be done for the real-world implementation of MIMO relaying networks, especially MIMO two-way relaying networks. The future work on this research subject may include the following categories.

For the analysis of coverage in cellular MIMO relaying networks, the impact of intracell and intercell interference may be considered in the future work. In this case, neighboring cells or different relay stations in the same cell may use the same frequency channel to serve their mobile users at the same time. Under this circumstance, multiple base stations or relay stations may transmit to the same user simultaneously, and coordination between those base or relay stations may be necessary. This will make the coverage analysis more complicated.

For the transmission techniques in MIMO two-way relaying system, the following topics may be important for further research:

- When the network coding scheme is applied in the BRC phase of two-way DF relaying systems, we only require the combined bit information after the XOR operation. It may be suboptimum to separately decode the two sets of data from the two user stations in the MAC phase as revealed by [161]. The lattice coding scheme proposed in [161] is of theoretical value. Practical coding design combining the network coding in the MAC phase, especially for wireless fading channels, may be the future direction of research.
- For two-way DF relaying systems, the relay has the soft bit information available after decoding in the MAC phase is completed. Such information is lost if only the XOR-ed bits after hard decision are transmitted back to the two user stations in the BRC phase. Transmission schemes that can retain the soft decoding information for the BRC phase transmission may further improve the system performance.
- For practical implementation of two-way relaying systems, synchronization for the transmission of the user stations in the MAC phase is an important issue. However, most previous work only considers perfect synchronization. Practical synchronization schemes for the two-way relaying systems are called for in future research.

Acronyms

3GPP	Third Generation Partnership Project
4G	Fourth Generation
AF	Amplify-and-Forward
APP	A Posteriori Probability
AWGN	Additive White Gaussian Noise
BCJR	Bahl-Cocke-Jelinek-Raviv
BER	Bit Error Rate
BICM-ID	Bit-Interleaved Coded Modulation with Iterative Decoding
BLAST	Bell Laboratories Layered Space-Time
BPSK	Binary Phase-Shift Keying
BRC	Broadcast
BS	Base Station
BSA	Binary Switching Algorithm
CDMA	Code Division Multiple Access
CF	Compress-and-Forward
CRC	Cyclic Redundancy Check
CSI	Channel State Information
CSIT	Channel State Information at the Transmitter
D-BLAST	Diagonal-Bell Laboratories Layered Space-Time
DAF	Decode-Amplify-Forward
DemF	Demodulate-and-forward
DF	Decode-and-Forward
DFD	Division Free Duplex

DFE	Decision Feedback Equalizer
DMT	Diversity-Multiplexing Tradeoff
DPC	Dirty Paper Coding
DUSTM	Differential Unitary Space-Time Modulation
EASY-C	(Enablers for Ambient Services and Systems Part C
EDGE	Enhanced Data GSM Environment
EF	Estimate-and-Forward
FDD	Frequency-Division Duplexing
FIM	Fisher Information Matrix
GSM	Global System for Mobile Communications
i.i.d.	independent and identically distributed
IEEE	Institute of Electrical and Electronics Engineers
IST WINNER	Information Society Technologies Wireless World Initiative New Radio
LDPC	Low-Density Parity-Check
LLR	Log-Likelihood Ratio
LMMSE	Linear Minimum Mean Square Error
LS	Least-Square
LTE	Long Term Evolution
MAC	Multiple Access
MANET	Mobile Ad Hoc Network
MAP	Maximum A Posteriori Probability
MARC	Multi-Access Relay Channel
MED	Minimum Euclidean Distance
MIMO	Multiple-Input Multiple-Output
MISO	Multiple-Input Single-Output
ML	Maximum Likelihood
MMSE	Minimum Mean Square Error
MS	Mobile Station

MSE	Mean Square Error
MTS	Mobile Telephone Service
ODMA	Opportunity Driven Multiple Access
OFDM	Orthogonal Frequency Division Multiplex
OFDMA	Orthogonal Frequency Division Multiple Access
OSTBC	Orthogonal Space-Time Block Codes
PDF	Probability Density Function
PEP	Pairwise Error Probability
PHY	Physical Layer
PSK	Phase-Shift Keying
QAM	Quadrature Amplitude Modulation
QoS	Quality of Service
QPSK	Quadrature Phase-Shift Keying
RACooN	Radio Access with Cooperative Nodes
RS	Relay Station
SI	Self-Interference
SIMO	Single-Input Multiple-Output
SINR	Signal-to-Interference-plus-Noise Ratio
SISO	Single-Input Single-Output
SNR	Signal-to-Noise Ratio
SP	Set Partitioning
SPC	Superposition Coding
STBC	Space-Time Block Codes
STTC	Space-Time Trellis Codes
SVD	Singular Value Decomposition
TD	Time-Division
TDD	Time-Division Duplexing
TDMA	Time Division Multiple Access

UMTS	Universal Mobile Telecommunications System
V-BLAST	Vertical-Bell Laboratories Layered Space-Time
WF	Water-filling
WiMAX	Worldwide Interoperability for Microwave Access
WLAN	Wireless Local Area Network
XOR	Exclusive or
ZF	Zero-Forcing

Notation

$\mathbb{N}, \mathbb{R}, \mathbb{C}$	The set of all natural, real and complex numbers, respectively.
$\mathbb{R}_+^N$	The set of all N-dimensional nonnegative real numbers.
$\mathbb{C}^{M\times N}$	The set of $M \times N$ matrices with complex entries.
$\mathbf{x}$	Vector $\mathbf{x}$, boldface lowercase letters denote vectors.
$\|\mathbf{x}\|$	Euclidean norm of vector $\mathbf{x}$, i.e., $\|\mathbf{x}\| = \sqrt{\mathbf{x}^H\mathbf{x}}$.
$\mathbf{X}$	Matrix $\mathbf{X}$, boldface uppercase letters denote matrices.
$\mathbf{X}^T$	Transpose of the matrix $\mathbf{X}$.
$\mathbf{X}^H$	Complex conjugate and transpose (Hermitian) of the matrix $\mathbf{X}$.
$\mathbf{X}^*$	Element-wise conjugate of the matrix $\mathbf{X}$.
$\mathbf{X}^{-1}$	Inverse of the matrix $\mathbf{X}$.
$\mathbf{X}^\dagger$	Moore-Penrose pseudo-inverse of the matrix $\mathbf{X}$.
$\det(\mathbf{X})$	Determinant of the matrix $\mathbf{X}$.
$\mathrm{tr}(\mathbf{X})$	Trace of the matrix $\mathbf{X}$.
$\|\mathbf{X}\|_F$	Frobenius norm of matrix $\mathbf{X}$.
$[\mathbf{X}]_{ij}$	i, jth component of the matrix $\mathbf{X}$.
$\mathbf{X} \succeq 0$	Matrix $\mathbf{X}$ is positive semidefinite.
$\mathbf{I}$ or $\mathbf{I}_N$	Identity matrix and identity matrix of dimension $N \times N$, respectively.
$\mathrm{vec}(\mathbf{X})$	Vector obtained by stacking the columns of the matrix $\mathbf{X}$.
x	Scalar x.
$\{a\}$	The sequence composed of elements a.
$(x)^+$	Positive part of the real scalar x, i.e., $(x)^+ = \max\{0, x\}$.
$I(x; y)$	Mutual information between random variables x and y.
$H(x)$	Entropy of random variables x.

$\mathcal{CN}(\mathbf{m}, \mathbf{K})$	Circularly symmetric complex Gaussian vector distribution with mean $\mathbf{m}$ and covariance matrix $\mathbf{K}$.
$\mathrm{E}[\cdot]$	Mathematical expectation.
$\mathrm{E}_x\{y\}$	The expectation of y with respect to the random variable x.
$\mathrm{P}(\cdot)$	Probability.
$p(x)$	Probability density function of the random variable x.
$O(g(x))$	Landau symbol to denote that, if $f(x) = O(g(x))$, then for sufficiently large values of x, $f(x)$ is at most a constant multiplied by $g(x)$ in absolute value.
$Q(x)$	The Q-function, i.e., $Q(x) = \frac{1}{\sqrt{2\pi}} \int_x^\infty \exp(-t^2/2)\,\mathrm{d}t$.
$G(\cdot,\cdot)$	The regularized incomplete gamma function.
$\gamma(\cdot,\cdot)$	The lower incomplete Gamma function.
$\Gamma(\cdot)$	The Gamma function.
$\|\mathcal{A}\|$	Cardinality of the set $\mathcal{A}$, i.e., number of elements in $\mathcal{A}$.
$\arg$	Argument.
$\max$, $\min$	Maximum and minimum.
$\sup$, $\inf$	Supremum (lowest upper bound) and infimum (highest lower bound).
$\lim$	Limit.
$\log(\cdot)$	Logarithm (with base 2 unless otherwise stated).
$\ln(\cdot)$	Natural logarithm.
$\max^*$	The max-star operator, where $\max^*(x, y) = \max(x, y) + \ln(1 + \exp(-\|x - y\|))$.
$\mathrm{conv}\, C$	The convex hull of set C.
$\otimes$	Kronecker product.
$\oplus$	Exclusive OR.
$\approx$	Approximately equal to.
$\gg$	Much greater than.
$\sim$	Distributed according to or asymptotically equivalent to.
$\propto$	Equal up to a scaling factor (proportional).
$\mapsto$	Map to.
∇	Gradient.

Bibliography

[1] "Generalized linear precoder and decoder design for MIMO channels using the weighted MMSE criterion."

[2] "IEEE 802.11n pre-draft," IEEE 802.11n Task Group, pre-release.

[3] "IEEE Std 802.16e: Air interface for fixed and mobile broadband wireless access systems," IEEE WirelessMAN 802.16, Tech. Rep., Feb. 2006.

[4] "IEEE 802.11s project," http://www.open80211s.org, 2009.

[5] "IEEE Std 802.16j: Air interface for broadband wireless access systems – multihop relay specification," IEEE 802.16 Relay Task Group, Tech. Rep., June 2009.

[6] "Evolved universal terrestrial radio access (E-UTRA); LTE physical layer; general description," ETSI, Tech. Rep., Mar. 2010, v9.1.0.

[7] I. Abou-Faycal and M. Médard, "Optimal uncoded regeneration for binary antipodal signaling," in *Proc. IEEE Intl. Conf. on Commun. (ICC)*, vol. 2, Paris, France, Jun. 20–24, 2004, pp. 742–746.

[8] M. Abuthinien, S. Chen, A. Wolfgang, and L. Hanzo, "Joint maximum likelihood channel estimation and data detection for MIMO systems," in *Proc. IEEE Intl. Conf. on Commun. (ICC)*, Glasgow, UK, June 24–28, 2007.

[9] R. Ahlswede, "Multi-way communication channels," in *Proc. 2nd IEEE Int. Symposium on Inf. Theory*, Thakadsor, Armenian SSR, Sept. 1971, pp. 23–52.

[10] R. Ahlswede, N. Cai, S.-Y. R. Li, and R. W. Yeung, "Network information flow," *IEEE Trans. Inform. Theory*, vol. 46, no. 4, pp. 1204–1216, July 2000.

[11] J. Akhtman and L. Hanzo, "Maximum-likelihood enhanced sphere decoding for MIMO-OFDM," in *OFDM and MC-CDMA: A Primer.* John Wiley & Sons, 2006, pp. 253–302.

[12] S. Alamouti, "A simple transmit diversity technique for wireless communications," *IEEE J. Select. Areas Commun.*, vol. 16, pp. 1451–1458, Oct. 1998.

[13] J. B. Andersen, "Array gain and capacity for known random channels with multiple element arrays at both ends," *IEEE J. Select. Areas Commun.*, vol. 18, no. 11, pp. 2172–2178, Nov. 2000.

[14] L. Bahl, J. Cocke, F. Jelinek, and J. Raviv, "Optimal decoding of linear codes for minimizing symbol error rate," *IEEE Trans. Inform. Theory*, vol. 20, pp. 284–287, Mar. 1974.

[15] I.-J. Baik and S.-Y. Chung, "Network coding for two-way relay channels using lattices," in *Proc. IEEE Intl. Conf. on Commun. (ICC)*, Beijing, China, May 19–23, 2008.

[16] J. Baltersee, G. Fock, and H. Meyr, "Achievable rate of MIMO channels with data-aided channel estimation and perfect interleaving," *IEEE J. Select. Areas Commun.*, vol. 19, no. 12, pp. 2358–2368, Dec. 2001.

[17] X. Bao and J. Li, "Efficient message relaying for wireless user cooperation: decode-amplify-forward (DAF) and hybrid DAF and coded-cooperation," *IEEE Trans. Wireless Commun.*, vol. 6, no. 11, pp. 3975–3984, 2007.

[18] D. Baum *et al.*, "Final report on link level and system level channel models," Tech. Rep. IST-WINNER D 5.4, Nov. 2005.

[19] A. Behbahani, R. Merched, and A. Eltawil, "Optimizations of a MIMO relay network," *IEEE Trans. Signal Process.*, vol. 56, no. 10, pp. 5062–5073, Oct. 2008.

[20] T. Beniero, S. Redana, J. Hamalainen, and B. Raaf, "Effect of relaying on coverage in 3gpp lte-advanced," in *69th IEEE Veh. Tech. Conf. (VTC)*, Barcelona, Spain, April 26–29, 2009.

[21] S. Berger and A. Wittneben, "Cooperative distributed multiuser MMSE relaying in wireless ad-hoc networks," in *Proc. Asilomar Conf. Signals, Syst., Comput.*, Pacific Grove, CA, Oct. 30–Nov. 3, 2005.

[22] ——, "Experimental performance evaluation of multiuser zero forcing relaying in indoor scenarios," in *Proc. 61th IEEE Veh. Tech. Conf. (VTC)*, May 2005.

[23] D. P. Bertsekas, *Nonlinear Programming*, 2nd ed. Athena Scientific, 1999.

[24] E. Biglieri, R. Calderbank, and A. Constantinides, *MIMO wireless communications*. Cambridge University Press, 2007.

[25] E. Biglieri, J. Proakis, S. Shamai, and D. di Elettronica, "Fading channels: Information-theoretic and communications aspects," *IEEE Trans. Inform. Theory*, vol. 44, no. 6, pp. 2619–2692, Oct. 1998.

[26] M. Biguesh and A. B. Gershman, "Training-based MIMO channel estimation: A study of estimator tradeoffs and optimal training signals," *IEEE Trans. Signal Process.*, vol. 54, no. 3, pp. 884–893, Mar. 2006.

[27] H. Bölcskei, D. Gesbert, and A. J. Paulraj, "On the capacity of OFDM-based spatial multiplexing systems," *IEEE Trans. Commun.*, vol. 50, no. 2, pp. 225–234, Feb. 2002.

[28] H. Bölcskei, R. U. Nabar, O. Oyman, and A. J. Paulraj, "Capacity scaling laws in MIMO relay networks," *IEEE Trans. Wireless Commun.*, vol. 5, no. 6, pp. 1433–1444, June 2006.

[29] H. Bölcskei, R. W. Heath Jr., and A. J. Paulraj, "Blind channel identification and equalization in OFDM-based multi-antenna systems," *IEEE Trans. Signal Process.*, vol. 50, no. 1, pp. 96–109, Jan. 2002.

[30] H. Bölcskei and A. J. Paulraj, *"The Communications Handbook"*, 2nd ed. CRC Press, 2002, ch. Multiple-input multiple-output (MIMO) wireless systems, pp. 90.1 – 90.14.

[31] J. Bonnet and G. Auer, "Optimized iterative channel estimation for OFDM," in *Proc. 64th IEEE Veh. Tech. Conf. (VTC)*, Montreal, Canada, Sept. 25–28, 2006.

[32] M. Borgmann and H. Bölcskei, "On the capacity of noncoherent wideband MIMO-OFDM systems," in *Proc. IEEE Int. Symposium on Inf. Theory*, Adelaide, Australia, Sept. 2005, pp. 651–655.

[33] S. Boyd and L. Vandenberghe, *Convex Optimization*. Cambridge University Press, 2004.

[34] D. Brennan, "Linear diversity combining techniques," *Proceedings of the IRE*, vol. 47, no. 6, pp. 1075–1102, June 1959.

[35] A. G. C. Berrou and P. Thitimajshima, "Near shannon limit error-correcting coding and decoding: Turbo-codes."

[36] G. Caire, G. Taricco, and E. Biglieri, "Bit-interleaved coded modulation," *IEEE Trans. Inform. Theory*, vol. 44, no. 3, pp. 927–945, May 1998.

[37] A. Chakrabarti, A. De Baynast, A. Sabharwal, and B. Aazhang, "Low density parity check codes for the relay channel," *IEEE J. Select. Areas Commun.*, vol. 25, no. 2, pp. 280–290, 2007.

[38] R. W. Chang, "Synthesis of band-limited orthogonal signals for multichannel data transmission," *Bell System Techn. J.*, vol. 45, pp. 1775–1796, Dec. 1966.

[39] L. Chebli, C. Hausl, G. Zeitler, and R. Koetter, "Cooperative uplink of two mobile stations with network coding based on the WiMAX LDPC code," in *Proc. IEEE Global Commun. Conf. (GLOBECOM)*, Honolulu, USA, Nov. 30 – Dec. 4, 2009.

[40] C.-J. Chen and L.-C. Wang, "Coverage and capacity enhancement in multiuser MIMO systems with scheduling," in *Proc. IEEE Global Commun. Conf. (GLOBECOM)*, Dallas, TX, Nov. 29 – Dec. 3, 2004.

[41] D. Chen and J. Laneman, "The diversity-multiplexing tradeoff for the multiaccess relay channel," in *Proc. Conf. on Information Sciences and Systems (CISS)*, March 22–24, 2006, pp. 1324–1328.

[42] ——, "Modulation and demodulation for cooperative diversity in wireless systems," *IEEE Trans. Wireless Commun.*, vol. 5, no. 7, p. 1785, July 2006.

[43] S. Chen, M. A. Beach, and J. P. McGeehan, "Division-free duplex for wireless applications," *Electron. Lett.*, vol. 34, no. 2, pp. 147–148, Jan. 1998.

[44] W. Chen, L. Dai, K. Letaief, and Z. Cao, "A unified cross-layer framework for resource allocation in cooperative networks," *IEEE Trans. Wireless Commun.*, vol. 7, no. 8, pp. 3000–3012, Aug. 2008.

[45] M. Coldrey and P. Bohlin, "Training-based MIMO systems – part I: Performance comparison," *IEEE Trans. Signal Process.*, vol. 55, no. 11, pp. 5464 – 5476, Nov. 2007.

[46] ——, "Training-based MIMO systems: Part II – improvements using detected symbol information," *IEEE Trans. Signal Process.*, vol. 56, no. 1, pp. 296 – 303, Jan. 2008.

[47] M. Costa, "Writing on dirty paper," *IEEE Trans. Inform. Theory*, vol. 29, no. 3, pp. 439–441, May 1983.

[48] T. M. Cover and A. El Gamal, "Capacity theorems for the relay channel," *IEEE Trans. Inform. Theory*, vol. 25, pp. 572–584, Sept. 1979.

[49] T. M. Cover and J. A. Thomas, *Elements of Information Theory.* John Wiley & Sons, 1991.

[50] CTIA, "CTIA semi-annual wireless industry survey," http://www.ctia.org, 2009.

[51] T. Cui, T. Ho, and J. Kliewer, "Memoryless relay strategies for two-way relay channels," *IEEE Trans. Commun.*, vol. 57, no. 10, pp. 3132–3143, Oct. 2009.

[52] T. Cui, F. Gao, T. Ho, and A. Nallanathan, "Distributed space¨ctime coding for two-way wireless relay networks," *IEEE Trans. Signal Process.*, vol. 57, no. 2, pp. 658–671, Feb. 2009.

[53] T. Cui, F. Gao, and C. Tellambura, "Differential modulation for two-way wireless communications: a perspective of differential network coding at the physical layer," *IEEE Trans. Commun.*, vol. 57, no. 10, pp. 2977–2987, Oct. 2009.

[54] O. Damen, A. Chkeif, and J.-C. Belfiore, "Lattice code decoder for space-time codes," *IEEE Commun. Lett.*, vol. 4, no. 5, pp. 161–163, May 2000.

[55] Z. Dawy, S. Davidović, and I. Oikonomidis, "Coverage and capacity enhancement of CDMA cellular systems via multihop transmission," in *Proc. IEEE Global Commun. Conf. (GLOBECOM)*, San Francisco, CA, Dec. 1–5, 2003.

[56] X. Deng, A. M. Haimovich, and J. Garcia-Frias, "Decision directed iterative channel estimation for MIMO systems," in *Proc. IEEE Intl. Conf. on Commun. (ICC)*, Anchorage, AK, May 11–15, 2003.

[57] A. Dinnis and J. Thompson, "Increasing high data rate coverage in cellular systems using relaying," in *60th IEEE Veh. Tech. Conf. (VTC)*, vol. 5, Los Angeles, CA, Sep. 26–29, 2004, pp. 3424–3428.

[58] M. Dohler and Y. Li, *Cooperative Communications: Hardware, Channel and PHY*. John Wiley & Sons, 2010.

[59] M. Dong, L. Tong, and B. Sadler, "Optimal insertion of pilot symbols for transmissions over time-varying flat fading channels," *IEEE Trans. Signal Process.*, vol. 52, no. 5, pp. 1403–1418, May 2004.

[60] M. Effros, T. Ho, and S. Kim, "A tiling approach to network code design for wireless networks," in *Proc. Inform. Theory Workshop (ITW)*, Punta del Este, Uruguay, Mar. 13 – 17, 2006, pp. 62–66.

[61] P. Elia, K. Kumar, S. Pawar, P. Kumar, and H.-F. Lu, "Explicit space-time codes achieving the diversity-multiplexing gain tradeoff," *IEEE Trans. Inform. Theory*, vol. 52, no. 9, pp. 3869–3884, 2006.

[62] ETSI, "Universal mobile telecommunications system (UMTS): Multiplexing and channel coding (TDD)," Tech. Rep. 3GPP TS 25.222 version 8.4.0, Mar. 2009.

[63] K. Fan, "Minimax theorems," *Proceedings of the National Academy of Sciences of the United States of America*, vol. 39, no. 1, pp. 42–47, Jan. 15, 1953.

[64] R. F. H. Fischer, C. Windpassinger, A. Lampe, and J. B. Huber, "Space-time transmission using Tomlinson-Harashima precoding," in *Proc. 4th ITG Conference on Source and Channel Coding*, pp. 139–147.

[65] G. J. Foschini, "Layered space-time architecture for wireless communication in a fading environment when using multi-element antennas," *Bell System Techn. J.*, vol. 1, no. 2, pp. 41–59, Autumn 1996.

[66] G. J. Foschini and M. J. Gans, "On limits of wireless communications in a fading environment when using multiple antennas," *Wireless Personal Communications*, vol. 6, no. 3, pp. 311–335, Mar. 1998.

[67] G. Foschini, G. Golden, R. Valenzuela, and P. Wolniansky, "Simplified processing for high spectral efficiency wireless communication employing multi-element arrays," *IEEE J. Select. Areas Commun.*, vol. 17, no. 11, pp. 1841–1852, Nov. 1999.

[68] A. Fujiwara, S. Takeda, H. Yoshino, and T. Otsu, "Area coverage and capacity enhancement by multihop connection of CDMA cellular network," in *Proc. 56th IEEE Veh. Tech. Conf. (VTC)*, Vancouver, Canada, Sept. 24–28, 2002.

[69] K. Fukuda, "The CDD solver," http://www.cs.mcgill.ca/~fukuda/soft/cdd_home/cdd.html, Aug. 2005.

[70] R. G. Gallager, "Low density parity check codes," MIT Press, Cambridge, Mass, 1963, monograph.

[71] F. Gao, T. Cui, and A. Nallanathan, "On channel estimation and optimal training design for amplify and forward relay networks," *IEEE Trans. Wireless Commun.*, vol. 7, no. 5, pp. 1907–1916, May 2008.

[72] ——, "Optimal training design for channel estimation in decode-and-forward relay networks with individual and total power constraints," *IEEE Trans. Signal Process.*, vol. 56, no. 12, pp. 5937–5949, Dec. 2008.

[73] F. Gao, R. Zhang, and Y.-C. Liang, "On channel estimation for amplify-and-forward two-way relay networks," in *Proc. IEEE Global Commun. Conf. (GLOBECOM)*, New Orleans, LA, Nov. 30 – Dec. 4, 2008.

[74] M. Gastpar and M. Vetterli, "On the capacity of wireless networks: The relay case," in *Proc. IEEE INFOCOM*, New York, USA, Jun. 23–27 2002, pp. 1577–1586.

[75] D. Gesbert, M. Kountouris, R. Heath, C. Chae, and T. Salzer, "From single user to multiuser communications: Shifting the MIMO paradigm," *IEEE Signal Processing Mag.*, vol. 24, no. 5, pp. 36–46, 2007.

[76] G. Golden, G. Foschini, R. Valenzuela, and R. Wolniansky, "Detection algorithm and initial laboratory results using V-BLAST space-time communication architecture," *Electron. Lett.*, vol. 35, no. 1, pp. 14–15, Jan. 1999.

[77] A. Goldsmith, *Wireless Communications*. New York, NY, USA: Cambridge University Press, 2005.

[78] A. Goldsmith, S. A. Jafar, N. Jindal, and S. Vishwanath, "Capacity limits of MIMO channels," *IEEE J. Select. Areas Commun.*, vol. 21, no. 5, pp. 684–702, June 2003.

[79] J. S. Gomadam, K.S., "Duality of MIMO multiple access channel and broadcast channel with amplify-and-forward relays," *IEEE Trans. Commun.*, vol. 58, no. 1, pp. 211–217, Jan. 2010.

[80] K. Gomadam and S. Jafar, "Optimal relay functionality for SNR maximization in memoryless relay networks," *IEEE J. Select. Areas Commun.*, vol. 25, no. 2, pp. 390–401, 2007.

[81] D. Gore, A. Gorokhov, and A. Paulraj, "Joint MMSE versus V-BLAST and antenna selection," in *Proc. Asilomar Conf. Signals, Syst., Comput.*, vol. 1, Pacific Grove, CA, Nov. 3–6, 2002, pp. 505–509.

[82] D. Gore, J. Heath, R.W., and A. Paulraj, "On performance of the zero forcing receiver in presence of transmit correlation," in *Proc. IEEE Int. Symposium on Inf. Theory*, Lausanne, Switzerland, June 30 – July 5, 2002.

[83] M. Grossglauser and D. Tse, "Mobility increases the capacity of ad-hoc wireless networks," *IEEE/ACM Transactions on Networking*, vol. 10, no. 4, pp. 477–486, Aug. 2002.

[84] D. Gündüz, E. Tuncel, and J. Nayak, "Rate regions for the separated two-way relay channel," in *46th Allerton Conf. Comm., Contr. and Comp.*, Allerton House, Monticello, Illinois, Sept. 23–26, 2008.

[85] D. Gündüz, A. Goldsmith, and H. V. Poor, "MIMO two-way relay channel: Diversity-multiplexing trade-off analysis," in *Proc. Asilomar Conf. Signals, Syst., Comput.*, Pacific Grove, CA, 2008.

[86] D. Gündüz, A. Yener, A. Goldsmith, and H. V. Poor, "The multi-way relay channel," in *Proc. IEEE Int. Symposium on Inf. Theory*, Seoul, South Korea, June 28 – July 3, 2009, pp. 339–343.

[87] P. Gupta and P. Kumar, "The capacity of wireless networks," *IEEE Trans. Inform. Theory*, vol. 46, no. 2, pp. 388–404, Mar. 2000.

[88] I. Hammerstroem, M. Kuhn, C. Esli, J. Zhao, A. Wittneben, and G. Bauch, "MIMO two-way relaying with transmit CSI at the relay," in *Proc. IEEE Int. Workshop on Signal Process. Advances for Wireless Comm. (SPAWC)*, Helsinki, Finland, Jun. 17–20, 2007.

[89] I. Hammerstroem and A. Wittneben, "Power allocation schemes for amplify-and-forward MIMO-OFDM relay links," *IEEE Trans. Wireless Commun.*, vol. 6, no. 8, pp. 2798–2802, 2007.

[90] Y. Han, S. H. Ting, C. K. Ho, and W. H. Chin, "Performance bounds for two-way amplify-and-forward relaying," *IEEE Trans. Wireless Commun.*, vol. 8, no. 1, pp. 432–439, Jan. 2009.

[91] H. Harashima and H. Miyakawa, "Matched-transmission technique for channels with intersymbol interference," *IEEE Trans. Commun.*, vol. 20, no. 4, pp. 774–780, 1972.

[92] M. Hartl, C. Rauch, C. Sattler, and A. Baier, "Trial of a hybrid DVB-H / GSM mobile broadcast system," in *14th IST Summit on Mob. and Wirel. Comm.*, Dresden, Germany, Jun. 19–23, 2005.

[93] M. O. Hasna and M. S. Alouini, "Harmonic mean and end-to-end performance of transmission systems with relays," *IEEE Trans. Commun.*, vol. 52, no. 1, pp. 130–135, Jan. 2004.

[94] M. Hasna and M.-S. Alouini, "End-to-end performance of transmission systems with relays over rayleigh-fading channels," *IEEE Trans. Wireless Commun.*, vol. 2, pp. 1126–1131, Nov. 2003.

[95] M. Hasna and M. Alouini, "A performance study of dual-hop transmissions with fixed gain relays," *IEEE Trans. Wireless Commun.*, vol. 3, no. 6, pp. 1963–1968, Nov. 2004.

[96] B. Hassibi and B. M. Hochwald, "High-rate codes that are linear in space and time," *IEEE Trans. Inform. Theory*, vol. 48, no. 7, pp. 1804–1824, July 2002.

[97] ——, "How much training is needed in multiple-antenna wireless links," *IEEE Trans. Inform. Theory*, vol. 49, no. 4, pp. 951–963, Apr. 2003.

[98] C. Hausl, "Joint network-channel coding for the multiple-access relay channel based on turbo codes," *Europ. Trans. Telecommun.*, vol. 20, no. 2, pp. 175–181, 2009.

[99] C. Hausl and J. Hagenauer, "Iterative network and channel decoding for the two-way relay channel," in *Proc. IEEE Intl. Conf. on Commun. (ICC)*, vol. 4, Istanbul, Turkey, June 11–15, 2006, pp. 1568–1573.

[100] S. He, J. K. Tugnait, and X. Meng, "On superimposed training for MIMO channel estimation and symbol detection," *IEEE Trans. Signal Process.*, vol. 55, no. 6, pp. 3007–3021, June 2007.

[101] X. He and A. Yener, "On the role of feedback in two-way secure communication," in *Proc. Asilomar Conf. Signals, Syst., Comput.*, Pacific Grove, CA, Oct. 26 – Oct. 29, 2008.

[102] R. W. Heath and A. J. Paulraj, "Linear dispersion codes for MIMO systems based on frame theory," *IEEE Trans. Signal Process.*, vol. 50, no. 10, pp. 2429–2441, Oct. 2002.

[103] C. K. Ho, B. Farhang-Boroujeny, and F. Chin, "Added pilot semi-blind channel estimation scheme for OFDM in fading channels," in *Proc. IEEE Global Commun. Conf. (GLOBECOM)*, vol. 5, San Antonio, TX, Nov. 25–29, 2001, pp. 3075–3079.

[104] T. Ho, M. Medard, R. Koetter, D. Karger, M. Effros, J. Shi, and B. Leong, "A Random Linear Network Coding Approach to Multicast," *IEEE Trans. Inform. Theory*, vol. 52, no. 10, pp. 4413–4430, 2006.

[105] B. M. Hochwald, C. B. Peel, and A. L. Swindlehurst, "A vector-perturbation technique for near-capacity multiantenna multiuser communication–part ii: Perturbation," *IEEE Trans. Commun.*, vol. 53, no. 3, pp. 537–544, 2005.

[106] B. M. Hochwald and W. Sweldens, "Differential unitary space-time modulation," *IEEE Trans. Commun.*, vol. 48, no. 12, pp. 2041–2052, Dec. 2000.

[107] B. M. Hochwald and S. ten Brink, "Achieving near-capacity on a multiple-antenna channel," *IEEE Trans. Commun.*, vol. 51, no. 3, pp. 389–399, Mar. 2003.

[108] P. Hoeher and J. Lodge, "Turbo DPSK: Iterative differential PSK demodulation and channel decoding," *IEEE Trans. Commun.*, vol. 47, no. 6, pp. 837–843, June 1999.

[109] P. Hoeher and F. Tufvesson, "Channel estimation with superimposed pilot sequence," in *Proc. IEEE Global Commun. Conf. (GLOBECOM)*, Rio de Janeiro, Brazil, Dec. 5–9, 1999, pp. 2162–2166.

[110] H. Hu, H. Yanikomeroglu, D. D. Falconer, and S. Periyalwar, "Range extension without capacity penalty in cellular networks with digital fixed relays," in *Proc. IEEE Globecom'04*, Dallas, TX, Nov. 29 – Dec. 3, 2004.

[111] B. Hughes, "Differential space-time modulation," *IEEE Trans. Inform. Theory*, vol. 46, no. 7, pp. 2567–2578, Nov. 2000.

[112] T. E. Hunter and A. Nosratinia, "Diversity through coded cooperation," *IEEE Trans. Wireless Commun.*, 2004, submitted.

[113] ——, "Cooperative diversity through coding," in *Proc. IEEE Int. Symposium on Inf. Theory*, Lausanne, Switzerland, June 30 – July 5, 2002, p. 220.

[114] M. T. Ivrlač and J. A. Nossek, "MIMO eigenbeamforming in correlated fading," in *Proc. IEEE International Conference on Circuits and Systems for Communications (ICCSC)*, June 26 – 28, 2002, pp. 212 – 215.

[115] S. A. Jafar, K. S. Gomadam, and C. Huang, "Duality and rate optimization for multiple access and broadcast channels with amplify-and-forward relays," *IEEE Trans. Inform. Theory*, vol. 53, no. 10, pp. 3350–3370, Oct. 2007.

[116] W. C. Jakes, Ed., *Microwave Mobile Communications*. John Wiley & Sons, 1975.

[117] M. Janani, A. Hedayat, T. E. Hunter, and A. Nosratinia, "Coded cooperation in wireless communications: Space-time transmission and iterative decoding," *IEEE Trans. Signal Process.*, vol. 52, no. 2, Feb. 2004.

[118] C. Jandura, P. Marsch, A. Zoch, and G. Fettweis, "A testbed for cooperative multi cell algorithms," in *Proc. 4th International Conference on Testbeds and Research Infrastructures for the Development of Networks & Communities (TRIDENTCOM'08)*, Innsbruck, Austria, Mar. 18–20, 2008.

[119] S. Jayaweera and H. Poor, "Capacity of multiple-antenna systems with both receiver and transmitter channel state information," *IEEE Trans. Inform. Theory*, vol. 49, no. 10, pp. 2697–2709, Oct. 2003.

[120] M. Jiang, J. Akhtman, and L. Hanzo, "Iterative joint channel estimation and multi-user detection for multiple-antenna aided OFDM systems," *IEEE Trans. Wireless Commun.*, vol. 6, no. 8, pp. 2904–2914, Aug. 2007.

[121] Y. Jiang, J. Li, and W. W. Hager, "Joint transceiver design for MIMO communications using geometric mean decomposition," *IEEE Trans. Signal Process.*, vol. 53, no. 10, pp. 3791–3803, 2005.

[122] N. Jindal and Z.-Q. Luo, "Capacity limits of multiple antenna multicast," in *Proc. IEEE Int. Symposium on Inf. Theory*, Seattle, WA, July 2006.

[123] ——, "Capacity limits of multiple antenna multicast," in *Proc. IEEE Int. Symposium on Inf. Theory*, Seattle, WA, July 9–14, 2006, pp. 1841–1845.

[124] Y. Jing and B. Hassibi, "Distributed space-time coding in wireless relay networks," *IEEE Trans. Wireless Commun.*, vol. 5, no. 12, pp. 3524–3536, Dec. 2006.

[125] K. Jitvanichphaibool, R. Zhang, and Y.-C. Liang, "Optimal resource allocation for two-way relay-assisted ofdma," *IEEE Trans. Veh. Technol.*, vol. 58, no. 7, pp. 3311–3321, Sept. 2009.

[126] K. Josiam and D. Rajan, "Bandwidth efficient channel estimation using super-imposed pilots in OFDM systems," *IEEE Trans. Wireless Commun.*, vol. 6, no. 6, pp. 2234–2245, June 2007.

[127] S. Katti, S. Gollakota, and D. Katabi, "Embracing wireless interference: analog network coding," in *Proc. ACM SIGCOMM*, 2007, pp. 397–408.

[128] S. Katti, D. Katabi, H. Balakrishnan, and M. Medard, "Symbol-level network coding for wireless mesh networks," in *ACM SIGCOMM*, Seattle, WA, Aug. 17–22, 2008.

[129] S. M. Kay, *Fundamentals of Statistical Signal Processing, Volume I: Estimation Theory*. Prentice Hall PTR, 1993.

[130] T. Koike-Akino, P. Popovski, and V. Tarokh, "Optimized constellations for two-way wireless relaying with physical network coding," *IEEE J. Select. Areas Commun.*, vol. 27, no. 5, pp. 773–787, June 2009.

[131] G. Kramer, M. Gastpar, and P. Gupta, "Cooperative strategies and capacity theorems for relay networks," *IEEE Trans. Inform. Theory*, vol. 51, no. 9, pp. 3037–3063, Sept. 2005.

[132] G. Kramer and A. van Wijngaarden, "On the white gaussian multiple-access relay channel," in *Proc. IEEE Int. Symposium on Inf. Theory*, Sorrento, Italy, June 25–30, 2000, p. 40.

[133] H. Krim and M. Viberg, "Two decades of array signal processing research: the parametric approach," *IEEE Signal Processing Mag.*, vol. 13, no. 4, pp. 67–94, July 1996.

[134] J. N. Laneman, D. N. Tse, and G. W. Wornell, "Cooperative diversity in wireless networks: Efficient protocols and outage behavior," *IEEE Trans. Inform. Theory*, vol. 50, no. 12, pp. 3062–3080, Dec. 2004.

[135] J. N. Laneman and G. W. Wornell, "Distributed space-time-coded protocols for exploiting cooperative diversity in wireless networks," *IEEE Trans. Inform. Theory*, vol. 49, no. 10, pp. 2415–2425, Oct. 2003.

[136] P. Larsson, N. Johansson, and K.-E. Sunell, "Coded bidirectional relaying," in *Proc. IEEE Veh. Tech. Conf. (VTC)*, vol. 2, Melbourne, Australia, May 7–10, 2006, pp. 851–855.

[137] V. K. Lau and Y.-K. R. Kwok, *Channel-Adaptive Technologies and Cross-Layer Designs for Wireless Systems with Multiple Antennas: Theory and Applications*. Wiley-Interscience, 2006.

[138] J. Lee and N. Jindal, "Symmetric capacity of MIMO downlink channels," in *Proc. IEEE Int. Symposium on Inf. Theory*, Seattle, WA, July 2006, pp. 1031–1035.

[139] K.-J. Lee, K. W. Lee, H. Sung, and I. Lee, "Sum-rate maximization for two-way MIMO amplify-and-forward relaying systems," in *69th IEEE Veh. Tech. Conf. (VTC)*, Barcelona, Spain, April 26–29, 2009.

[140] K. Lee and L. Hanzo, "MIMO-assisted hard versus soft decoding-and-forwarding for network coding aided relaying systems," *IEEE Trans. Wireless Commun.*, vol. 8, no. 1, pp. 376–385, Jan. 2009.

[141] N. Lee, H. Park, and J. Chun, "Linear precoder and decoder design for two-way AF MIMO relaying system," in *Proc. 67th IEEE Veh. Tech. Conf. (VTC)*, Singapore, May 11–14, 2008, pp. 1221–1225.

[142] S.-Y. R. Li, R. W. Yeung, and N. Cai, "Linear network coding," *IEEE Trans. Inform. Theory*, vol. 49, no. 2, pp. 371–381, Feb. 2003.

[143] Y. Li and B. Vucetic, "On the performance of a simple adaptive relaying protocol for wireless relay networks," in *67th IEEE Veh. Tech. Conf. (VTC)*, Singapore, May 11–14, 2008.

[144] Y. Li, B. Vucetic, T. Wong, and M. Dohler, "Distributed turbo coding with soft information relaying in multihop relay networks," *IEEE J. Select. Areas Commun.*, vol. 24, no. 11, pp. 2040–2050, 2006.

[145] Y. Li, J. Winters, and N. Sollenberger, "MIMO-OFDM for wireless communications: signal detection with enhanced channel estimation," *IEEE Trans. Commun.*, vol. 50, no. 9, pp. 1471–1477, Sept. 2002.

[146] Y. Li, "Distributed coding for cooperative wireless networks: An overview and recent advances," *IEEE Commun. Mag.*, vol. 47, no. 8, pp. 71–77, Aug. 2009.

[147] T.-J. Liang, W. Rave, and G. Fettweis, "Iterative joint channel estimation and decoding using superimposed pilots in OFDM-WLAN," in *Proc. IEEE Intl. Conf. on Commun. (ICC)*, Istanbul, Turkey, June 11–15, 2006.

[148] H. Liao, "Multiple access channels," Ph.D. dissertation, University of Hawaii, 1972.

[149] J. Liu and Y. T. Hou, "Maximum weighted sum rate of multi-antenna broadcast channels," in *Proc. IEEE Intl. Conf. on Commun. (ICC)*, Beijing, China, May 19–23, 2008.

[150] M. P. Llisterri, "Impact of channel state information on the analysis and design of multi-antenna communication systems," Ph.D. dissertation, Universitat Politècnica de Catalunya (UPC), Barcelona, Spain, Dec. 2006.

[151] E. S. Lo and K. B. Letaief, "Network coding versus superposition coding for two-way wireless communication," in *Proc. IEEE Wirel. Comm. and Netw. Conf. (WCNC)*, Budapest, Hungary, Apr. 5–8, 2009.

[152] X. Ma and G. Giannakis, "Full-Diversity Full-Rate Complex-Field Space-Time Coding," *IEEE Trans. Signal Process.*, vol. 51, no. 11, pp. 2917–2930, 2003.

[153] D. MacKay and R. Neal, "Near shannon limit performance of low density parity check codes," *Electron. Lett.*, vol. 33, no. 6, pp. 457–458, Mar 1997.

[154] M. R. McKay and I. B. Collings, "Error performance of MIMO-BICM with zero-forcing receivers in spatially-correlated rayleigh channels," *IEEE Trans. Wireless Commun.*, vol. 6, no. 3, pp. 787–792, Mar. 2007.

[155] W. Mohr, "Spectrum demand for systems beyond IMT-2000 based on data rate estimates," *Wireless Communications and Mobile Computing*, vol. 3, no. 7, pp. 817–835, Nov. 2003.

[156] W. Mohr, R. Lüder, and K.-H. Möhrmann, "Data rate estimates, range calculations and spectrum demand for new elements of systems beyond IMT-2000," in *The 5th International Symposium on Wireless Personal Multimedia Communications*, vol. 1, Oct. 2002, pp. 37–46.

[157] O. Muñoz, A. Agustín, and J. Vidal, "Cellular capacity gains of cooperative MIMO transmission in the downlink," in *Int. Zurich Seminar on Communication (IZS)*, Zurich, Switzerland, Feb. 2004.

[158] O. Muñoz, J. Vidal, and A. Agustín, "Linear transceiver design in non regenerative relays with channel state information," *IEEE Trans. Signal Process.*, vol. 55, no. 6, pp. 2593–2604, June 2007.

[159] R. U. Nabar, H. Bölcskei, and F. W. Kneubühler, "Fading relay channels: Performance limits and space-time signal design," *IEEE J. Select. Areas Commun.*, vol. 22, no. 6, pp. 1099–1109, June 2004.

[160] W. Nam, S.-Y. Chung, and Y. H. Lee, "Capacity of the gaussian two-way relay channel to within 1/2 bit," *IEEE Trans. Inform. Theory*, 2009, submitted.

[161] K. Narayanan, M. P. Wilson, and A. Sprintson, "Joint physical layer coding and network coding for bi-directional relaying," in *Proc. 45th Allerton Conf. Comm., Contr. and Comp.*, Allerton House, Monticello, Illinois, Sept. 26–28, 2007.

[162] A. Narula, M. Lopez, M. Trott, and G. Wornell, "Efficient use of side information in multiple-antenna data transmission over fading channels," *IEEE J. Select. Areas Commun.*, vol. 16, no. 8, pp. 1423–1436, Oct. 1998.

[163] B. Nazer and M. Gastpar, "Compute-and-forward: Harnessing interference through structured codes," *IEEE Trans. Inform. Theory*, 2009, submitted.

[164] M. C. Necker and G. L. Stüber, "Totally blind channel estimation for OFDM on fast varying mobile radio channels," *IEEE Trans. Wireless Commun.*, vol. 3, no. 5, pp. 1514–1525, Sept. 2004.

[165] T. C.-Y. Ng and W. Yu, "Joint optimization of relay strategies and resource allocations in cooperative cellular networks," *IEEE J. Select. Areas Commun.*, vol. 25, no. 2, pp. 328–339, Feb. 2007.

[166] V. Nguyen, H. D. Tuan, H. H. Nguyen, and N. N. Tran, "Optimal superimposed training design for spatially correlated fading MIMO channels," *IEEE Trans. Wireless Commun.*, vol. 7, no. 8, pp. 3206–3217, Aug. 2008.

[167] A. Nosratinia, T. E. Hunter, and A. Hedayat, "Cooperative communication in wireless networks," *IEEE Commun. Mag.*, Oct. 2004.

[168] T. Oechtering, "Spectrally efficient bidirectional decode-and-forward relaying for wireless networks," Ph.D. dissertation, Technical University Berlin, Oct. 2007.

[169] T. J. Oechtering and H. Boche, "Optimal time-division for bidirectional relaying using superposition encoding," *IEEE Commun. Lett.*, vol. 12, no. 4, Apr. 2008.

[170] T. J. Oechtering, C. Schnurr, I. Bjelakovic, and H. Boche, "Broadcast capacity region of two-phase bidirectional relaying," *IEEE Trans. Inform. Theory*, vol. 54, no. 1, pp. 454–458, Jan. 2008.

[171] T. J. Oechtering, R. F. Wyrembelski, and H. Boche, "Optimal time-division of two-phase decode-and-forward bidirectional relaying," in *Proc. IEEE Int. Symposium on Inf. Theory and its Applications*, Auckland, New Zealand, Dec. 7–10, 2008, pp. 829–834.

[172] ——, "Multiantenna bidirectional broadcast channels - optimal transmit strategies," *IEEE Trans. Signal Process.*, vol. 57, no. 5, pp. 1948–1958, May 2009.

[173] A. Ozgur, O. Leveque, and D. N. C. Tse, "Hierarchical cooperation achieves optimal capacity scaling in ad hoc networks," *IEEE Trans. Inform. Theory*, vol. 53, no. 10, pp. 3549–3572, Oct. 2007.

[174] R. Pabst, B. Walke, D. Schultz, P. Herhold, H. Yanikomeroglu, S. Mukherjee, H. Viswanathan, M. Lott, W. Zirwas, M. Dohler, H. Aghvami, D. Falconer, and G. Fettweis, "Relay-based deployment concepts for wireless and mobile broadband radio," *IEEE Commun. Mag.*, vol. 42, no. 9, pp. 80–89, Sept. 2004.

[175] D. P. Palomar, A. Agustín, O. Muñoz, and J. Vidal, "Decode and forward protocol for cooperative diversity in multi-antenna wireless networks," in *Proc. Conference on Information Sciences and Systems (CISS)*, Feb. 2004.

[176] D. P. Palomar, M. Bengtsson, and B. Ottersten, "Minimum BER linear transceivers for MIMO channels via primal decomposition," *IEEE Trans. Signal Process.*, vol. 53, no. 8, pp. 2866–2882, Aug. 2005.

[177] D. P. Palomar, J. M. Cioffi, and M. A. Lagunas, "Joint Tx-Rx beamforming design for multicarrier MIMO channels: a unified framework for convex optimization," *IEEE Trans. Signal Process.*, vol. 51, no. 9, pp. 2381–2401, Sept. 2003.

[178] ——, "Uniform power allocation in MIMO channels: A game-theoretic approach," *IEEE Trans. Inform. Theory*, vol. 49, no. 7, pp. 1707–1727, July 2003.

[179] D. P. Palomar, "A unified framework for communications through MIMO channels," Ph.D. dissertation, Technical University of Catalonia (UPC), Barcelona, Spain, May 2003.

[180] D. Palomar, M. Lagunas, and J. Cioffi, "Optimum linear joint transmit-receive processing for MIMO channels with QoS constraints," *IEEE Trans. Signal Process.*, vol. 52, no. 5, pp. 1179–1197, May 2004.

[181] A. Papoulis and S. U. Pillai, *Probability, Random Variables and Stochastic Processes*, 4th ed. McGraw-Hill, 2002.

[182] C. S. Patel, S. Member, and G. L. Stüber, "Channel estimation for amplify and forward relay based cooperation diversity systems," *IEEE Trans. Wireless Commun.*, vol. 6, no. 6, pp. 2348–2356, June 2007.

[183] A. J. Paulraj and T. Kailath, "Increasing capacity in wireless broadcast systems using distributed transmission/directional reception," U. S. Patent 5 345 599, 1994.

[184] A. Paulraj, R. Nabar, and D. Gore, *Introduction to Space-Time Wireless Communications*. Cambridge University Press, 2003.

[185] C. B. Peel, B. M. Hochwald, and A. L. Swindlehurst, "A vector-perturbation technique for near-capacity multiantenna multiuser communication–part I: channel inversion and regularization," *IEEE Trans. Commun.*, vol. 53, no. 1, pp. 195–202, 2005.

[186] T.-H. Pham, Y.-C. Liang, A. Nallanathan, and G. Krishna, "On the design of optimal training sequence for bi-directional relay networks," *IEEE Signal Processing Lett.*, vol. 16, no. 3, pp. 200–203, Mar. 2009.

[187] P. Popovski and H. Yomo, "The anti-packets can increase the achievable throughput of a wireless multi-hop network," in *Proc. IEEE Intl. Conf. on Commun. (ICC)*, Istanbul, Turkey, June 11–15, 2006.

[188] ——, "Physical network coding in two-way wireless relay channels," in *Proc. IEEE Intl. Conf. on Commun. (ICC)*, Glasgow, UK, June 24–28, 2007.

[189] V. Protassov, G. Frenk, and G. Kassay, "Extended minimax theorems," May 2001, monograph.

[190] B. Rankov and A. Wittneben, "Achievable rate regions for the two-way relay channel," in *Proc. IEEE Int. Symposium on Inf. Theory*, Seattle, (WA), 2006, submitted.

[191] ——, "Distributed spatial multiplexing in a wireless network," in *Proc. of the Asilomar Conference on Signals, Systems, and Computers 2004*, Pacific Grove, CA, Nov. 2004.

[192] ——, "Spectral efficient signaling for half-duplex relay channels," in *Asilomar Conference on Signals, Systems, and Computers 2005*, Pacific Grove, CA, Nov. 2005.

[193] ——, "Spectral efficient protocols for half-duplex fading relay channels," *IEEE J. Select. Areas Commun.*, vol. 25, no. 2, pp. 379–389, Feb. 2007.

[194] R. Roy and B. Ottersten, "Spatial division multiple access wireless communication systems," U. S. Patent 5 515 378, May 1996.

[195] L. Sankaranarayanan, G. Kramer, and N. B. Mandayam, "Capacity theorems for the multiple-access relay channel," in *Allerton Conf. Comm., Contr. and Comp.*, Allerton House, Monticello, Illinois, Sep. 29 – Oct. 1, 2004.

[196] F. Sanzi, S. Jelting, and J. Speidel, "A comparative study of iterative channel estimators for mobile OFDM systems," *IEEE Trans. Wireless Commun.*, vol. 2, no. 5, pp. 849–859, Sept. 2003.

[197] H. Sato, "Information transmission through a channel with relay," *The ALOHA System, University of Hawaii, Honolulu, Tech. Rep. B*, vol. 7, 1976.

[198] A. Scaglione, P. Stoica, S. Barbarossa, G. Giannakis, and H. Sampath, "Optimal designs for space-time linear precoders and decoders," *IEEE Trans. Signal Process.*, vol. 50, no. 5, pp. 1051–1064, 2002.

[199] S. V. Schell, "An overview of sensor array processing for cyclostationary signals," in *Cyclostationarity in Communications and Signal Processing*, W. A. Gardner, Ed. New York: IEEE, 1994, pp. 168–239.

[200] F. Schreckenbach, N. Görtz, J. Hagenauer, and G. Bauch, "Optimization of symbol mappings for bit-interleaved coded modulation with iterative decoding," *IEEE Commun. Lett.*, vol. 7, no. 12, pp. 593–595, Dec. 2003.

[201] ——, "Optimized symbol mappings for bit-interleaved coded modulation with iterative decoding," in *Proc. IEEE Global Commun. Conf. (GLOBECOM)*, San Francisco, CA, Dec. 1–5, 2003.

[202] M. Sellathurai and S. Haykin, "TURBO-BLAST for wireless communications: Theory and experiments," *IEEE Trans. Signal Process.*, vol. 50, no. 10, pp. 2538–2546, Oct. 2002.

[203] A. Sendonaris, E. Erkip, and B. Aazhang, "Increasing uplink capacity via user cooperation diversity," in *Proc. IEEE Int. Symposium on Inf. Theory*, Cambridge, MA, Aug. 16–21, 1998, p. 156.

[204] ——, "User cooperation diversity-Part I: System description," *IEEE Trans. Commun.*, vol. 51, no. 11, pp. 1927–1938, Nov. 2003.

[205] ——, "User cooperation diversity-Part II: Implementation aspects and performance analysis," *IEEE Trans. Commun.*, vol. 51, no. 11, pp. 1939–1948, Nov. 2003.

[206] S. Serbetli and A. Yener, "Transceiver optimization for multiuser MIMO systems," *IEEE Trans. Signal Process.*, vol. 52, no. 1, pp. 214–226, 2004.

[207] S. Shahbazpanahi, A. B. Gershman, and G. B. Giannakis, "Semiblind multiuser MIMO channel estimation using Capon and MUSIC techniques," *IEEE Trans. Signal Process.*, vol. 54, no. 9, pp. 3581–3591, Sept. 2006.

[208] C. E. Shannon, "Communication theory of secrecy systems," *Bell System Techn. J.*, vol. 28, no. 4, pp. 656–715, Oct. 1949.

[209] ——, "Two-way communication channels," in *Proc. 4th Berkeley Symp. Math. Stat. and Prob.*, vol. 1, 1961, pp. 611–644.

[210] M. Sharif and B. Hassibi, "On the capacity of MIMO broadcast channels with partial side information," *IEEE Trans. Inform. Theory*, vol. 51, no. 2, pp. 506–522, 2005.

[211] Z. Shen, R. W. Heath Jr., J. G. Andrews, and B. L. Evans, "Comparison of space-time water-filling and spatial water-filling for MIMO fading channels," in *Proc. IEEE Global Commun. Conf. (GLOBECOM)*, vol. 1, Dallas, TX, Nov. 29 – Dec. 3, 2004.

[212] S. Shi, M. Schubert, and H. Boche, "Downlink MMSE transceiver optimization for multiuser MIMO systems: Duality and sum-MSE minimization," *IEEE Trans. Signal Process.*, vol. 55, no. 11, pp. 5436–5446, 2007.

[213] ——, "Downlink MMSE transceiver optimization for multiuser MIMO systems: MMSE balancing," *IEEE Trans. Signal Process.*, vol. 56, no. 8 Part 1, pp. 3702–3712, 2008.

[214] N. D. Sidiropoulos, T. N. Davidson, and Z.-Q. T. Luo, "Transmit beamforming for physical-layer multicasting," *IEEE Trans. Signal Process.*, vol. 54, no. 6, pp. 2239–2251, June 2006.

[215] Q. H. Spencer, A. L. Swindlehurst, and M. Haardt, "Zero-forcing methods for downlink spatial multiplexing in multiuser MIMO channels," *IEEE Trans. Signal Process.*, vol. 52, no. 2, pp. 461–471, 2004.

[216] A. Stefanov and E. Erkip, "Cooperative coding for wireless networks," *IEEE Trans. Commun.*, vol. 52, no. 9, pp. 1470–1476, Sept. 2004.

[217] C. Studer, A. Burg, and H. Bölcskei, "Soft-output sphere decoding: Algorithms and VLSI implementation," *IEEE J. Select. Areas Commun.*, vol. 26, no. 2, pp. 290–300, Feb. 2008.

[218] J. F. Sturm, "Using SeDuMi 1.02, a Matlab toolbox for optimization over symmetric cones," *Optimization Methods and Software*, vol. 11–12, pp. 625–653, 1999.

[219] X. Tang and Y. Hua, "Optimal design of non-regenerative MIMO wireless relays," *IEEE Trans. Wireless Commun.*, vol. 6, no. 4, pp. 1398–1407, Apr. 2007.

[220] V. Tarokh, H. Jafarkhani, and A. Calderbank, "Space-time block codes from orthogonal designs," *IEEE Trans. Inform. Theory*, vol. 45, no. 5, pp. 1456–1467, July 1999.

[221] ——, "Space-time block coding for wireless communication: Performance results," *IEEE J. Select. Areas Commun.*, vol. 17, no. 3, pp. 451–460, Mar. 1999.

[222] V. Tarokh, N. Seshadri, and A. Calderbank, "Space-time codes for high data rate wireless communication: Performance criterion and code construction," *IEEE Trans. Inform. Theory*, vol. 44, no. 2, pp. 744–765, Mar. 1998.

[223] S. Tavildar and P. Viswanath, "Approximately universal codes over slow-fading channels," *IEEE Trans. Inform. Theory*, vol. 52, no. 7, pp. 3233–3258, July 2006.

[224] I. E. Telatar, "Capacity of multi-antenna Gaussian channels," *Europ. Trans. Telecommun.*, vol. 10, no. 6, pp. 585–595, Nov. 1999.

[225] S. ten Brink, F. Sanzi, and J. Speidel, "Two-dimensional iterative APP channel estimation and decoding for OFDM systems," in *Proc. IEEE Global Commun. Conf. (GLOBECOM)*, San Francisco, CA, Nov. 27 – Dec. 1, 2000.

[226] B. Timus, "A coverage analysis of amplify-and-forward relaying schemes in outdoor urban environment," Bucharest, Romania, July 29–31 2006.

[227] M. Tomlinson, "New automatic equaliser employing modulo arithmetic," *Electron. Lett.*, vol. 7, no. 5–6, pp. 138–139, 1971.

[228] L. Tong, B. M. Sadler, and M. Dong, "Pilot assisted wireless transmissions," *IEEE Signal Processing Mag.*, vol. 21, no. 6, pp. 12–25, Nov. 2004.

[229] D. Tse and P. Viswanath, *Fundamentals of Wireless Communication*. Cambridge University Press, 2005.

[230] R. Tütüncü, K. Toh, and M. Todd, "Solving semidefinite-quadratic-linear programs using SDPT3," *Mathematical Programming*, no. 95, pp. 189–217, 2003.

[231] G. Ungerboeck, "Channel coding with multilevel/phase signals," *IEEE Trans. Inform. Theory*, vol. 28, no. 1, pp. 55–67, Jan. 1982.

[232] H. Y. V. Sreng and D. Falconer, "Coverage enhancement through two-hop relaying in cellular radio systems," in *IEEE Wirel. Comm. and Netw. Conf. (WCNC)*, Orlando, FL, March 17–21 2002.

[233] M. C. Valenti and B. D. Woerner, "Iterative channel estimation and decoding of pilot symbol assisted turbo codes over flat-fading channels," *IEEE J. Select. Areas Commun.*, vol. 19, no. 9, pp. 1697–1705, Sept. 2001.

[234] E. C. Van der Meulen, "A survey of multi-way channels in information theory," *IEEE Trans. Inform. Theory*, vol. 23, pp. 1–37, 1977.

[235] E. Van Der Meulen, "Three-terminal communication channels," *Advances in Applied Probability*, vol. 3, no. 1, pp. 120–154, 1971.

[236] B. Van Veen and K. Buckley, "Beamforming: a versatile approach to spatial filtering," *IEEE ASSP Mag.*, vol. 5, no. 2, pp. 4–24, Apr. 1988.

[237] L. Vandenberghe and S. Boyd, "Semidefinite programming," vol. 38, pp. 49–95, Mar. 1996.

[238] U. Varshney, "Multicast over wireless networks," *Commun. ACM*, vol. 45, no. 12, pp. 31–37, 2002.

[239] R. Vaze and R. W. Heath, "Capacity scaling for MIMO two-way relaying," in *Proc. IEEE Int. Symposium on Inf. Theory*, Nice, France, June 24 – 29, 2007.

[240] S. Verdú, *Multiuser Detection.* Cambridge University Press, Aug. 1998.

[241] S. Vishwanath, N. Jindal, and A. Goldsmith, "Duality, achievable rates, and sum-rate capacity of Gaussian MIMO broadcast channels," *IEEE Trans. Inform. Theory*, vol. 49, no. 10, pp. 2658–2668, Oct. 2003.

[242] H. Viswanathan, S. Venkatesan, and H. Huang, "Downlink capacity evaluation of cellular networks with known-interference cancellation," *IEEE J. Select. Areas Commun.*, vol. 21, no. 5, pp. 802–811, June 2003.

[243] A. Viterbi, "CDMA: principles of spread spectrum communication," *Prentice Hall*, 1995.

[244] E. Viterbo and J. Boutros, "A universal lattice code decoder for fading channels," *IEEE Trans. Inform. Theory*, vol. 45, no. 5, pp. 1639–1642, July 1999.

[245] S. Vorobyov, A. Gershman, and Z. Luo, "Robust adaptive beamforming using worst-case performance optimization: A solution to the signal mismatch problem," *IEEE Trans. Signal Process.*, vol. 51, no. 2, pp. 313–324, 2003.

[246] A. Vosoughi and A. Scaglione, "Everything you always wanted to know about training: Guidelines derived using the affine precoding framework and the CRB," *IEEE Trans. Signal Process.*, vol. 54, no. 3, pp. 940–954, Mar. 2006.

[247] M. Vu and A. Paulraj, "MIMO wireless linear precoding," *IEEE Signal Processing Mag.*, vol. 24, no. 5, pp. 86–105, 2007.

[248] J. Wang, D. J. Love, and M. D. Zoltowski, "Improved space-time coding for multiple antenna multicasting," in *Proc. of IEEE International Waveform Diversity and Design Conference*, Kauai, HA, Jan. 2006.

[249] T. Wang, A. Cano, G. Giannakis, and J. Laneman, "High-performance cooperative demodulation with decode-and-forward relays," *IEEE Trans. Commun.*, vol. 55, no. 7, pp. 1427–1436, July 2007.

[250] H. Weingarten, Y. Steinberg, and S. Shamai, "The capacity region of the Gaussian multiple-input multiple-output broadcast channel," *IEEE Trans. Inform. Theory*, vol. 52, no. 9, pp. 3936–3964, Sept. 2006.

[251] C. Windpassinger, R. F. H. Fischer, and J. B. Huber, "Lattice-reduction-aided broadcast precoding," *IEEE Trans. Commun.*, vol. 52, no. 12, pp. 2057–2060, Dec. 2004.

[252] C. Windpassinger, R. F. H. Fischer, T. Vencel, and J. B. Huber, "Precoding in multi-antenna and multi-user communications," *IEEE Trans. Wireless Commun.*, vol. 3, no. 4, pp. 1305–1316, July 2004.

[253] J. Winters, J. Salz, and R. Gitlin, "The impact of antenna diversity on the capacity of wirelesscommunication systems," *IEEE Trans. Commun.*, vol. 42, no. 2/3/4, pp. 1740–1751, Feb. 1994.

[254] T. Wirth, V. Venkatkumar, T. Haustein, E. Schulz, and R. Halfmann, "LTE-advanced relaying for outdoor range extension," in *70th IEEE Veh. Tech. Conf. (VTC)*, Anchorage, USA, Sept. 20–23, 2009.

[255] A. Wittneben, "Basestation modulation diversity for digital SIMULCAST," in *Proc. 41st IEEE Veh. Tech. Conf. (VTC)*, St. Louis, MO, May 19–22, 1991, pp. 848–853.

[256] A. Wittneben and I. Hammerstroem, "Multiuser zero forcing relaying with noisy channel state information," in *Proc. IEEE Wirel. Comm. and Netw. Conf. (WCNC)*, Mar. 13–17,, New Orleans, LO 2005.

[257] A. Wittneben and B. Rankov, "Impact of cooperative relays on the capacity of rank-deficient MIMO channels," in *Proc. 12th IST Summit on Mob. and Wirel. Comm.*, Aveiro, Portugal, Jun. 15–18, 2003, pp. 421–425.

[258] ——, "Distributed antenna systems and linear relaying for gigabit MIMO wireless," in *Proc. 60th IEEE Veh. Tech. Conf. (VTC)*, Los Angeles, CA, Sep. 26–29, 2004.

[259] P. W. Wolniansky, G. J. Foschini, G. D. Golden, and R. A. Valenzuela, "V-BLAST: An architecture for realizing very high data rates over the rich-scattering wireless channel," in *Proc. Int. Symp. on Sig., Sys. and Elec.*, Pisa, Italy, Sep 29, 1998, pp. 295–300.

[260] Y. Wu, P. Chou, and S.-Y. Kung, "Information exchange in wireless networks with network coding and physical-layer broadcast," in *Proc. of 39th Annual Conf. on Information Sciences and Systems (CISS)*, Baltimore, MD, Mar. 2005.

[261] Y. Wu, "Broadcasting when receivers know some messages a priori," in *Proc. IEEE Int. Symposium on Inf. Theory*, Nice, France, June 24 – 29, 2007, pp. 1141–1145.

[262] R. F. Wyrembelski, T. J. Oechtering, I. Bjelakovic, C. Schnurr, and H. Boche, "Capacity of Gaussian MIMO bidirectional broadcast channels," in *Proc. IEEE Int. Symposium on Inf. Theory*, Toronto, Canada, July 2008, pp. 584–588.

[263] R. F. Wyrembelski, T. J. Oechtering, and H. Boche, "Decode-and-forward strategies for bidirectional relaying," in *19th IEEE Int. Symp. on Pers., Ind. and Mob. Rad. Comm. (PIMRC)*, Cannes, France, Sep. 15–18, 2008, pp. 1–6, invited.

[264] L.-L. Xie, "Network coding and random binning for multi-user channels," in *Proc. of the 10th Canadian Workshop on Information Theory*, Edmonton, Alberta, Canada, June 6-8, 2007, pp. 85–88.

[265] F. Xue, C.-H. Liu, and S. Sandhu, "MAC-layer and PHY-layer network coding for two-way relaying: Achievable regions and opportunistic scheduling," in *45th Allerton Conf. Comm., Contr. and Comp.*, Allerton House, Monticello, Illinois, Sept. 26–28, 2007.

[266] J. Yang and S. Roy, "On joint transmitter and receiver optimization for multiple-input-multiple-output (MIMO) transmission systems," *IEEE Trans. Commun.*, vol. 42, no. 12, pp. 3221–3231, 1994.

[267] S. Yang and R. Koetter, "Network coding over a noisy relay: a belief propagation approach," in *Proc. IEEE Int. Symposium on Inf. Theory*, Nice, France, June 24 – 29, 2007.

[268] H. Yao and G. Wornell, "Achieving the full MIMO diversity-multiplexing frontier with rotation-based space-time codes," in *41st Allerton Conf. Comm., Contr. and Comp.*, Allerton House, Monticello, Illinois, 2003.

[269] S. Yao and M. Skoglund, "Analog network coding mappings for the gaussian multiple-access relay channel," in *Proc. IEEE Int. Symposium on Inf. Theory*, Seoul, South Korea, June 28 – July 3, 2009, pp. 104–108.

[270] T. Yu, S. H. Han, S. Jung, J. Son, Y. Chang, H. Kang, and R. Taori, "Proposal for full duplex relay," IEEE 802.16 Broadband Wireless Access Working Group, May 2008.

[271] W. Yu, W. Rhee, S. Boyd, and J. M. Cioffi, "Iterative water-filling for Gaussian vector multiple access channels," *IEEE Trans. Inform. Theory*, vol. 50, pp. 145–152, Jan. 2004.

[272] J. Yuan, Z. Chen, Y. Li, and J. Chu, "Distributed space-time trellis codes for a cooperative system," *IEEE Trans. Wireless Commun.*, vol. 8, no. 10, pp. 4897–4905, Oct. 2009.

[273] K. Zeger and A. Gersho, "Pseudo-gray coding," *IEEE Trans. Commun.*, vol. 38, no. 12, pp. 2147–2158, Dec. 1990.

[274] G. Zeitler, R. Koetter, G. Bauch, and J. Widmer, "Design of network coding functions in multihop relay networks," in *Proc. 5th International Symposium on Turbo Codes and Related Topics*, Lausanne, Switzerland, Sept. 1–5 2008.

[275] ——, "On quantizer design for soft values in the multiple-access relay channel," in *Proc. IEEE Intl. Conf. on Commun. (ICC)*, Dresden, Germany, June 14–18, 2009.

[276] J. Zhan, B. Nazer, M. Gastpar, and U. Erez, "MIMO compute-and-forward," in *Proc. IEEE Int. Symposium on Inf. Theory*, Seoul, South Korea, June 28 – July 3, 2009.

[277] R. Zhang, Y.-C. Liang, C. C. Chai, and S. Cui, "Optimal beamforming for two-way multi-antenna relay channel with analogue network coding," *IEEE J. Select. Areas Commun.*, vol. 27, no. 5, pp. 699–712, June 2009.

[278] B. Zhao and M. C. Valenti, "Distributed turbo coded diversity for relay channel," *Electron. Lett.*, vol. 39, pp. 786–787, May 2003.

[279] J. Zhao, M. Kuhn, A. Wittneben, and G. Bauch, "Self-interference aided channel estimation in two-way relaying systems," in *Proc. IEEE Global Commun. Conf. (GLOBECOM)*, New Orleans, LA, Nov. 30 – Dec. 4, 2008.

[280] L. Zheng and D. N. C. Tse, "Communication on the grassmann manifold: A geometric approach to the noncoherent multiple-antenna channel," *IEEE Trans. Inform. Theory*, vol. 48, no. 2, pp. 359–383, Feb. 2002.

[281] ——, "Diversity and multiplexing: A fundamental tradeoff in multiple-antenna channels," *IEEE Trans. Inform. Theory*, vol. 49, no. 5, pp. 1073–1096, May 2003.

[282] H. Zhu, R.-R. Chen, and B. Farhang-Boroujeny, "Capacity of pilot-aided MIMO communication systems," in *Proc. IEEE Int. Symposium on Inf. Theory*, Chicago, IL, June 27 – July 2, 2004, p. 544.

[283] H. Zhu, B. Farhang-Boroujeny, and C. Schlegel, "Pilot embedding for joint channel estimation and data detection in MIMO communication systems," *IEEE Commun. Lett.*, vol. 7, no. 1, pp. 30–32, Jan. 2003.

Curriculum Vitae

Name: **Jian Zhao**
Birthday: May 7, 1979
Birthplace: Nanjing, China

Education

2004/09– 2010/07	**Swiss Federal Institute of Technology (ETH) Zurich, Switzerland** Department of Information Technology and Electrical Engineering Degree awarded: Dr. sc. ETH Zürich
2001/10– 2004/07	**Hamburg University of Technology, Germany** Information and Communication Systems Degree awarded: M.Sc. Degree
1997/09– 2001/07	**Nanjing University, China** Department of Electronic Science and Engineering Degree awarded: B.S. Degree Graduated with TOP 1 GPA out of 148 students

Awards

2009	**Chinese Government Award for Outstanding Students Abroad**
2008	**Best Paper Award** in IEEE Global Communications Conference
2001–2003	**DAAD-Siemens Asia 21st Century Scholarship**
1997–2001	**Academic Excellence Scholarship**, Nanjing University

Experience

2004/09– 2010/07	**Research and Teaching Assistant** Communication Technology Laboratory, ETH Zurich, Switzerland ▲ Responsible for the joint project with DOCOMO Euro-labs; investigating and proposing new transmission schemes for relaying technologies in the next generation cellular wireless communication systems; ▲ Responsible for teaching exercises classes, preparing course materials and correcting students' homework and exams.
2002/07– 2002/10	**Intern** Information and Communication Mobile (ICM), Siemens AG, Germany ▲ Analyzed and implemented a simulation chain for BLAST type MIMO wireless systems in the simulation environment Cossap using C++.

Patents

1. **J. Zhao**, M. Kuhn, A. Wittneben, and G. Bauch, "Method, apparatus and system for channel estimation in two-way relaying networks", *Patent number(s): EP2079209-A1; CN101483622-A; JP2009171576-A; US2009190634-A1; EP2079209-B1*
2. **J. Zhao**, M. Kuhn, A. Wittneben, and G. Bauch, "Coding and modulation techniques for two-way relaying systems with asymmetric channel quality", *Patent pending*

Publications

1. **J. Zhao**, M. Kuhn, A. Wittneben, and G. Bauch, "Asymmetric Data Rate Transmission in Two-Way Relaying Systems With Network Coding", *IEEE International Conference on Communications (ICC)*, Cape Town, South Africa, May 2010.
2. **J. Zhao**, M. Kuhn, A. Wittneben, and G. Bauch, "Achievable Rates of MIMO Bidirectional Broadcast Channels With Self-Interference Aided Channel Estimation", *IEEE Wireless Communications & Networking Conference (WCNC)*, Budapest, Hungary, Apr. 2009.
3. **J. Zhao**, M. Kuhn, A. Wittneben, and G. Bauch, "Self-Interference Aided Channel Estimation in Two-Way Relaying Systems", *IEEE Global Communications Conference (GLOBECOM)*, New Orleans, LA, Nov. 2008 **(BEST PAPER AWARD)**
4. **J. Zhao**, M. Kuhn, A. Wittneben, and G. Bauch, "Optimum Time-Division in MIMO Two-Way Decode-and-Forward Relaying Systems", *The 42nd Annual Asilomar Conference on Signals, Systems, and Computers*, Pacific Grove, CA, Oct. 2008.
5. **J. Zhao**, M. Kuhn, A. Wittneben, and G. Bauch, "Cooperative Transmission Schemes for Decode-and-Forward Relaying", *The 18th Annual IEEE International Symposium on Personal, Indoor and Mobile Radio Communications (PIMRC)*, Athens, Greece, Sept. 2007.
6. I. Hammerström, M. Kuhn, C. Esli, **J. Zhao**, A. Wittneben, and G. Bauch, "MIMO Two-Way Relaying with Transmit CSI at the Relay", *The 8th IEEE International Workshop on Signal Processing Advances in Wireless Communications (SPAWC)*, Helsinki, Finland, pp. 5, June 2007, (invited paper).
7. **J. Zhao**, I. Hammerström, M. Kuhn, A. Wittneben, M. Herdin, and G. Bauch, "Coverage Analysis for Cellular Systems With Multiple Antennas Using Decode-and-Forward Relays", *IEEE Vehicular Technology Conference (VTC)*, Dublin, Ireland, Apr. 2007.
8. I. Hammerström, **J. Zhao**, S. Berger, and A. Wittneben, "Experimental Performance Evaluation of Joint Cooperative Diversity and Scheduling", *IEEE Vehicular Technology Conference (VTC)*, Dallas, Sept. 2005.
9. **J. Zhao** and A. Wittneben, "Cellular Relaying Networks: State of the Art and Open Issues", *The 2nd COST 289 Workshop*, Kemer, Antalya, Turkey, July 2005.

10. I. Hammerström, **J. Zhao**, and A. Wittneben, "Temporal Fairness Enhanced Scheduling for Cooperative Relaying Networks in Low Mobility Fading Environments", *The 6th IEEE International Workshop on Signal Processing Advances in Wireless Communications (SPAWC)*, New York, June 2005.

Bisher erschienene Bände der Reihe

Series in Wireless Communications

ISSN 1611-2970

1	Dirk Benyoucef	Codierte Multiträgerverfahren für Powerline– und Mobilfunkanwendungen ISBN 978-3-8325-0137-2 40.50 EUR
2	Marc Kuhn	Space-Time Codes und ihre Anwendungen in zukünftigen Kommunikationssystemen ISBN 978-3-8325-0159-4 40.50 EUR
3	Frank Althaus	A New Low-Cost Approach to Wireless Communication over Severe Multipath Fading Channels ISBN 978-3-8325-0158-7 40.50 EUR
4	Ingmar Hammerström	Cooperative Relaying and Adaptive Scheduling for Low Mobility Wireless Access Networks ISBN 978-3-8325-1466-2 40.50 EUR
5	Thomas Zasowski	A System Concept for Ultra Wideband (UWB) Body Area Networks ISBN 978-3-8325-1715-1 40.50 EUR
6	Boris Rankov	Spectral Efficient Cooperative Relaying Strategies for Wireless Networks ISBN 978-3-8325-1744-1 40.50 EUR
7	Stephan Sand	Joint Iterative Channel and Data Estimation in High Mobility MIMO-OFDM Systems ISBN 978-3-8325-2385-5 39.00 EUR
8	Florian Troesch	Design and Optimization of Distributed Multiuser Cooperative Wireless Networks ISBN 978-3-8325-2400-5 42.50 EUR
9	Celal Esli	Novel Low Duty Cycle Schemes: From Ultra Wide Band to Ultra Low Power ISBN 978-3-8325-2484-5 42.00 EUR

10	Stefan Berger	Coherent Cooperative Relaying in Low Mobility Wireless Multiuser Networks ISBN 978-3-8325-2536-1 42.00 EUR
11	Christoph Steiner	Location Fingerprinting for Ultra-Wideband Systems. The Key to Efficient and Robust Localization ISBN 978-3-8325-2567-5 36.00 EUR